Chitin and Collagen: Isolation, Purification, Characterization, and Applications

Chitin and Collagen: Isolation, Purification, Characterization, and Applications

Azizur Rahman

MDPI

Basel • Beijing • Wuhan • Barcelona • Belgrade • Novi Sad • Cluj • Manchester

Azizur Rahman
Centre for Climate Change Research
University of Toronto
(ONRamp at UTE)
Toronto
Canada

Editorial Office
MDPI AG
Grosspeteranlage 5
4052 Basel, Switzerland

This is a reprint of articles from the Special Issue published online in the open access journal *Polysaccharides* (ISSN 2673-4176) (available at: www.mdpi.com/journal/polysaccharides/special_issues/chitin_collagen).

For citation purposes, cite each article independently as indicated on the article page online and using the guide below:

Lastname, A.A.; Lastname, B.B. Article Title. *Journal Name* **Year**, *Volume Number*, Page Range.

ISBN 978-3-7258-1808-2 (Hbk)
ISBN 978-3-7258-1807-5 (PDF)
https://doi.org/10.3390/books978-3-7258-1807-5

Contents

About the Editor

Azizur Rahman

Azizur Rahman, Ph.D., is a distinguished research lead, president, and CEO of the Centre for Climate Change Research, AR Environmental Solutions, and AR Biotech Canada at the University of Toronto Entrepreneurship program. Dr. Rahman has earned numerous accolades for his pioneering research and contributions to science innovation. Dr. Rahman's exemplary work has earned him the Presidential Honorary Award for Scientists from the University of Ryukyus in Japan, the institution's highest honor for novel research findings. Additionally, he has received the prestigious Global Peer Review Award from the Web of Science Group in both 2018 and 2019. Throughout his career, he has been honoured with over 20 major national and international awards, including multiple best paper awards and a gold medal. Recently, in 2024, Dr. Rahman was acknowledged as one of the most cited and viewed authors of MDPI.

Dr. Rahman's interdisciplinary research spans a broad array of critical fields, including the following: biotechnology and human health, global climate change, organic agriculture, biodiversity, ocean and marine biology, urban air pollution, and food waste management.

Dr. Rahman is passionate about leveraging science to promote community advancement through innovation and entrepreneurship. He applies his diverse expertise to drive impactful change, supporting the intersection of science and social innovation. A serial entrepreneur, he has founded multiple companies and non-profit organizations dedicated to fostering scientific and social progress.

Dr. Azizur Rahman's career reflects a commitment to excellence in research and a dedication to utilizing science for the advancement of society. His work continues to inspire and drive significant advancements across multiple disciplines.

Preface

Chitin and collagen are the most important biopolymers in nature. Interestingly, these two biopolymers exhibit similar hierarchical structural organizations. Chitin can also be enzymatically deacetylated to chitosan, a more flexible and soluble biopolymer. It has many applications, including in the medical, environmental, and agricultural sectors. Similarly, nature is a source of massive quantities of collagen, especially in marine organisms. Collagen is the main fibrous structural protein in animals' extracellular matrix and connective tissue. It contributes greatly to biotechnology products and medical applications.

In this reprint, the latest innovations of chitin and collagen have been integrated and could potentially be applied in various multidisciplinary fields. This reprint covers recent trends in applied scientific research on polysaccharides and two biopolymers of chitin and collagen.

Azizur Rahman
Editor

polysaccharides

Editorial

Two Important Biopolymers: The Transformative Power of Chitin and Collagen in Multidisciplinary Applications

Azizur Rahman [1,2,3]

[1] Centre for Climate Change Research, University of Toronto, ONRamp at UTE, Toronto, ON M5G1L5, Canada; mazizur.rahman@utoronto.ca or aziz@climatechangeresearch.ca
[2] A.R. Environmental Solutions, ICUBE-University of Toronto, Mississauga, ON L5L1C6, Canada
[3] AR Biotech Canada, Toronto, ON M2H 3P8, Canada

Biopolymers are natural polymers produced by living organisms' cells, and have promising multidisciplinary applications. Chitin and collagen are the most important biopolymers in nature. Interestingly, these two biopolymers exhibit similar hierarchical structural organizations. After cellulose, chitin is the world's second most important natural polymer, and has been identified in bacteria, fungi, plants, and marine invertebrates. Chitin can also be enzymatically deacetylated to chitosan, a more flexible and soluble biopolymer. It has many applications, including in the medical, environmental, and agricultural sectors [1–5]. Similarly, nature is a source of massive quantities of collagen, especially in marine organisms. Collagen is the main fibrous structural protein in animals' extracellular matrix and connective tissue. It contributes greatly to biotechnology products and medical applications [6–9].

This Editorial is cover the Special Issue "Chitin and Collagen: Isolation, Purification, Characterization, and Applications". This Special Issue of *Polysaccharides* about chitin and collagen aimed at their isolation, purification, characterization, and promising applications, contains 10 high-quality original articles and two review articles. At this point, an overview of the research results and reviews of existing public literature by the authors is provided, which could help readers find appropriate articles for their field and findings of interest. The contributions are listed below:

Contribution 1 reported that the electrochemical determination of adsorption thermodynamics on chitin is directly linked to its applications in environmental monitoring and technology. There is a strong adsorption of metal ions and their complexes to chitin, which depends on both the oxidation and complexation states of many of the said elements. With chitin forming the outer hull of mobile organisms (animals), this biopolymer is expected to take part in metal distribution in aquatic (limnetic and riverine) ecosystems. The authors demonstrated how chitin properties can control element transport in aquatic ecosystems.

Contribution 2 developed a method for glucose oxidase (GOx) immobilization onto graphite rod electrodes, avoiding enzyme denaturation through adsorption by incubation in chitosan solutions. The suitability of this method in keeping enzyme activity, which is the key aspect regarding immobilization procedures for bioelectrodes development, is evaluated in this paper through cyclic voltammetry in oxygen-free solutions, in order to assess electron transfer between the graphite surface and the active centre of GOx. Furthermore, the catalytic response of the enzyme for energy harvesting in a glucose BFC has been evaluated to confirm the proper stability of the enzyme using this immobilization procedure.

Contribution 3 established a new electrostatic complex-based lecithin and chitosan crosslinked with sodium tripolyphosphate, its use for the encapsulation of orange essential oil, and a subsequent evaluation of its cytotoxicity in zebrafish liver cells (ZFL cell line).

Contribution 4 explored chitosan- and collagen-based materials containing curcumin as a bioactive compound for wound-healing targets. The effects of incorporating curcumin and increasing its concentration on both the rheological properties of the formed solutions

Citation: Rahman, A. Two Important Biopolymers: The Transformative Power of Chitin and Collagen in Multidisciplinary Applications. *Polysaccharides* **2024**, *5*, 96–99. https://doi.org/10.3390/polysaccharides5020007

Received: 30 January 2024
Accepted: 21 March 2024
Published: 17 April 2024

and the morphological and thermal properties of the three-dimensional scaffolds obtained from them were evaluated. The authors show the appearances of the scaffolds obtained, as well as their surface and cross-sectional surface morphologies. The incorporation of curcumin into the polymeric matrix of chitosan and collagen led to changes in the coloration of the scaffolds. Concerning the surface morphology of the scaffolds, all presented interconnected pores in the polymeric network, which is advantageous and desirable for the proposed application, since scaffolds must allow for the transport of nutrients, absorb fluids and moisture, and allow for cell migration and proliferation. Regarding the cross-sectional surface micrographs of the scaffolds, the polymeric fibres were channelled-distributed, and the addition of curcumin in its highest concentration led to greater compaction of these channels, which was probably due to the cross-linking effect caused by the polyphenol. Therefore, the materials developed in this study were revealed as promising biomaterials for their biological evaluation in tissue regeneration.

Contribution 5 demonstrated that collagen, hyaluronic acid, and chitosan materials can be modified with laser treatment. Scanning electron microscopy (SEM) imaging and infrared spectroscopy (FTIR-ATR) were applied to evaluate laser beam effects on the surface structure. SEM images revealed important changes in the biopolymer film structure. The treatment performed in this article could be used for material modification of potential biomedical purposes.

Contribution 6 reported that chitin is an effective sorbent that can be used in environmental monitoring, without the applications in withholding metal-containing pollutants from wastewater or nuclear fuel reprocessing flows, since the background levels in chitin are very low, with the exception of a few metals.

Contribution 7 assessed the effects of pomegranate peel extract, which is an agro-industrial residue with active properties. The extracts were examined for the rheological properties of potential coatings based on chitosan and gelatine. The rheological properties of the polymeric solutions were also investigated, as was its incorporation order into the system.

Contribution 8 explored how to control the surface properties of chitosan/hyaluronan multilayered coatings for a tumour cell capture. The authors looked for a marker for the development of prostate cancer. In this scenario, films composed of hyaluronic acid and chitosan have demonstrated significant capture potential of prostate tumour cells. This study shows that surface chemistry and morphology are critical factors for the development of biomaterials designed for several cell adhesion applications, such as rapid diagnostic, cell signalling, and biosensing mechanisms.

The application of chitosan from edible cricket species was demonstrated as a Hypolipidemic and Antimicrobial Agent in Contribution 9. The authors successfully converted cricket chitin to chitosan with an approximately 72%, 76% and 80% degree of deacetylation, achieved by varying deacetylation times using concentrated alkaline treatments. This study concludes that chitosan derived from U.S.-reared edible crickets has physicochemical and bioactive properties, similar to commercial crustacean (e.g., shrimp) chitosan. The results established that there is huge potential for the mass production of cricket-based chitosan as the consumer acceptability for arthropods widens outside the traditional source of crustaceans.

Toncheva-Moncheva N. et al. [10] have discovered the conversion of electrospun chitosan into chitin (Figure 1). This innovative procedure allows for tuning and modifying the thermal and mechanical properties and, more importantly, the biodegradation of the prepared nanofibrous mats. This is a reproducible method that offers the unique advantage of modulating the membrane properties leading to stable 2D biomimetic CsU and/or chitin (CsE) scaffolds tailor-made for specific purposes in the field of tissue engineering. This is a robust strategy to tune the properties of 2D biomimetic nanofibers and to transfer chitosan to chitin.

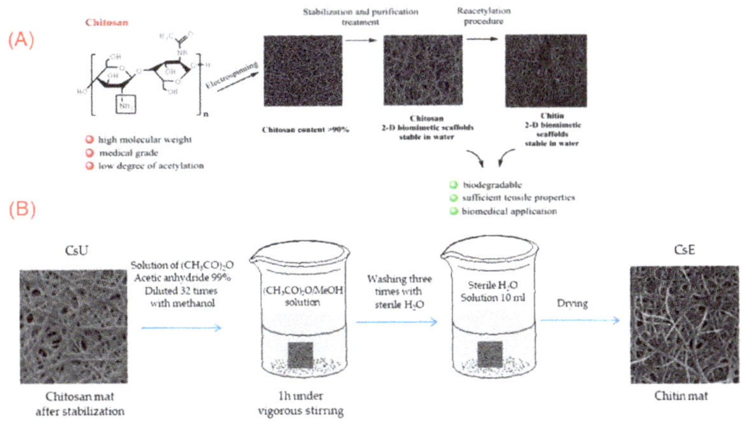

Figure 1. (**A**) Conversion of electrospun chitosan into chitin. (**B**) Reacetylation reaction procedure of chitosan into chitin [10].

Contributions 11 and 12 reviewed the polysaccharides stalk of *Didymosphenia geminata*, which is a species of freshwater diatom seen as invasive and which propagates quickly around the world. Although invasive species are generally considered a nuisance, the authors explored some useful applications for *D. geminata* in the biomedical field and wastewater remediation. Here, the authors highlight the polysaccharide-based stalks of *D. geminata* that enable versatile potential applications and uses as a biopolymer, in drug delivery and wound healing, and as biocompatible scaffolding in cell adhesion and proliferation. Moreover, this article focuses on how the polysaccharide nature of stalks and their metal-adsorption capacity allow them to have excellent wastewater remediation potential. They also aim to assess the economic impact of *D. geminata*, as an invasive species, on its immediate environment. Overall, the authors described both the benefits and environmental concerns of this invasive species, which could help researchers and communities around the world understand the facts.

As a guest editor, I highly appreciate the efforts provided by all the authors who contributed their excellent results to this Special Issue. I thank all the reviewers who carefully evaluated the submitted manuscripts. I would also like to thank the editorial board, managing editors and editorial assistants of *Polysaccharides* for their support and kind help.

Institutional Review Board Statement: Not applicable.

Data Availability Statement: No new data was created or analyzed in this study.

Conflicts of Interest: The author declares no conflict of interest.

List of Contributions

1. Fränzle, S.; Blind, F. Reversible Metal Ion/Complex Binding to Chitin Controlled by Ligand, Redox, and Photochemical Reactions and Active Movement of Chitin on Aquatic Arthropods. *Polysaccharides* **2022**, *3*, 515–543. https://doi.org/10.3390/polysaccharides3030031.
2. Buaki-Sogó, M.; García-Carmona, L.; Gil-Agustí, M.; García-Pellicer, M.; Quijano-López, A. Low-Denaturazing Glucose Oxidase Immobilization onto Graphite Electrodes by Incubation in Chitosan Solutions. *Polysaccharides* **2022**, *3*, 388–400. https://doi.org/10.3390/polysaccharides3020023.
3. De Toledo, A.M.N.; Machado, A.R.; de Souza-Soares, L.A. Development and In Vitro Cytotoxicity of *Citrus sinensis* Oil-Loaded Chitosan Electrostatic Complexes. *Polysaccharides* **2022**, *3*, 347–355. https://doi.org/10.3390/polysaccharides3020020.
4. Milan, E.P.; Bertolo, M.R.V.; Martins, V.C.A.; Bogusz Junior, S.; Plepis, A.M.G. Chitosan and Collagen-Based Materials Enriched with Curcumin (*Curcuma longa*): Rheological and Morphological Characterization. *Polysaccharides* **2022**, *3*, 236–249. https://doi.org/10.3390/polysaccharides3010013.

5. Grabska-Zielińska, S.; Sionkowska, A. Surface Property Modification of Collagen, Hyaluronic Acid, and Chitosan Films with the Neodymium Laser. *Polysaccharides* **2022**, *3*, 178–187. https://doi.org/10.3390/polysaccharides3010008.
6. Blind, F.; Fränzle, S. Chitin as a Sorbent Superior to Other Biopolymers: Features and Applications in Environmental Research, Energy Conversion, and Understanding Evolution of Animals. *Polysaccharides* **2021**, *2*, 773–794. https://doi.org/10.3390/polysaccharides2040047.
7. Romanelli Vicente Bertolo, M.; Leme, R.; da Conceição Amaro Martins, V.; de Guzzi Plepis, A.M.; Bogusz Junior, S. Rheological Characterization of the Influence of Pomegranate Peel Extract Addition and Concentration in Chitosan and Gelatin Coatings. *Polysaccharides* **2021**, *2*, 648–660. https://doi.org/10.3390/polysaccharides2030039.
8. Lima, G.G.; Rocha Neto, J.B.M.; Carvalho, H.F.d.; Beppu, M.M. Control of Surface Properties of Hyaluronan/Chitosan Multilayered Coatings for Tumor Cell Capture. *Polysaccharides* **2021**, *2*, 387–399. https://doi.org/10.3390/polysaccharides2020025.
9. Malm, M.; Liceaga, A.M. Physicochemical Properties of Chitosan from Two Commonly Reared Edible Cricket Species, and Its Application as a Hypolipidemic and Antimicrobial Agent. *Polysaccharides* **2021**, *2*, 339–353. https://doi.org/10.3390/polysaccharides2020022.
10. Toncheva-Moncheva, N.; Aqil, A.; Galleni, M.; Jérôme, C. Conversion of Electrospun Chitosan into Chitin: A Robust Strategy to Tune the Properties of 2D Biomimetic Nanofiber Scaffolds. *Polysaccharides* **2021**, *2*, 271–286. https://doi.org/10.3390/polysaccharides2020019.
11. Somanader, E.; Sreenivas, R.; Siavash, G.; Rodriguez, N.; Gao, T.; Ehrlich, H.; Rahman, M.A. Polysaccharide Stalks in *Didymosphenia geminata* Diatom: Real World Applications and Strategies to Combat Its Spread. *Polysaccharides* **2022**, *3*, 83–94. https://doi.org/10.3390/polysaccharides3010004.
12. Ejaz, H.; Somanader, E.; Dave, U.; Ehrlich, H.; Rahman, M.A. Didymo and Its Polysaccharide Stalks: Beneficial to the Environment or Not? *Polysaccharides* **2021**, *2*, 69–79. https://doi.org/10.3390/polysaccharides2010005.

References

1. Kurita, K. Chitin and Chitosan: Functional Biopolymers from Marine Crustaceans. *Mar. Biotechnol.* **2006**, *8*, 203. [CrossRef] [PubMed]
2. Rahman, M.A.; Halfar, J. First evidence of chitin in calcified coralline algae: New insights into the calcification process of Clathromorphum compactum. *Sci. Rep.* **2014**, *4*, 6162. [CrossRef] [PubMed]
3. Ehrlich, H. Chitin and collagen as universal and alternative templates in biomineralization. *Int. Geol. Rev.* **2010**, *52*, 661. [CrossRef]
4. Da Sacco, L.; Masotti, A. Chitin and Chitosan as Multipurpose Natural Polymers for Groundwater Arsenic Removal and As_2O_3 Delivery in Tumor Therapy. *Mar. Drugs* **2010**, *8*, 1518–1525. [CrossRef] [PubMed]
5. El Hadrami, A.; Adam, L.R.; El Hadrami, I.; Daayf, F. Chitosan in Plant Protection. *Mar. Drugs* **2010**, *8*, 968–987. [CrossRef] [PubMed]
6. Benayahu, D.; Sharabi, M.; Pomeraniec, L.; Awad, L.; Haj-Ali, R.; Benayahu, Y. Unique Collagen Fibers for Biomedical Applications. *Mar. Drugs* **2018**, *16*, 102. [CrossRef] [PubMed]
7. Ehrlich, H.; Wysokowski, M.; Żółtowska-Aksamitowska, S.; Petrenko, I.; Jesionowski, T. Collagens of Poriferan Origin. *Mar. Drugs* **2018**, *16*, 79. [CrossRef] [PubMed]
8. Chen, J.; Gao, K.; Liu, S.; Wang, S.; Elango, J.; Bao, B.; Dong, J.; Liu, N.; Wu, W. Fish Collagen Surgical Compress Repairing Characteristics on Wound Healing Process In Vivo. *Mar. Drugs* **2019**, *17*, 33. [CrossRef] [PubMed]
9. Carvalho, A.M.; Marques, A.P.; Silva, T.H.; Reis, R.L. Evaluation of the Potential of Collagen from Codfish Skin as a Biomaterial for Biomedical Applications. *Mar. Drugs* **2018**, *16*, 495. [CrossRef] [PubMed]
10. Toncheva-Moncheva, N.; Aqil, A.; Galleni, M.; Jérôme, C. Conversion of Electrospun Chitosan into Chitin: A Robust Strategy to Tune the Properties of 2D Biomimetic Nanofiber Scaffolds. *Polysaccharides* **2021**, *2*, 271–286. [CrossRef]

polysaccharides

MDPI

Article

Chitin as a Sorbent Superior to Other Biopolymers: Features and Applications in Environmental Research, Energy Conversion, and Understanding Evolution of Animals

Felix Blind and Stefan Fränzle *

International School Zittau (IHI), Department of Biological and Environmental Sciences, Dresden Technical University, Markt 23, D-02763 Zittau, Germany; felix.blind@gmx.de

* Correspondence: Stefan.fraenzle@tu-dresden.de

Abstract: Chitin is an effective sorbent which can be used in environmental monitoring, beyond obvious applications in withholding metal-containing pollutants from wastewater- or nuclear fuel reprocessing flows, since background levels in (purified) chitin are very low except for a few metals (Fe, Cu, Al, Ti, and Zn). Since retention of M^{x+} and their complexes on chitin depend on an oxidation state, and to a lesser extent the presence of possible ligands or co-ligands, partition between chitin samples exposed to sediment and those exposed to water can be changed by environmental factors such as local biota producing or absorbing/metabolizing effective ligands such as citrate or oxalate and by changes of redox potential. Thermodynamics are studied via log P, using calibration functions log P vs. 1/r or log P vs. $\Sigma\sigma$ (sum of Hammett parameters of ligand donor groups) for di- and trivalent elements not involved in biochemical activity (not even indirectly) and thus measuring "deviations" from expected values. These "deviations" can be due to input as a pollutant, biochemical use of certain elements, precipitation or (bio-induced reduction of $SO_4{}^{2-}$ or CO_2) dissolution of solids in sediment. Biochemical processes which occur deep in sediment can be detected due to this effect. Data from grafted chitin (saturation within $\leq$ 10 min) and from outer surfaces of arthropods caught at the same site do agree well. Log P is more telling than total amounts retrieved. Future applications of these features of chitin are outlined.

Keywords: adsorption to chitin; complex formation; biomonitoring; charge (redox) fractionation; comparison of isolated chitin and arthropod covers; photoredox chemistry on chitin

Citation: Blind, F.; Fränzle, S. Chitin as a Sorbent Superior to Other Biopolymers: Features and Applications in Environmental Research, Energy Conversion, and Understanding Evolution of Animals. *Polysaccharides* **2021**, 2, 773–794. https://doi.org/10.3390/polysaccharides2040047

Academic Editor: Azizur Rahman

Received: 16 July 2021
Accepted: 9 September 2021
Published: 9 October 2021

1. Introduction

1.1. Chitin and Chitosan—A Comparison of Chemical and Sorptive Properties

Although chitosan, that is chitin which has undergone hydrolysis of acetamido side chains, is commonly used in medical applications and as a glue known to be inert (or get slowly dissolved and metabolized) in physiological terms [1,2], it

(a) does not occur in natural conditions and
(b) is much less robust against thermal or oxidative stress than the native chitin.

Chitin does adsorb all electrophilic agents such as metal ions [3], hydrocarbons [4] and anions such as arsenate, sulfate [5] or hexacyanoferrate which produce strong H bridges with OH groups of sugars. Even though the acetylation extent of chitin may vary (some 80–95% [6]), it proved to display reproducible features as a sorbent (for quantitative terms see below). Chitin is difficult to oxidize both by an anode and by chemical agents such as the common agents for saccharide analytic cleavage; chitin itself is completely silent in electrochemical experiments (when dissolved by adding Li salts in dimethyl formamide) even though it does contain trace amounts of redox-active ions (Fe, Ti, Cu). It can be oxidized by certain staining agents such as Ag^+ or OsO_4 only after deacetylation [6] while not being attacked by common reagents used in saccharide cleavage and analysis, such as

IO_4^- or Pb(IV) acetate [7], both of which do react with the monomer, N-acetyl glucosamine. However, alkaline oligoperiodate, identified as $H_2I_2O_{10}^{4-}$, does react, leaving behind chitin nanocrystals [8]. None of the said oxidants will occur in natural surroundings, however. Chitosan and monomers glucosamine and N-acetyl glucosamine are degraded by both IO_4^- and $Pb(CH_3COO)_4$. Pb(IV) acetate does not react with either chitin or (potato) starch, but with chitosan in DMSO/glacial acetic acid [7].

Thus, chitin is the material of choice if one wants to study sorbent properties concerning, e.g., trace metals in environmental analytics. Moreover, many different organisms, including such of meaningful operative surface sizes such as spiders [9], lichens [3,10] or many insects, crabs, and crayfish are fully or partially covered by chitin which can in turn be sampled without necessarily killing the animals or lichens. The surface of grafted chitin surface used for adsorbing metal ions has a diameter of 1 cm and thus an active surface of some 0.78 cm^2. Much smaller areas would suffice in a trade-off for sensitivity. Legs, wings, or mandibles of bigger arthropods will readily exceed this area. Chitin is much more robust towards pyrolysis than other biopolymers [11]. Degradation of natural chitin, containing proteins and metal carbonates, in moist environments, is distinguished by proteins removed first [12] while chitin may persist over >> 10^8 years in some cases [12,13].

1.2. Chemical Peculiarities Which Turn up at Ecotones and Similar Interfaces

Generally speaking, chitin behaves as a very effective sorbent which even retains most metal ions at aq. levels < 1 nM/L ([14] and literature quoted there). Chitin can be dissolved in various solvents with (slow esterification by neat formic or dichloroacetic acids) or without permanent chemical modifications (hexafluoroisopropanol [15], carboxamides or lactams containing high levels of Li salts [14]), permitting production of fibers or films upon addition of water (all these solvents are completely miscible with H_2O). Different oxidation states may differ in their retention properties, such as with di- and trivalent Eu (Eu^{2+} being eluted from chitin by slowly running water or DMF). Thus, photo- or electroredox reactions taking place next to a chitin layer in flowing water will produce a voltage due to a concentration gradient which exceeds the electrokinetic flow potential (zeta potential) by a factor of 10 at least (the pzzp of chitin is pH $\approx$ 4.6 [14], considerably lower than the pH values obtained in water with carboxylate ions or even ethanol amines [$C_{substr.}$ = 0.1 mol/L]). Aq. arthropods other than water mites, caddisfly larvae [15] or *Asellus aquaticus* would not withstand pH < 3.5 for long either [16]).

Chitin does report on non-equilibrium phenomena next to water-sediment interfaces in a unique manner. These are caused by biological uptake or precipitation of certain elements in sediment [14], distinguishing part of benthic animals exposed to water only or penetrating the sediment-water interface regularly (i.e., antennae and leg tips of crayfish [17]). A rigorous formal analysis of partition data calibrated against "inert" di- and trivalent cations, respectively, yields much more information (see below). In addition, adsorption of M to chitin was studied depending on presence of such ligands which can be expected in soil liquids, ambient (fresh) waters or mining leachates. None of these features is present in chitosan or any other native or modified biopolymer, including cellulose (moss-based biomonitoring) or proteins. Hence, it is worthwhile to describe methods and results concerning chitin within this framework, too.

Although chitin does undergo slow surface hydrolysis, significant change of metal ion adsorption properties is limited to a few elements (Y, Pb, Bi). "Normal" (crustacean) chitin (non-digging crayfish species *Orconectes limosus*) compares well to grafted chitin samples once exposed to boiling water; this was conducted either to euthanize crayfish caught alive for studying adsorption from either water or sediment depending on body parts/organs and their position/function in situ, or for comparison of data using grafted chitin. This does imply hydrolysis is negligible insofar as changes of complexation are concerned. Metals forming very stable ammin-, amine complexes such as Ni and Cu actually display a slight effect on adsorption after heat treatment of chitin samples. To put it the other way round, we still deal with chitin rather than chitosan after the chitin surface or animal had been (briefly) exposed to boiling water (see below). Apart from this, chitin is exceptionally robust in terms of thermal treatment, unlike other biopolymers [3,12].

Using this property of chitin takes one beyond simply deriving some relationship between adsorbed ion quantity and environmental concentrations. In fact, non-equilibrium states can be detected by comparing adsorption at adjacent sites (partition) and using transport. This feature is unique in biomonitoring. This is shown in the following diagrams. This can be observed at a water/inundated sediment interface. When biological activity or precipitation of solids cause "downward" (water→sediment) diffusion, the partition coefficient will increase, meaning more is adsorbed on chitin exposed to sediment (green arrow in Figure 1):

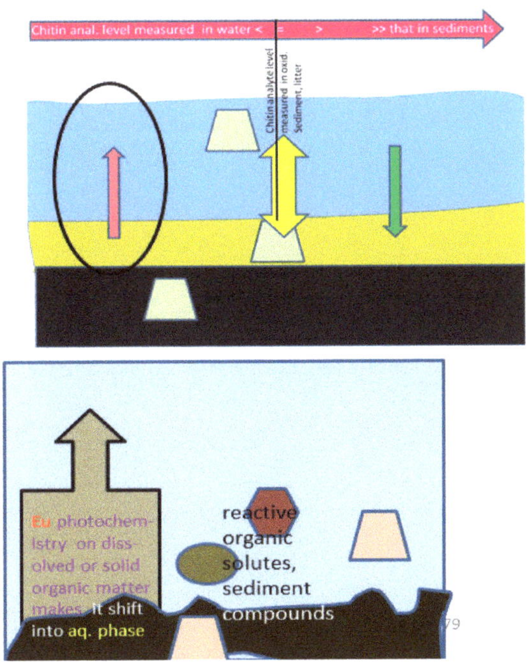

Figure 1. Partition of metals on chitin flakes embedded in water (**top**) and sediment (**below**) and corresponding effects concerning europium (Eu) co-illuminated with organic matter and photo-oxidizing it. Violet upward-pointing arrow refers to removal of Ba and Pb from sediment and depletion on the chitin flakes located there. The green downward arrow may be due to either input from above (limnetic or aerosol) or uptake deeper in the sediment. By this method one can detect processes which occur at least 40 cm from the interface, including methanogenesis linked to Ni which otherwise is too scarce to effect it without aerial input in ombrotrophic bogs. It does not matter whether this is the bottom of some water pool or its beach.

2. Materials and Methods

The procedure of sample processing and extraction of data was described before ([14,17,18]). Surface dissolution is conducted using $LiClO_4$ solutions in DMF, and absorption of this solute by cation exchanger resins kept in home-built Nylon baglets [17]. Backward ion exchange is achieved by flushing the bag several times by 1% aq. nitric acid subsequent to ion exchange which permits direct ICP-MS measurements (and re-use of the ion exchanger bag if wanted [17,18]). Amounts of reagent and periods of time in the workup steps were conducted according to the optimization procedures given by [14,17,18]. ICP-MS was undertaken by a Perkin-Elmer DR-Ce, not using the reaction cell. Detection limits were exceeded by far also for most REEs and other elements. A total of 26 elements was studied, omitting those which are abundant in biomass, yet do not bind to chitin (alkali metals, Mg, Ca) [14]. Solutions of chitin prepared in this manner do also bind metal ions up to a comparable amount. When doing electrochemistry with such a solution, metal ions (if exceeding chitin saturation), chitin adducts and complex chitin adducts all do contribute to an electrochemical signal which can be determined by means of cyclic voltammetry. An example is given in Figure 2:

Whereas both metal (e.g., Fe like in this example) ions and complexes can be adsorbed to chitin, the latter may or may not adsorb without exchange of ligands other than halides. This enhancement effect was demonstrated recently in a simple experiment by these authors: after dissolving (anhydrous) $FeCl_3$ in 40 mL DMF (2.7 mM/L), same amount (i.e., 100 μmol = 7.5 mg) of solid acetohydroxamic acid $CH_3CON(H)$-OH (causing immediate color change, meant to represent exudates from soil fungi or bacteria) and finally 20 mL of a saturated solution of shrimp chitin in DMF (1.6 M $LiClO_4$) were added.

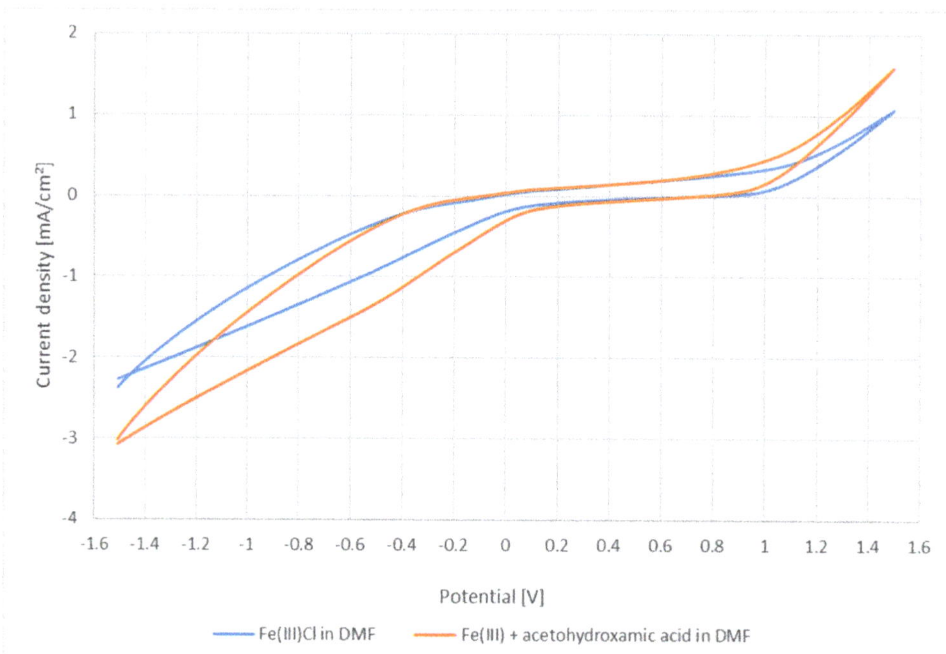

(a)

Figure 2. *Cont.*

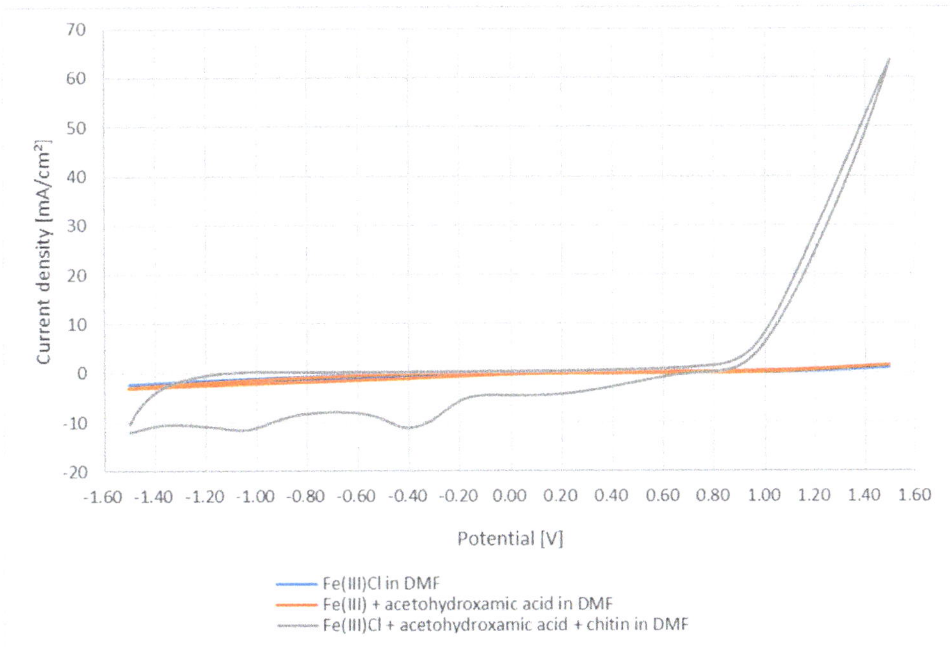

(b)

Figure 2. (**a**): Electrochemical behavior of $FeCl_3$ dissolved in DMF (cp. [18,19]) as such and after adding acetohydroxamic acid. (**b**): Superposition of all three redox systems between −1.5 and + 1.5 V vs. SCE. Chitin adsorption does markedly increase measured current by organizing the redox system near electrode. Smell of solution indicates that a terminal ($\varepsilon \geq 0.9$ V vs. SCE) increase in irreversible oxidation current is due to decay product dimethyl amine (for its oxidation properties cp. [20]).

The main redox transition of $[Fe(CH_3CO\text{-}N(H)O)]^{2+}$ at −0.4 V remains at its position but becomes both stronger and irreversible, indicating desorption of Fe^{2+} from chitin. In the original solution there are just two species, namely partially hydrated $FeCl_2^+$ and $[FeCl_4]^-$ solvated by DMF [21]. The effect of enhancing electrochemical activity may be important for both fuel cell applications and electroanalytical sensor methods (see below).

2.1. Technique of Sampling

Sampling was conducted by either grafted chitin [14] and/or collecting arthropods [17,22], lichens [10]. The former method is preferred for reasons of both

(a) protection of possibly endangered species and
(b) full control of exposition times while

The chitin sample can be located in either sediment or open water (it is recommended to mark chitin flakes placed in gravel- or other sediment by colored plastic strips or the like to improve the chances of retrieving them). Grafted chitin thus duplicates exposition features of "live" chitin whether or not the animal is a benthic one. Additionally, grafted chitin can be used in conditions by far exceeding those which would permit animals to survive, like excessive heat, cold, insecticides, radioactivity, etc., for most aquatic animals pH < 3.5 [16]. The latter fact allows detection of traces of contaminants, as well as determination of non-equilibrium states due to "uncommon" metabolic activities (i.e., such catalyzed by elements less common in biology). Typical examples are Ni (methanogenesis) or V (haloperoxidases, N_2 fixation when Mo is depleted) or precipitation/dissolution of insoluble salts. The latter may include biochemical activity, too, particularly when Ba or Pb sulfates are dissolved by reduction consuming H_2, CH_4, or other organics in inundated sediment (see below).

The sampling is least invasive; in fact, the technique was designed to sample the same individuals of living arthropods (cricket *Gryllus assimilis*) several times, just etching the surface a little bit without intoxicating or killing the animal (both for protection concerns and to obtain a baseline for a single animal before beginning active biomonitoring) [14,22]. The DMF/Li^+ solution (employing the most simple and convenient tertiary carboxamide available) can reach/suspend chitin concentrations of some 30 g/L solvent at $\geq$ 2 M/L Li^+ (Li salt = nitrate, chloride, bromide, perchlorate, trifluoromethane sulfonate but not acetate or other carboxylates which do not suspend chitin in DMF even at similar Li levels). This method was selected and optimized (times, concentrations) after experiments conducted in my workgroup to obtain films of materials dissolving surface chitin and its sorbates without working in a liquid medium gave disappointing results: f.e., when melting ε-caprolactam or δ-valerolactam (m.p. = 40 °C) and adding Li salts decomposition and carbonization invariably occurred when adding solid Li salts to the lactam melts before even making contact with chitin. The end was to finally remove the solidified lactam/nylon film containing chitin and the analytes from the surface. Hence the experiment could not be completed into testing the retention to some organic film ripped from the chitin support [23].

Such solutions of chitin in DMF ($LiClO_4$ as a conducting salt) are completely silent in electrochemical terms (ε in between −2.6 up to +1.73 V vs. SCE (onset potential of oxidative solvent breakdown in DMF and similar carboxamides) although there are traces of metals in native chitin which might undergo redox transitions within this range (10–20 ppm Fe, some 15 ppm Ti, 1.5 ppm Cu). Redox transitions do only occur after addition of metal salts or complexes and can be used to study adsorption to (dissolved) polymer strands by adding little by little metal salt until the original transition in DMF turns up (saturation; effects of added or present ligands can be studied in the same manner).

2.2. Processing of Data

Thus, meaningful partition coefficients are defined using sets of elements (di- or tri- or tetravalent metals, As, Sb, Se, Ge, etc.) which are not involved in biology. Partition coefficients P of such elements are plotted against functions which relate to complex formation constants β of "similar" complexes vs. Hammett's σ constant or reciprocal ion radii 1/r. The equations depend on the assumption that

$$\log P \sim \log \beta(ML) = a \times \log \beta + b \quad (1)$$

for some "test" or "equivalent" ligand L. a and b are given in a table below; the sign a

Then, corresponding data for "active" elements are introduced (e.g., Cu, Zn) to calculate differences vs. the expected value. The same kind of shift happens when equilibrium is perturbed by photochemical processes (mediated Eu (bottom of Figure 1), or U, then changing partition of either element) or by external redox processes, e.g., responding to strongly reducing conditions (Figure 6). These effects can be correlated with observations of enzyme-related product formation/substrate consumption along the interface. Examples from our work include CH_4 formation vs. Ni distribution (open field: a bog located right on the Polish–Czech border W of Mt. Śniezka (Sudety Mountains)) and nitrate reduction vs. uptake of Mo (model study done in our lab; Figure 3):

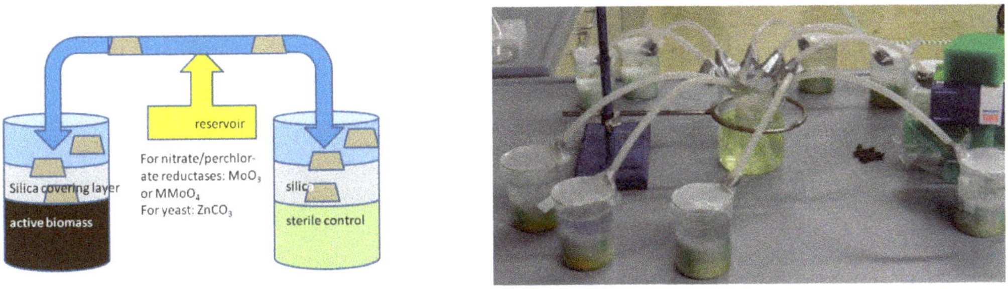

(central beaker with Mo stock solution connected to both active and inert [control] vessels)

(active [left] and inert [right] vessels)

Figure 3. Transfer of Mo from central solution (pale yellow solution in central beaker in top right picture) via filled tubes to beakers where there is glucose, nitrate and nitrate reductase (bacterial pellets) or the latter is missing (sterile control). When there is biological activity, nitrate will be depleted while the levels of Mo inside silica layer are 2.5–2.9 times those on chitin placed in aq. supernatant. In controls (right), $P_{chitin;sedim./chitin;H2O} < 1.5$. Once nitrate is depleted, glucose keeps on reducing Mo turning the supernatant blue (bottom picture, left). Chitin flakes within tubes were omitted, transport occurring simply by diffusion, the "pump effect" of evaporation (and oxidation by O_2 rather than NO_3^-) being limited by Parafilm® covers.

Possible reasons for non-equilibrium according to Figure 1 include photochemistry, biological/biochemical activity involving less-common and less-abundant metals and redox reactions which may cut linkage between chitin strands in either solid polymers or dissolved chitin.

The retention by chitin does also depend on oxidation state of an element. Thus, it will respond to redox or photoredox processes by changes of partition. Accordingly, some reduced (Eu^{2+}) or oxidized ($HCrO_4^-$) forms of an element can be removed/extracted from the chitin layer by some slow flow of solvent, sometimes facilitated by light (reduction of Eu, U) or presence of a natural catalyst such as MnO_2 that promotes air oxidations of Ce, Cr [24], Cl^-, phenols [25], alkyl benzenes, etc., with both solid Mn phases [26] and Mn^{2+}, although weakly adsorbing to chitin [22]). Then, two electrodes placed on either side of the percolated chitin pack will record different potentials due to the concentration drop (scheme Figure 4, practical setup Figure 5 below):

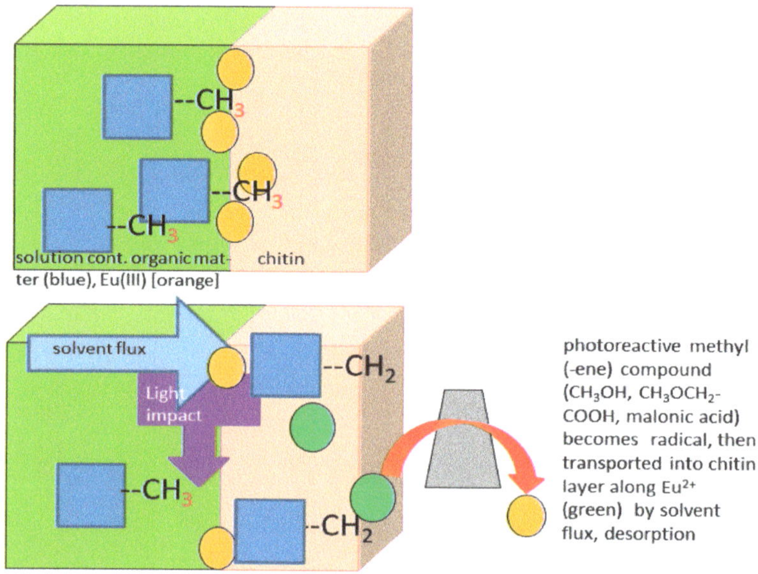

Figure 4. Eu can only pass the solvent/chitin-layer interface if reduced, otherwise it is retained by chitin surface.

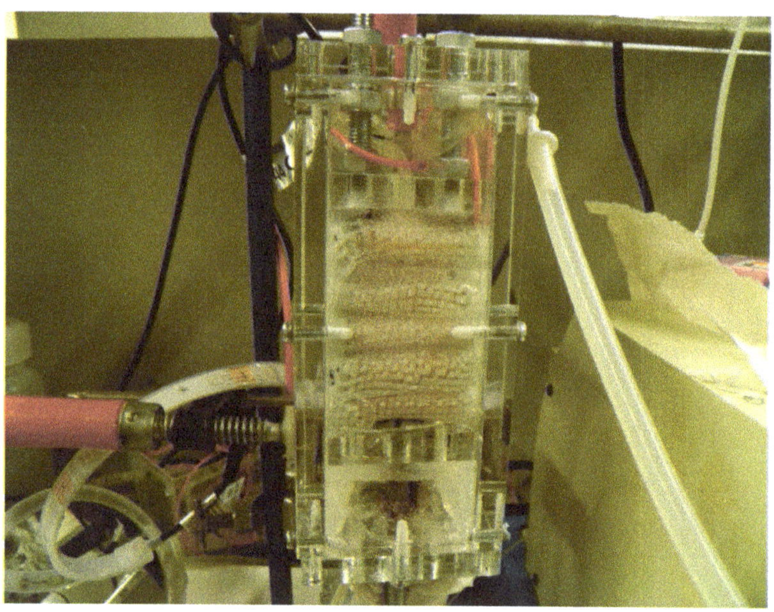

Figure 5. Design of a photo-assisted fuel-cell where excited Eu(III) does abstract hydrogen atoms from organic matter while the metalliferous product Eu^{2+} loses contact with/adsorption to chitin and is eluted "downward" to create an electrochemical potential like shown in Figure 3. When a reductant like ethanolamine spreads by diffusion in an open cell to reach Eu(III) at different times, initial potentials can be >> 100 mV implying that > 99% of Eu are reduced by LED-light before the diffusion front reaches the counter-electrode also. The folded material is glass fiber fabric which was selected to be both inert and transparent for visible light (UV radiation is not required).

Here, sensor technology and energy conversion—meant to induce fuel cell conversion of organic compounds containing CH and NH bonds while not requiring Pt group metals for CH bond activation and storing some light energy by making Eu^{2+} and H_2 (in a secondary step)—do merge.

Beyond the statement that there is retention down to aq. M^{x+} levels < 1 nM/L, it is also feasible to compare chitin to a small ligand molecule the dative structure of which is very similar to that of chitin. One such ligand for which many complex formation constants are known is antibiotic agent chloramphenicol. Except for very simplified models of bidentate C-hydroxo/carboxamide (probably O-) binding such as N-formyl ethanolamine or N-acetylethanolamine, the best-investigated ligand for which a reasonable number of complex formation constants are available is the antibiotic agent chloramphenicol ([27], introduced into clinical applications back in 1949), cp. the key structural motifs which are relevant for M^{n+} complexation. The similarity of binding modes is shown in Figure 6a,b.

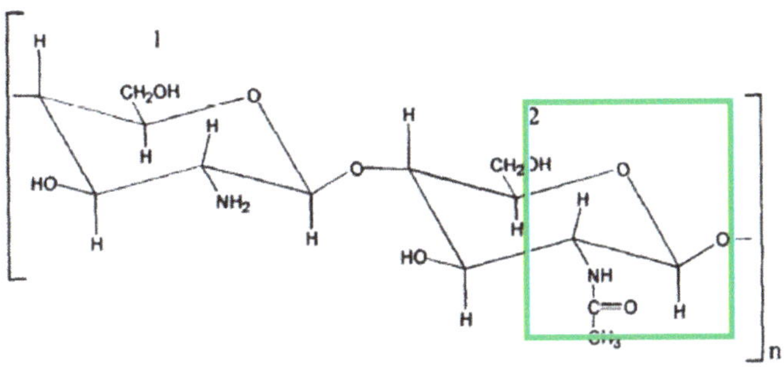

(**a**)

vs. chloramphenicol (bottom)

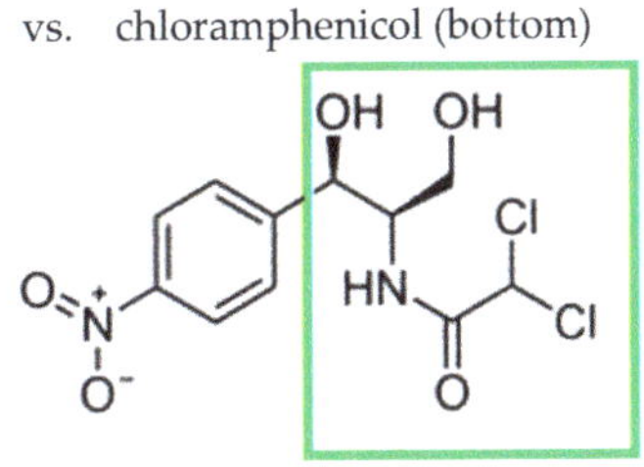

(**b**)

Figure 6. (**a**): Chitin, displaying a free amino group at left saccharide ring (<15% of total N in natural chitin samples, almost 100% in chitosan). (**b**) Chloramphenicol: bidentate binding of metal ions via OH and carboxamide. For complex stabilities see Table 1, note similarity of binding motif. Dichloroacetamide will display a slightly increased NH acidity as compared with acetamide groups, thus possibly enhancing binding of metal ions like Cu, Pt (cp. complexation properties of substituted hydantoins).

Table 1. List of chemicals and materials used.

Substance	Supplier	Purpose	Remarks
chitin	Sigma-Aldrich, Taufkirchen, Germany	sorbent	purified, background levels see text. No electrochemical signal unless metal salts were added
Lithium perchlorate	Sigma-Aldrich, Taufkirchen, Germany	Dissolution of chitin, conducting salt in CV	"battery-grade" purity
Cation exchanger resin	Amberlite H-120, Supelco, Bellefonte, PA, USA	Fixation and transfer of analytes	Checked for trace metal background, nothing detected. Can be re-used several times after leaching ions bound from chitin solution. Retrieval rate about 70% [18]
Nitric acid	Merck Suprapur, Darmstadt, Germany		
Crayfish *Orconectes limosus*			Caught in local waters, also exuvia were studied
Eu(III) trifluoromethanesulfonate	Sigma-Aldrich, Taufkirchen, Germany	Photooxidation reagent	
Solacor glue	Boldt & Co., Wermelskirchen, Germany	Fixation of chitin flakes	Photo-hardening, metal-free glue, rather resistant towards organic solvents including DMF

For the latter monomeric ligand the following complex formation constants (aq., I = 0.1 M, T = 25 °C) do hold (Table 2) ([27], except for lead and iron):

Table 2. Complex formation constants of M^{2+} ions, Fe(III) with chloramphenicol ([24] and other sources).

M^{2+}	Log k	Remarks
Pb	3.26	Pb does accumulate on chitin more than other metals [9,13]
Cd	3.93	background on chitin essentially zero
Mn	3.54	
Ni	3.33	Detection of methanogenesis
Zn	3.16	
Fe(III)	4.13	different oxidation state

The corresponding calculated electrochemical ligand parameter is—0.154 V (bidentate binding). This value would correspond to log k ≈ 5.7 for binding of Cu(II) to chitin using the formula introduced by this author [28]:

$$\text{Log } \beta = x \times E_L(L) + c \tag{2}$$

with x and c tabulated in [28,29]. Equation (2) is formally equivalent to (1) since there is some linear relationship between Hammett´s constant σ and the electrochemical ligand parameter [30,31], as defined by Lever [31] (there are also other scales/benchmark redox systems but these comprehend far less ligands), whereas x and c are linearly correlated and thus can be traced to Equation (1) by corresponding coefficients (suggested in [30] without actually revealing correlation equations). Understanding LREE binding to chitin is most important because it offers the only sizable dataset for trivalent ions (besides of Al, Cr, Bi) unless there is biocatalytic activity associated with some of them [32–35], and because of its application in fuel cells and sensor devices (see below).

However, both x and the slope of the latter equation (not explicitly given by [30])

$$E_L(L) = e_{nd} \times \Sigma\sigma + f_{nd}\ [V] \quad (3)$$

(parameters calculated by SF [36]) depend on denticity d and charge of ligand. Thus, a couple of parameters is required rather than a single set, given in the following Table 3:

Table 3. Relationship between σ and electrochemical ligand parameter for different kinds of ligands.

Kind of Ligands	No. of Ligands in Regression	e [Σσ]	f [V]	Remarks
Monodentate, neutral	7	0.278	0.124	N-, P-, As-, S- and C- (isocyanides) donors, which all are π-bonding (NH_3, aliphatic amines do not correlate properly)
	7	0.265	0.133	omitting pyridine, plus $AsPh_2(CH_3)$
Monodentate anions	15	0.593	−0.402	Halides, pseudohalides including CN^-, OH^-, carboxylates, etc., not H^-
Bidentate, different charges	6	−0.185	−0.176	Negative slope, also includes neutral L en (ethylene diamine)

The advantage of (1) is that many more Hammett constants σ are available (e.g., [28]) than electrochemical ligand parameters [10,29,31].

Most Hammett parameters were taken from literature [31] while complex stability constants for benzene-1,2-dicarboxylic (ortho-phthalic) acid anion, salicylate, and maleate (M = La, Nd, and Eu [37]) gave the following additional values for σ′ (Table 4):

Table 4. Hammett constants of different ligands (calculated using [13]).

Ligand (Bidentate)	Σσ	σ′	Σ of "Normal" Ligands	
o-phthalate	−0.36	−0.18	Carboxylate 0, OH −0.37	
salicylate	−0.40	OH (aromatic) = −0.22		
maleate	−0.30	Acrylic carboxylate −0.15		Cannot be broken up as definition applies to bidentate ligands

An actual chitin polymer strand has M* ≈ 400.000, and the structural motif binding M^{x+} is present about 2000 times. Hence, actual complex formation constants (log) $\beta_{M;chitin}$ are much higher than for complexation by chloramphenicol, agreeing to the fact that even pMol/l levels cause measurable adsorption of element ions.

Chitin-based biomonitoring does require reliable and fast adsorption. In our experiments, equilibrium was usually reached within 10 min. Earlier work had shown that even traces of 'heavy metals', including radionuclides [38] could be removed from water flows and slurries, causing us to explore chitin application chances in environmental analytics. Background levels of metals in commercial chitin obtained by shrimp peeling (*Pandalus borealis*; Sigma-Aldrich, Taufkirchen, Germany) are very low, permitting detections of traces of contaminants in ambient matter, too: levels of Al, Ti, and Fe are some 10–20 μg/g chitin; Cu and Zn come with about 1.5 μg/g regardless of the significant role of either metal in arthropods, while important toxic metals [22] such as Be, Cd, and U cannot easily be detected in the native material, implying almost any trace will be pinpointed. Binding capacity is about 40 μmol/g for most metals, while values for Al [22], Ce or Eu [39], V or Pb [14,22,40] are considerably higher. Most metals for which some biological role was established recently (e.g., [41]) bind rather well, so do Bi and certain others for which no (catalytic) biological role was detected so far.

3. Results

3.1. Photoredox Processes and Retention

Of course, complexation can alter the extent/equilibrium of metal ion retention. Many ligands, however, contain "free" terminal oxygen or nitrogen (carboxylate-, sulfate groups, unbound nitriles in AAN or cyanoanilines, etc.) which form H bonds with OH groups of the polysaccharide. Then, complexation will rather enhance retention of M^{x+} to chitin. Since retention of metal ions by chitin is not just "all (most metal ions, -complexes, down to < 1 mM/L) or nothing (alkali metals, Mg, Ca)", binding rather depends on oxidation states, too; there can be both energy conversion (subsequent to redox- or photochemical changes of oxidation state) and some effect—not generally a perturbation—on adsorption for analytical purposes. Because levels of these ligands used to be low, this does not compromise results on M in normal lakes or ponds. However, it should be noted that the above change of partition might also extend to LREEs as the latter, especially La and Nd, are involved in oxidation of methanol [33,34] and other primary alcohols run by aerobic proteobacteria [35]; that is, this will happen right in the uppermost sediment layer. Except for this case, REEs can be taken—such as and adding to Al, Cr, Bi—to represent physiologically inert M^{3+} for production of such a calibration function. With M^{2+}, it can be Sr, Ba, Co, Ni (at sufficiently high potentials which preclude methanogenesis [40]), Pb, or Cd. Mg and Ca do not bind to chitin, whereas Mn, Cu, and Zn are involved in all life-forms (and thus shall generally "deviate" from the regression obtained in the above Equation (1) in favor of a stronger adsorption within sediment). However, Ba and Pb can be mobilized once there is sulfate reduction. There are no difficulties with coarse, gravel-like sediments or beaches, but fine-grained sediments are more difficult to understand if inundated. Elements commonly occurring in oxidation states +IV and higher normally do not take part in biochemistry (except for Mo, V, and sometimes W [40]).

3.2. Relationship between Donor Groups and Log σ_{para} or Log P Constants for Metal Coordination to Chitin

It is an interesting case to compare chitin in its "physiological place"—covering aquatic or moisture-exposed arthropods (like crayfish, shrimps, ants or bumblebees building nests in moist soil)—to isolated and grafted chitin samples: will adsorption patterns match insofar as measurements on living beings can be attributed to sample either the aq. phase or sediment (which means crayfish to be studied must not dig tunnels in which they live, hide or even reproduce). As can be seen from Table 4, differences against the simple aq. system are small enough to analyze adsorption chemistry/selectivity along the common way of reasoning but large enough to observe effects due to local chemistry. There are no indications for isotopic fractionation of multi-isotopic elements during chitin adsorption; for example, determinations of Sm content on chitin surfaces using isotopes ^{147}Sm, ^{149}Sm, and ^{152}Sm, respectively, provide the same result assuming the normal isotopic composition of samarium [18]; (this does likely permit to reconstruct the whereabouts of chitin-covered organisms by means of Pb [41] isotopic signature). Solvent impact may alter chitin structure [42]; acidity values of potential ligands (protonated forms thereof) in DMF are given in [43] and were used besides own measurements to derive Equations (9) and (10).

Given the binding groups of chitin (or that in any other polymeric ligand) are either obvious (say, in polyacrylonitrile, all being -CN) or can (must) be inferred from complex formation studies, the following equation (4) can be used to accomplish, with (5), referring to partition and assumed to be equivalent to (4), polymer-metal ion interactions which are taken to happen by complexation, with or without additional ligands involved:

$$\text{Log } \beta = a \times \sigma_{para} + b \text{ and} \quad (4)$$

$$\text{Log P} = \text{a} \times \Delta\sigma_{para} \quad (5)$$

where σ_{para} is Hammett´s constant (for para substitution and log P denotes the partition between chitin samples immersed into sediment and adjacent water (e.g., at bottom of some pond or creek or at a beach or bog pool lining), tacitly assuming metal binding to chitin occurs by complexation as with ligands dissolved in some solvent (this is to account for the large differences in metal retention by chitin and by either chitosan or cellulose). Data for meta substitution (e.g., [31]) are linearly correlated with σ_{para}; however, considerably less values are available for σ_{meta}. Some σ_{para} values must be modified as metal ions may bind to ligands in manners not available to withhold the aromatic groups such as phenyl or naphthyl. This also holds for carboxamide or 3-ketoenolate ligands which interact via O with metal ions, rather than forming CC bond, as well as carboxylates, nitrate, sulfate, etc. which would form esters or acid containing a C-O-E or HO-E single bond chain whereas REE^{3+} will bind via two O atoms, which is why there are different descriptions for REE ions.

Equation (3) can be rearranged to calculate an effective value of σ or $\Sigma\sigma$ even without knowing the binding functions, or in cases where M ligation differs from binding modes to aromatics, such as carboxamide or 3-ketoenolate ligands:

$$\text{Log } \beta - \text{b} = \text{a} \times \sigma_{para} \text{ and thus} \quad (6)$$

$$\sigma_{eff} = (\log \beta - \text{b})/\text{a} \quad (7)$$

Hence, log P represents the difference of local chemical conditions between the two phases connected by an ecotone while log β refers to complexation of the ion in the same solvent. This does mean the equilibrium state shown in Figure 1 (yellow double arrow) can exist for log P $\neq 0$ for respective parameters and ion sizes.

3.3. *Effects of Boiling Water on M Retention*

Generally speaking, chitin is very robust against both thermolysis (wet or dry) [12,13] and oxidation [7,14]. This holds for both chemical, using the colloquial reagents for poly- or monosaccharide analysis (which, however, do readily react with chitosan, see above) such as periodate or Pb(IV) acetate, and electrochemical processes (own experiments in DMF solvent containing Li^+). Yet, acetamide groups may be labile towards (hot) water hydrolysis (chitosan is prepared by rough (e.g., warm NaOH) hydrolysis from chitin [2,6]). Such hydrolysis might occur when chitin is exposed to hot (boiling) water [17]. Effects from this were studied in both grafted chitin and crayfish (this is the regular method of euthanizing them, rather than applying chemicals). As a rule, the layer from which metal ions are mobilized by DMF/Li^+ becomes thicker, and thus more metal ions are leached. Certain metals, however, show the opposite behavior, namely when amino groups produced by acetamide hydrolysis do better to bind metal ions, e.g., with Y, Ni or Cu. Complexation of trace metals essential to microorganisms (too) by chitin or chitosan apparently do block growth/multiplication of pathogenic bacteria [44] or yeasts [45].

3.4. *Effects of Ligands on M Retention*

While effects of non-specified ligands were considered before in this paper, adsorption depending on ligand additions was studied for a couple of elements also [18] in the test medium (water/DMF 50/50 *v*/*v*, buffered to $H_0 = 5.8$ by adding anilinium salts. We did study this for M = Mn^{2+}, Ni^{2+} [46], and six REEs (i.e., La, Ce, Sm, Eu, Dy, and Yb [18], also covering Mn). The only ion which generally does lose contact with chitin interfaces upon addition of ligands known or expected to be present in the soil solution (delivered by plant roots, fungal mycelia, or soil bacteria) is Sm(III) introduced as trifluoromethane sulfonate salt, like the other REEs [18]. The buffer was chosen to circumvent the problem that most buffers either do precipitate REEs or else undergo photooxidation by Eu(III)* while retention by chitin depends on pH or H_0 [14,22]. La and Eu do better bind to chitin when glycine was added, while retention of both Ce and HREEs Dy, Yb is increased by adding anyone of many ligands, including humic acids. Some of them are photoactive with Eu(III) while others are not. This will change results of partition measurements (Figure 1 bottom), but it is included in calibration as these measurements were conducted in very dim light.

3.5. Comparison to Chitin on Living Benthic Animals and Their Exuvia

Data on antennae (comparing to open water) and other parts of benthic, non-digging crayfish *Orconectes limosus* showed highest accumulations (BCF values) for elements such as Bi, V, and light REEs from water [17] while sediment/water partition on both chitin flakes and crayfish cover parts which penetrate into sediment (leg tips, telson) is most pronounced for LREEs La, Ce. Knowing that adsorption of "integer" complexes prevails over that of M^{x+} for $M \neq Sm$ in many/most cases [18], $\Sigma\sigma_{para} = -0.44$ for chitin in pure water and calculations from log P of "bioinert" metals in different local waters are feasible. They give an idea of the size of changes caused by presence of (possibly ligating) organics in water and/or sediment of local waters. Samples were obtained in upper Lusatia, Saxony, FR Germany, representing mainly small ponds, lakes, and (lignite, gravel) quarry ponds, and beside small rivers (data from [17], here Table 5); water depth in sampling was 1 m or less:

Table 5. Partition factor P for chitin flakes and comparison telson/antennae, leg tips/antennae for calculation of effective $\Sigma\sigma$ (calculation from a, b for Al, La and Ce. $\Sigma\sigma$ does refer to aq. phase while $\Sigma\sigma'$ denotes conditions in top sediment layers (leg tips, chitin flakes). La^{3+} does accumulate on flakes (FA) much more rapidly than in (leg tips of) living animals (LA) likewise immersed in sediment except when there is lignite. Animals were caught and exuvia collected next to the shore at water depths ≤ 1 m [17]. Exuvia were not "boiled", thus site VII gives a direct comparison on the effect of thermal treatment of chitin.

Site	Description of Site	$\Sigma\sigma$	$\Sigma\sigma'$	Remarks	La Partitions (P) [–]
I	Pond, former lignite pit (12 m max. depth), exuvia	−0.574	−0.532		LA 1.27 FA 0.39
II	Small river	−0.522	−0.475		LA 1.45 FA 6.00
III	Small lake inhabited by tench (strongly digging fishes, introducing O_2 and NO_3^- into sediment); exuvia	−0.566	−0.412	highest value of $\Sigma\sigma'$	LA 0.09 FA 1.32
IV	Shallow pond, very rich in SO_4^{2-}; many exuvia could be sampled	−0.621	−0.508	REEs form sulfatocomplexes; lowest value of $\Sigma\sigma$	LA 0.63 FA 10.93
V	Small river	−0.401	−0.441	pH > 8; highest value of $\Sigma\sigma$, hard sediment	LA 2.19 FA 9.83
VI	Lake, former lignite pit (41 m deep)	−0.470	−0.573	pH > 8	LA 2.60 FA 0.42
VIIa	Flooded former basalt quarry, exuvia	−0.611	−0.446		LA 0.36 FA 0.295
VIIb	Flooded former basalt quarry, living crayfish	−0.582	−0.459	Either value of $\Sigma\sigma$ is very similar to those of VIIa	LA 0.29 FA 4.47

Antennae (*Orconectes limosus*) are distinguished by $\Sigma\sigma$ somewhat lower than for both pure water-chitin and leg tips penetrating top sediment layers; the shift generally is negative and of order −0.1 to −0.2. It is difficult to explain this value into much detail, yet this does mean local organic contributions will hardly change or limit application of the method. The two sets of effective $\Sigma\sigma$ are related according to Equation (8)

$$\Sigma\sigma = -\ 0.321\ \Sigma\sigma' - 0.666 \quad (8)$$

obtained from linear regression of the above values. Accordingly, partition between (a) leg tips and (b) telson (getting contacted with sediment while normally not penetrating it) on one side and antennae (exclusively exposed to water given the living habits of this crayfish species) is correlated but in a negative way, indicating that telsons of *O. limosus* are kept in water for most of the time, unlike leg tips.

4. Discussion

Table 6 gives the parameters required for calculation/prediction of complex formation constants (empirical data: [47,48]) of several trivalent metals, including Al, Cr and many REEs. For Eu [48], but in theory also other REEs such as Ce, Yb, Sm, or Nd using appropriate solvents, electrochemical data from [47,49] can be used to calculate complex formation constants in DMF. However, there are some difficulties: the reduction signal of $Yb^{3+/2+}$ [48] in DMF does match that of the polysaccharide protons of chitin in same solvent at −1.5 V whereas Ce gives poorly reproducible data on adsorption unless the redox state is tightly controlled. Reduction of Sm is feasible when there are ligands, though a and b in Equations (3) and (7) are empirical parameters, which in turn were calculated by this author using tables of complex formation constants (data from [42,43]). The sum of Hammett parameters (taken from [31]) for the ligand donor sites (e.g., 0 [carboxylate]–0.66 [-NH_2 group] for glycinate or anthranilate, −0.75 for sarcosinate or −0.37 for glycolate or lactate, 2 × [−0.66] = −1.32 for ethylene diamine, etc.) can be compared to partition for environmental sites. For HOX acids (except of water, H_2O_2, HSO_4^-) (e.g., nitric, formic, benzoic, methanesulfonic, hydroxamic acids, phenol, glyconitrile, or HCO_3^-) the following equation applies (own calculation):

$$pKa = 9.97 - 12.48\ \sigma_X \tag{9}$$

Table 6. a and b parameters for mono- to tridentate ligands and certain trivalent ions Al, Cr, several REEs. The values in square brackets for a_{1d} and b_{1d} hold for OX^- donors excluding nitrate (carboxylates, sulfate, carbonate, phenolate), for bidentate ligands complex formation constants for glycinate, glycolate, oxalate, or hydroxamates were used. For chitin, $\Sigma\sigma \approx -0.44$ and bidentate binding do hold.

M(III)	a_{1d}	b_{1d}	a_{2d}	b_{2d}	a_{3d}	b_{3d}
Al			12.48	6.92		
Cr			−8.50	4.52		
Y			2.57	5.67		
La	[−10.08]	[6.90]	8.32	6.21	−2.52	3.09
Ce	−8.00 [−11.50]	2.89 [7.58]	3.70	2.56	−2.58	3.31
Pr			3.95	2.57	−2.74	3.35
Nd			4.02	2.60	−2.77	3.43
Sm	[−11.07]	[7.54]	8.08	6.02		
Eu	−8.98 [−11.92]	3.44 [7.97]	9.45	7.02		
Gd	[−11.55]	[7.71]	9.91	7.05		
Dy	[−14.55]	[8.90]	9.65	7.04		
Tm	[−12.58]	[8.02]				
Yb	[−13.22]	[8.29]				
Lu	[−12.23]	[7.86]	10.59	7.30		

CN = 6; CN = 9; variable CN; CN = 8, orange: tentative values.

This does describe adsorption performance of chitin somewhat influenced by ambient ligands which may bind to REEs and other metal ions. This can either produce some complex withhold in its dissolved state (e.g., with Sm) and/or (on the contrary) facilitate/increase adsorption by stronger adsorption of large complex molecules. Ligands may even act as anchor groups connecting metal ions and chitin, such as with glycine and other amino acids [18,22]. Values used for calibration and in Equations (4), (6)–(8) were calculated and most of them (the pertinent ones) are given in Table 6 (additional values can be obtained from Equation (3), Table 4 and data in [28]).

These data then are used to identify binding sites from observed sequences of binding preferences, e.g., BCF values following Equation (2).

This gives an account for differences in binding of REEs in different organs of a plant or animal or fungus; here, we limit our focus of attention to chitin. Chitin is one of the ligands where the modified formula must be used for its containing carboxamide ligands. Interestingly, metal retention to chitin does produce an additional reductive half-wave in DMF/Li^+ solutions regardless of the kind of metal which was bound [14,17]. It is attributed to an increased acidity of sugar OH groups caused by complexation. This transition is located at -1.50 V vs. SCE which translates into -1.07 V vs. NHE in DMF (the larger difference vs. SCE owing to the higher proton affinity of this solvent) and thus $pK_{a(M^{X+}\text{-}OH);\ dmf} \approx 18$, comparable to anion acids (HSO_4^-, $HC_2O_4^-$) [43] or glycine (own measurements) in this solvent. Using the equations

$$pK_{a(dmf)} = 23.48 - 22.91\ \sigma_{para} \tag{10}$$

$$pK_{a(dmf)} = 10.82 - 9.12\ \sigma_R \text{ for RCOOH} \tag{11}$$

derived by us for acidities of oxoacid solutions in DMF, one can estimate σ_{para} for the metal-loaded OH groups to be + 0.24 (comparing to metal-free OH = -0.37). By now, there are no σ_{para}-based regressions for all the metals studied concerning their interactions with chitin, but there is a focus on those which are most important in our own work.

4.1. Photoassisted Redox Reactions Occurring When Adsorbed to Chitin

Eu(III) can promote photooxidation (cleavage of CH bonds) of organics and become reduced also when it is adsorbed to chitin [50]. Besides being injected by vegetation (oxalic acid also by lichens [10]), chemical species enhancing or bridging adsorption of Eu can be subject to degradation by bacteria (humic acids also by earthworms). Potential changes between the two chitin-modified electrodes thus must be anticipated and are going to be studied in our continuing experiments (also see outlook). Applications are concerned with sensors rather than fuel cells.

Among established substrates of photooxidation by Eu(III), the only one to enhance adsorption is glycine while humic acid, caffeic acid, citric and malonic acid all do reduce binding of Eu towards chitin. Generally speaking, chitin´s adsorption capacity for Eu (and Ce, Dy, Pb) is considerably higher than for other ions including other REEs [19,37]. Hence, one would predict that glycine photo-processing to afford ethanolamine, HCN and CO_2 [51] would mainly occur at the chitin/solvent interface while the photooxidations of the other substrates take place in solution mainly. However, retention means an electrode placed at the top of chitin or right in it will maintain a rather positive potential, then acting as an anode while the leached Eu^{2+} gives a downstream electrode potential more positive than standard (cp. [48,50]) and thus makes the downstream electrode the cathode. We use Ta (foil or wire) as an electrode material [50] to avoid both

(1) heterogeneous-catalytic electrotransformations of organic compounds (to be anticipated on Pt) and
(2) electrode corrosion (sizable with Cu).

4.2. Change of Chitin Functions during Evolution

The distribution of chitin among taxa [52] does strongly suggest it was present in many animals or their predecessor taxa in late Precambrian much before any hard structures other than mouthpieces (of putative mollusk *Kimberella*) were formed from it, omitting open issues on taxonomy of late Precambrian organisms capable of active movement and on details of the Cambrian explosion. When there was chitin way back in the Precambrian with common antecessors of certain pairs of taxa separating then (i.e., 550–575 mio. y BP) or even earlier (fungi and animals) while totally or mainly absent in others, genetic data of either can be used to determine when it came into being even though chitin biosynthetic pathways in fungi and in animals differ somewhat while those in animals are conserved. However, the sorbent properties of chitin and the origins of mouthpieces combine to yield a meaningful interpretation: in the very beginnings, namely feeding, or retrieval of trace mineral nutrients partially relied on placing chitin fibers/appendices into the sediment. Covering structures, antennae, or legs would form only later (i.e., during Cambrian at the earliest [12,13,52–55]). Yet, given the data on adsorption extent [3,14,22] and kinetics [10,18,22], combined with the chance to leach them from some chitin surface by either acid or complexation, that is, by licking. Besides organisms such as *Hallucigenia sparsa*, it is telling that gill baskets of lancelets also contain chitin, which otherwise is rare in "genuine" vertebrates [55].

In previous and further course of evolution, chitin-bearing organisms developed many modes of interactions with those displaying cell walls that consist of other materials. Later, symbioses rendered chitin and its adsorption features available to organisms uncapable of making it themselves, as in plants (algae, cyanobacteria) teaming with fungi. Plants (*Archaeoplastida*) and cyanobacteria now make use of chitin´s unique properties by way of (mostly obligate) symbiosis with fungi (lichens). In arboreal mykorrhiza there is another version of the same effect, with mykorrhiza known to be involved in transport of metals, too, implying adsorption and desorption of ambient metal ions by chitin in mycelia. Different kinds/materials of cell walls are also shown in the diagram (Figure 7):

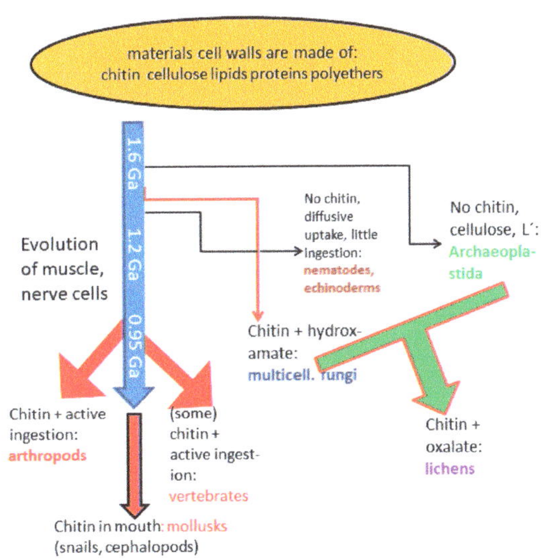

Figure 7. Chitin acts as a nutrient- and metal-retrieving sorbent within soft sediment (Ediacaran creatures purported to represent early animals used to be benthic or dwell within the sediment rather than swimming). (Multicellular) fungi (yeasts do not make chitin) use the same effect in mycelia. Here, uptake, transport, and release are likewise controlled by redox processes. REE contents in lichens may be rather high. Other kinds of cell walls existing in bacteria (*B. subtilis*) can discriminate and enrich terminal HREEs Tm, Yb, and Lu against all lighter REEs and other trivalent ions.

5. Conclusions

The prevalent role of chitin in terms of forming the outer covers, locomotive organs, and mouthpieces in a number of species which exceeds any other group by far (there are >1.4 mio. recognized species of beetles alone, even disregarding other arthropods) and other biopolymers by orders of magnitude in quantity can be traced back to its sorbent properties [3,5,34,54]. These can be exploited far beyond simple purification of wastewater flows [5,38,54]. Since adsorption depends on the oxidation state of some metal ions, such as Eu, Mn, and U, charge separation producing electrochemical potentials can be achieved by passing a slow solvent flow through a chitin layer where chemical or photochemical reduction [50] or oxidation takes place. Because chitin is plentiful, optically transparent and readily available as scrap, and besides being very robust towards both oxidation and thermolysis, such devices should find their fields of application, e.g., in fuel cells and environmental sensors. It is anticipated that knowledge of such effects will improve understanding of peculiar effects seen in arthropod outer covers such as photovoltaic or thermoelectric activity in hornet *Vespa orientalis* [55].

6. Outlook: Additional Studies in Lab and Open-Field

Studies can be conducted in very different conditions, even without exceeding the limits of "extremophilic" arthropods (*Artemia salina*, crabs and shrimps inhabiting rather hot water pools) or lichens in the Antarctic given the chemical and thermal robustness of chitin. Certain uncommon biotopes and their adapted populations are particularly vulnerable, urging use of the chitin method for studies of local conditions, pollution inputs and metal mobilities, e.g., associated with formation of chloro-, sulfato-, or carbonatocomplexes in cold to mesothermal salinic media. These are inhabited by both arthropods and some fish (*Cyprinodon* spp.).

(a) Studying Salinic Conditions in Water Bodies

Among the vast number of living organisms covered by chitin, brine shrimp *Artemia salina* is easy to grow and keep (endures up to > 20% NaCl), enabling its use as a mobile electrode carrier in the following setup (Figure 8):

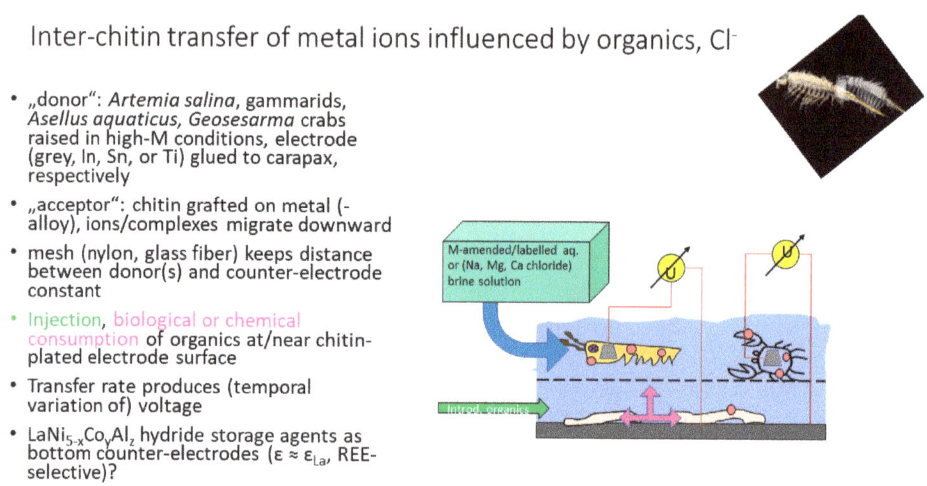

Figure 8. Animals swimming above (**left**) or running on (**right**) the separating mesh are kept at a given distance from the counter-electrode. This does control carry-over of metal ions from metal ions to chitin fixed to bottom electrode.

A "freshwater version" could be realized with gammarids or aq. woodlice (*Asellus aquaticus*) which are notorious for their robustness concerning both acidic pH and chemicals including dissolved chlorine. Other test organisms are crabs inhabiting extreme-condition tiny water pools such as phytotelmata (*Geosesarma* spp.) or—for moderately salinic counterparts —hermit crabs frequently inhabiting tidal rock pools.

(b) Adsorption vs. Transport around and Inside Wells, Springs, and Seepage Sites

Detection of non-equilibrium can be extended to artificial constructions as shown above (Figure 1) for Mo dynamics related to nitrate reductase and likely to wells (Figure 9). Microorganisms and excited Eu(III) compete for the same substrate to be oxidized (glucose, ethanolamine) and nitrate is either used by REE ions to induce (catalyze) nitrations of aromatics [56] such as benzyl alcohol, alkyl phenols or C,C-diphenylglycine. Besides REE catalysis of aromatic nitrations, nitrate can become reduced by nitrate-reducing bacteria (NRBs, chemolithotrophs) whereas chitin flakes in water (top and inside well) and sediment are used to measure partition. Controls are obtained in darkness or omitting living NRBs (Figure 9):

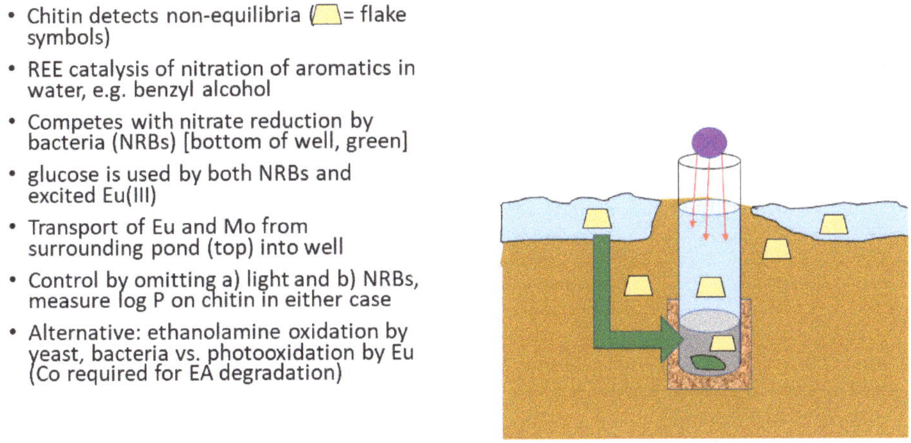

Figure 9. Model of a natural spring where biorelevant trace metals and Eu pass through a sediment layer (e.g., the support of a small island) before re-appearing in a spring or well. Transport and biochemical processes can be surveyed by both chitin adsorption/partition on flakes and photoreduction of Eu.

Solid matrices such as gelatin enable activation of compounds which do not undergo photooxidation in solutions, such as aniline, salicylic acid, and N,N-dimethyl glycine [57]. Both occurrence of this reaction and products may differ depending on the surroundings, including feedback to chitin flakes (Figures 8 and 9). We plan corresponding experiments in the near future.

Author Contributions: Data curation, F.B.; Formal analysis, S.F.; Writing manuscript, S.F.; Supervision, S.F. All authors have read and agreed to the published version of the manuscript.

Funding: Open Access Funding by the Publication Fund of the TU Dresden.

Institutional Review Board Statement: Not applicable.

Informed Consent Statement: There were no experiments on human beings or related to their medical treatment, so consent neither need nor could be obtained.

Data Availability Statement: Not applicable.

Acknowledgments: First of all, my appreciation and thanks go to my enthusiastic team involved in this research since 2013, also being aware of what this might mean in terms of biology and evolution (cp. the theses by A. Bauer, F. Blind, P. Budelmann, M. Erler, T. Gebauer, D. Retschke, and Y.J. Luoga), and of course to the people who did the analyses and helped to develop the corresponding methods, finally transferring chitin´s fascinating properties into the field of energy conversion topics. In addition, I owe much inspiration to colleagues from abroad, especially Italy, Mongolia and Russia (Stefano Loppi, Andrea Vannini, Marina Frontasyeva, Ariuna Lkhagva).

Conflicts of Interest: The authors declare no conflict of interest.

Abbreviations

aq.	aqueous, aquatic (refers to living animals)
DMF	N,N-dimethylformamide (solvent)
DMSO	dimethyl sulfoxide (solvent)
$E_L(L)$	electrochemical ligand parameter; denotes shift of redox potential by replacing member(s) of ligand sphere in $Ru^{II/III}$ system
FUV	far ultraviolet ($\lambda < 300$ nm)
H_0	"effective" pH measured in low-water or nonaqueous solvents. Range about–21 ("magic acid") to > + 30 (strong base solutions in DMSO or liquid ammonia)
HFIP	hexafluoroisopropanol (solvent, does dissolve chitin, extremely transparent in FUV, radiation passes down to some 170 nm)
HREE	heavy rare-earth elements (Z = 66–71 [Dy ... Lu])
ICP-MS	mass spectrometry using an inductively coupled plasma for sample ionization
LREE	light rare earth elements (Y and Z = 57–63 [La ... Eu])
M	metal
M*	molecular mass
M^{x+}	metal ion in an unspecified oxidation state x
NRB	nitrate-reducing bacteria
OX^-	oxoanion, e.g., nitrate ($X = NO_2$) or acetate ($X = CH_3CO$)
REE	rare earth elements (Z = 39; 57–71)
$\Sigma\sigma$	sum of Hammett constants for a chemical entity (ligand) interacting with some other via two or more functional groups described by their respective Hammett parameters σ.

References

1. Zhao, D.; Yu, S.; Sun, B.; Gao, S.; Guo, S.; Zhao, K. Biomedical Applications of Chitosan and Its Derivative Nanoparticles. *Polymers* **2018**, *10*, 462. [CrossRef] [PubMed]
2. Yamada, K.; Chen, T.; Kumar, G.; Vesnovsky, O.; Topoleski, L.D.T.; Payne, G.F. Chitosan Based Water-Resistant Adhesive. Analogy to Mussel Glue. *Biomacromolecules* **2000**, *1*, 252–258. [CrossRef] [PubMed]
3. Anastopoulos, I.; Bhatnagar, A.; Bikiaris, D.N.; Kyzas, G.Z. Chitin Adsorbents for Toxic Metals: A Review. *Int. J. Mol. Sci.* **2017**, *18*, 114. [CrossRef] [PubMed]
4. Barros, F.C.; Vasconcellos, L.C.G.; Carvalho, T.V.; do Nascimento, R.F. Adsorption of organics to chitin: Removal of Petroleum Spill in Water by Chitin and Chitosan. *Orbital-Electron. J. Chem.* **2014**, *6*, 70–74. [CrossRef]
5. Pinto, P.X.; Al-Abed, S.R.; McKernan, J. Comparison of the efficiency of chitinous and ligneous substrates in metal and sulfate removal from mining-influenced water. *J. Environ. Manag.* **2018**, *227*, 321–328. [CrossRef]
6. Pearlmutter, N.L.; A Lembi, C. Localization of chitin in algal and fungal cell walls by light and electron microscopy. *J. Histochem. Cytochem.* **1978**, *26*, 782–791. [CrossRef]
7. Zitko, V.; Bishop, C.T. Oxidation of polysaccharides by lead tetraacetate in dimethyl sulfoxide. *Canad. J. Chem.* **1966**, *44*, 1749–1756. [CrossRef]
8. Liu, P.; Liu, H.; Schäfer, T.; Gutmann, T.; Gibhardt, H.; Qi, H.; Tian, L.; Zhang, X.C.; Buntkowsky, G.; Zhang, K. Unexpected selective alkaline periodate oxidation of chitin for the isolation of chitin nanocrystals. *Green Chem.* **2020**, *23*, 745–751. [CrossRef]
9. Machałowski, T.; Amemiya, C.; Jesionowski, T. Chitin of Araneae origin: Structural features and biomimetic applications: A review. *Appl. Phys. A* **2020**, *126*, 1–17. [CrossRef]
10. Loppi, S.; Vannini, A.; Monaci, F.; Dagodzo, D.; Blind, F.; Erler, M.; Fränzle, S. Can Chitin and Chitosan Replace the Lichen *Evernia prunastri* for Environmental Biomonitoring of Cu and Zn Air Contamination? *Biology* **2020**, *9*, 301. [CrossRef]
11. Arthur Stankiewicz, B.; van Bergen, P.F.; Duncan, I.J.; Carter, J.F.; Briggs, D.E.; Evershed, R.P. Recognition of chitin and proteins in invertebrate cuticles using analytical pyrolysis/gas chromatography and pyrolysis/gas chromatography/mass spectrometry. *Rapid Comm. Mass Spectrom.* **1996**, *10*, 1747–1757. [CrossRef]
12. Ehrlich, H.; Rigby, J.K.; Botting, J.P.; Tsurkan, M.V.; Werner, C.; Schwille, P.; Geisler, T. Discovery of 505-million-year old chitin in the basal demosponge *Vauxia Gracilenta*. *Nat. Sci. Rep.* **2013**, *3*, 3497. [CrossRef]
13. Gupta, N.S.; Michels, R.; Briggs, D.E.G.; Evershed, R.P.; Pancost, R.P. The organic preservation of fossil arthropods: An experimental study. *Proc. Biol. Sci.* **2006**, *273*, 2777–2783. [CrossRef] [PubMed]

14. Fränzle, S.; Erler, M.; Blind, F.; Ariuntsetseg, L.; Narangarvuu, D. Chitin adsorption in environmental monitoring: Not an alternative to moss monitoring but a method providing (lots of) bonus information. *J. Sci. Arts. Univ. Valahia Chem.* **2019**, *19*, 659–674.
15. Hassanzadeh, P.; Sun, W.; De Silva, J.P.; Jin, J.; Makhnejia, K.; Cross, G.; Rolandi, M. Mechanical properties of self-assembled chitin nanofiber networks. *J. Mater. Chem. B* **2013**, *2*, 2461–2466. [CrossRef] [PubMed]
16. Bell, H.L.; Nebeker, A.V. Preliminary studies on the tolerance of aquatic insects to low pH. *J. Kans. Entomol. Soc.* **1969**, *42*, 230–236.
17. Budelmann, P. Verbreitung der Flusskrebse (Decapoda) in der südlichen Oberlausitz und die Eignung des invasiven Kamberkrebses (*Orconectes limosus*) für Chitin-Basiertes Monitoring von Schwermetallen in Limnischen Ökosystemen. Master's Thesis, IHI Zittau/TU Dresden, Zittau, Germany, 2021.
18. Gebauer, T. Methodische Optimierung des Übertrags von Metallionen aus Umweltprobenmodellen auf Chitinoberflächen und von Diesen zu Zwecken Analytischen Biomonitorings Sowie Untersuchungen zur Diffusion/Ausbreitung von Analyten in Chitinproben. Master's Thesis, IHI Zittau/TU Dresden, Zittau, Germany, 2016.
19. Retschke, D. Orientierende Untersuchungen zur Adsorption von Schwermetallen (Nickel) unter dem Einfluss Ausgewählter Komplexliganden Sowie in Arealen Potenzieller und Manifester Methanogenese. Master's Thesis, IHI Zittau/TU Dresden, Zittau, Germany, 2017.
20. Kim, Y.J.; Park, C.R. Analysis of problematic complexing behavior of ferric chloride with N,N-Dimethylformamide using combined techniques of FT-IR, XPS, and TGA/DTG. *Inorg. Chem.* **2002**, *41*, 6211–6216. [CrossRef] [PubMed]
21. Masui, M.; Sayo, H.; Tsuda, Y. Anodic Oxidation of Amines. Part 1. Cyclic Voltammetry of Aliphatic Amines at a Stationary Glassy-carbon Electrode. *J. Chem. Soc. B* **1968**, 973–976. [CrossRef]
22. Erler, M. Untersuchung des Bindungsverhaltens Ausgewählter Elemente und Ihrer Bodenrelevanten Komplexe an Chitin. Master's Thesis, IHI Zittau/TU Dresden, Zittau, Germany, 2020.
23. Fränzle, S.; Bauer, A. This observation was not published in print but reported in conference lectures. 2014.
24. Bauer, A. Orientierende Untersuchungen zur Bindung von Metallionen an Chitin und zur Davon Abhängigen Eignung von Arthropoden zur Bestimmung von Metallionenkonzentrationen in der Umwelt. Master's Thesis, IHI Zittau/TU Dresden, Zittau, Germany, 2014.
25. Chen, K.; Bocknek, L.; Manning, B. Oxidation of Cr(III) to Cr(VI) and Production of Mn(II) by Synthetic Manganese(IV) Oxide. *Crystals* **2021**, *11*, 443. [CrossRef]
26. Nico, P.S.; Zasoski, R.J. Mn(III) Center Availability as a Rate Controlling Factor in the Oxidation of Phenol and Sulfide on δ-MnO_2. *Environ. Sci. Technol.* **2001**, *35*, 3338–3343. [CrossRef]
27. Sharma, S.; Rai, O.P. Evaluation of stability constants of metal complexes with chloramphenicol. *Intern. J. Adv. Sci. Res.* **2017**, 6. [CrossRef]
28. Fränzle, S. *Chemical Elements in Plants and Soil*, Habilitation thesis of SF, extended and translated version, originally submittted to Vechta Univ. (Vechta, Germany) in 2007 which gave *venia legendi*; Springer: Berlin/Heidelberg, Germany, 2010.
29. Fränzle, S. Estimating and predicting chemical potentials, distributions, speciation modes and mobilities of radiometals in soil, water and biomass. *J. Environ. Radioact.* **2013**, *102*, 109–116. [CrossRef]
30. Masui, H.; Lever, A.B.P. Correlations between the ligand electrochemical parameter, $E_L(L)$, and the Hammett substituent parameter, sigma. *Inorg. Chem.* **1993**, *32*, 2199–2201. [CrossRef]
31. Hansch, C.; Leo, A.; Taft, R.W. A Survey of Hammett Substituent Constants and Resonance and Field Parameters. *Chem. Rev.* **1991**, *97*, 165–195. [CrossRef]
32. Lever, A.B.P. Electrochemical parametrization of metal complex redox potentials, using the ruthenium(III)/ruthenium(II) couple to generate a ligand electrochemical series. *Inorg. Chem.* **1990**, *29*, 1271–1285. [CrossRef]
33. Pol, A.; Barends, T.R.M.; Dietl, A.; Khadem, A.F.; Eygensteyn, J.; Jetten, M.S.M.; den Camp, H.J.M.O. Rare earth metals are essential for methanotrophic life in volcanic mudpots. *Environ. Microbiol.* **2013**, *16*, 255–264. [CrossRef] [PubMed]
34. Picone, N.; den Camp, H.J.M.O. Role of rare earth elements in methanol oxidation. *Curr. Opin. Chem. Biol.* **2019**, *4*, 39–44. [CrossRef]
35. Featherston, E.R.; Cotruvo, J.A., Jr. The biochemistry of lanthanide acquisition, trafficking, and utilization. *Biochim. Biophys. Acta (BBA)-Mol. Cell Res.* **2021**, *1868*, 118864. [CrossRef] [PubMed]
36. Fränzle, S. Hitherto unpublished work. Correlations between (sum of) Hammett constant(-s) of ligand donor groups and the electrochemical ligand parameter (which is defined by dividing total potential effect by denticity of ligand). 2020.
37. Vijayaraghavan, K.; Balasubramanian, R. Single and binary biosorption of cerium and europium onto crab shell particles. *Chem. Eng. J.* **2010**, *163*, 337–343. [CrossRef]
38. Muzzarelli, R.A.A.; Rocchetti, R.; Marangio, G. Separation of zirconium, niobium, cerium and ruthenium on chitin and chitosan columns for the determination of cesium in nuclear fuel solutions. *J. Radioanal. Chem.* **1972**, *10*, 17–25. [CrossRef]
39. Karraker, R.H. Stability Constants of Some Rare-Earth-Metal Chelates. Ph.D. Thesis, Iowa State University, Ames, IA, USA, 1961.
40. Dmytryk, A.; Tuhy, L.; Samoraj, M.; Chojnacka, K.W. Biological Functions of Cadmium, Nickel, Vanadium, and Tungsten. In *Recent Advances in Trace Elements*; Chojnacka, K.W., Saeid, A., Eds.; Wiley: Hoboken, NY, USA, 2018; pp. 219–234. [CrossRef]
41. Gonzalez-Davila, M.; Santana-Casiano, M.; Millero, F.J. The Adsorption of Cd(II) and Pb(II) to Chitin in Seawater. *J. Colloid Interf. Sci.* **2009**, *137*, 102–107. [CrossRef]

42. Ramirez-Wong, D.; Ramirez-Cardona, M.; Sánchez-Leija, R.J.; Rugerio, A.; Mauricio-Sánchez, R.A.; Hernández-Landaverde, M.A.; Carranza, A.; Pojman, J.A.; Garay-Tapia, A.M.; Prokhorov, E.; et al. Sustainable-solvent-induced polymorphism in chitin films. *Green Chem.* **2016**, *18*, 4303–4311. [CrossRef]
43. Kolthoff, I.M.; Chantooni, M.K.; Smagowski, H. Acid-base strengths in N,N-dimethylformamide. *Anal. Chem.* **1970**, *42*, 1622–1628. [CrossRef]
44. Abdel-Rahman, R.M.; Hrdina, R.; Abdel-Mohsen, A.-M.; Fouda, M.; Soliman, A.; Mohamed, F.; Mohsin, K.; Pinto, T.D. Chitin and chitosan from Brazilian Atlantic Coast: Isolation, characterization and antibacterial activity. *Int. J. Biol. Macromol.* **2015**, *80*, 107–120. [CrossRef] [PubMed]
45. Bernabé, P.; Becherán, L.; Cabrera, G.; Nesic, A.; Alburquenque, C.; Tapia, C.V.; Taboada, E.; Alderete, J.B.; Ríos, P.D.L. Chilean crab (Aegla cholchol) as a new source of chitin and chitosan with antifungal properties against Candida spp. *Int. J. Biol. Macromol.* **2020**, *149*, 962–975. [CrossRef] [PubMed]
46. Wood, S.A. The aqueous geochemistry of the rare-earth elements: Critical stability constants for complexes with simple carboxylic acids at 25 °C and 1 bar and their application to nuclear waste management. *Eng. Geol.* **1993**, *34*, 229–259. [CrossRef]
47. Couffin, F. *Potentiels d 'oxydo-reduction des elements lanthanides et actinides dans les solvants organiques*; CEA (Internal report (data collection) for CEA (Centre d´Energie Atomique), France. Does cover all lanthanides, many sol-vents including DMF, and actinides Th, Pa, U ... Cm, Cf in DMF, DMSO, ethylene diamine, or tributyl phosphate; CEA: Saclay, France, 1980.
48. Holleck, L. Zur Komplexchemie der Seltenen Erden. Die polarographischen Strompotentialkurven des Europiums als Nachweis-mittel komplexer Bindung. *Z. für Naturf. B* **1947**, *2*, 81–89. [CrossRef]
49. Blind, F. Orientierende Untersuchungen zur Platinmetall freien Aktivierung von CH-Bindungen für Europium basierte Brennstof-fzellenanwendungen. Master's Thesis, IHI Zittau/TU Dresden, Zittau, Germany, 2018.
50. Blind, F.; Kluge, M.; Fränzle. Unpublished GC/MS observation using silylation of photoredox products in DMF and alkylation of ions like cyanide.
51. Cody, G.D.; Gupta, N.S.; Briggs, D.E.; Kilcoyne, A.; Summons, R.E.; Kenig, F.; Plotnick, R.E.; Scott, A. Molecular signature of chitin-protein complex in Paleozoic arthropods. *Geology* **2011**, *39*, 255–258. [CrossRef]
52. Keleşoglu, S. Comparative Adsorption Studies of Heavy Metal Ions on Chitin and Chitosan Biopolymers. Master's Thesis, Izmir School of Technology, Izmir, Turkey, 2007.
53. Jaafarzadeh, N.; Mengelizadeh, N.; Takdastan, A.; Farsani, M.H.; Niknam, N.; Aalipour, M.; Hadei, M.; Bahrami, P. Biosorption of heavy metals from aqueous solutions onto chitin. *Int. J. Environ. Health Eng.* **2015**, *4*, 7. [CrossRef]
54. Sannasi, A.; Hermann, H.R. Chitin in the cephalochordata, Branchisotoma floridae. *Cell. Mol. Life Sci.* **1970**, *26*, 351–352. [CrossRef]
55. Ishay, J.; Benshalom-Shimony, T.; Ben-Shalom, A.; Kristianpoller, N. Photovoltaic effects in the Oriental hornet, Vespa orientalis. *J. Insect Physiol.* **1992**, *38*, 37–48. [CrossRef]
56. Shi, M.; Cui, S. Perfluorinated rare earth metals catalyzed nitration of aromatic compounds. *J. Fluor. Chem.* **2002**, *39*, 551–556. [CrossRef]
57. Pulsfort, J. Biopolymere als Potenzielle Substrate der Photooxidation Durch f-Block-Metalle. Master's Thesis, IHI Zittau/TU Dresden, Zittau, Germany, 2021.

polysaccharides

Article

Physicochemical Properties of Chitosan from Two Commonly Reared Edible Cricket Species, and Its Application as a Hypolipidemic and Antimicrobial Agent

Morgan Malm and Andrea M. Liceaga *

Protein Chemistry and Bioactive Peptides Laboratory, Department of Food Science, Purdue University, 745 Agriculture Mall Dr., West Lafayette, IN 47907, USA; mmeiser@purdue.edu
* Correspondence: aliceaga@purdue.edu; Tel.: +1-765-496-2460

Abstract: Insect-derived chitin and chitosan have gained interest as alternative sources to that derived from crustaceans; however, little information is available on chitin from the house cricket (*Acheta domesticus*) and tropical banded cricket (*Gryllodes sigillatus*), two cricket species commonly reared in the United States for human consumption. In this study, chitin was successfully isolated and purified from these two cricket species; using FTIR, chitins were found to be in alpha-crystalline form. Cricket chitosan was produced from both species with varying degrees of deacetylation (DDA) by varying alkaline conversion duration. *G. sigillatus* chitosan was larger (524 kDa) than *A. domesticus* chitosan (344 kDa). Both cricket chitosans showed similar ($p > 0.05$) lipid-binding capacity to that of shrimp chitosan. Both chitosans were as effective at inhibiting microbial growth of surrogate foodborne pathogens as the commercial shrimp chitosan. At a concentration of 0.50 mg/mL cricket chitosan, approximately 100% of *Listeria innocua* growth was inhibited, due to a contribution of both chitosan and the solvent-acetic acid. At the same concentration, growth of *Escherichia coli* was inhibited 90% by both cricket chitosan samples with ~80% DDA, where a decrease in the DDA led to decreased antimicrobial activity. However, varying the DDA had no effect on chitosan's lipid-binding capacity. As more edible insects become a normalized protein source in our diet, the use of by-products, such as chitin and chitosan, derived from insect protein processing, show promising applications for the pharmaceutical and food industries.

Keywords: insect chitin; cricket chitosan; lipid-binding activity; antimicrobial activity

Citation: Malm, M.; Liceaga, A.M. Physicochemical Properties of Chitosan from Two Commonly Reared Edible Cricket Species, and Its Application as a Hypolipidemic and Antimicrobial Agent. *Polysaccharides* **2021**, *2*, 339–353. https://doi.org/10.3390/polysaccharides2020022

Academic Editor: Azizur Rahman

Received: 9 April 2021
Accepted: 6 May 2021
Published: 12 May 2021

1. Introduction

Chitin is a polysaccharide consisting of *N*-acetyl-D-glucosamine units that form a polymer through covalent β-1,4-linkages; it is commonly converted to its counterpart chitosan through replacement of its acetyl group with an amino group [1]. Chitin occurs naturally in the exoskeletons of arthropods from the largest animal phylum Arthropoda, which accounts for 80% of the species in the animal kingdom. Crustacean (crab and shrimp) shells, a by-product of the food industry, are mainly used as commercial sources of chitin and chitosan [2–5].

Another subphylum of Arthropoda is the Hexapoda, which contains the class Insecta with over 1 million species. To date, research is available on chitin obtained from bumblebees, grasshoppers, crickets, hornets, wasps, centipedes, velvet worms and other species of cockroaches and beetles, among others [6–14]. In this context, edible insects have gained recent attention as emerging protein sources to help alleviate the demand of food in a growing world population [15]. In Europe, Canada and the United States, interest in edible insects has surged due to consumers' willingness to eat more sustainable and environmentally friendly proteins, driving insect-focused food product development [15]. For example, protein bars, baked goods and pasta products made with cricket flour (ground whole crickets) have launched in the western market. In addition to the consumption and product development of cricket-based products, technology has been developed for enzymatic hydrolysis of cricket protein [15–21]. These cricket protein hydrolysates provide concentrated protein powders rich in essential amino acids that can be incorporated into low or poor-quality protein foods to enhance the overall protein content/quality. Most importantly, a large by-product of this manufacturing technology is the chitin-rich exoskeletons of the crickets [15]. The giant cricket (*Brachytrupes portentosus*) and the African field cricket (*Gryllus bimaculatus*) have been used successfully as sources of chitin and chitosan [13,14,22]. However, there is currently no information on chitin isolated from edible cricket species commonly farmed in the United States, the house cricket (*Acheta domesticus*) and tropical banded cricket (*Gryllodes sigillatus*); as a result, little is known on the physicochemical properties of chitin derived from these two species.

Evidence suggests that crustacean (shellfish) chitosan may be an alternative treatment for obesity due to its impressive lipid-binding capacity [23], among other modes of action [24]. In addition, numerous studies are available detailing shellfish chitosan's antimicrobial activity, with potential applications in the food industry. Research suggests chitosan's physicochemical properties, such as degree of deacetylation (DDA) and molecular weight, have significant effects on its bioactivity [25]. Nevertheless, a research gap remains on the characteristics and bioactivity of edible insect chitin and chitosan. The aim of this study is to investigate chitin derived from two edible cricket species commercially reared in the United States and its conversion into chitosan with focus on its physicochemical properties. In addition to its characterization, this research will also explore the bioactive properties of cricket chitosan related to its entrapment of lipids under in vitro digestion and its ability to inhibit the growth of two bacteria important to the food industry.

2. Materials and Methods

All materials and chemical reagents were purchased from Fisher Scientific (Waltham, MA, USA) and Sigma Aldrich (St. Louis, MO, USA), unless specified. The two cricket species, *Acheta domesticus* (house cricket) and *Gryllodes sigillatus* (tropical banded cricket), were obtained from two edible cricket rearing facilities, Ovipost, Inc. (Labelle, FL, USA) and Three Cricketeers, LLC (St. Louis Park, MN, USA), respectively. Each cricket species was shipped frozen and stored in a −20 °C freezer until needed. Commercial chitosan sourced from shrimp shells and Alcalase® (protease from *Bacillus licheniformis*) was purchased from Sigma Aldrich. *Escherichia coli* ATCC 25922 and *Listeria innocua* ATCC 33090 were obtained from the American Type Culture Collection (Manassas, VA, USA).

2.1. Cricket Chitin Extraction

Chitin-rich pellets were obtained through enzymatic treatment of each cricket species, *Acheta domesticus* and *Gryllodes sigillatus*, using the procedure previously described [16,26] that was shown to most effectively separate chitin from the protein matrix. Briefly, frozen crickets (100 g) and 250 mL of water were homogenized for 2 min in a commercial blender (Waring Commercial, CT), followed by pasteurization in a 90 °C water bath to inactive endogenous enzymes. The slurry's pH was adjusted to 8.0 with 6 mol L^{-1} NaOH solution and equilibrated at 55 °C to produce optimum conditions for the enzyme. Proteolysis was performed with 3% (w/w) Alcalase® for 80 min and terminated by heating slurries at 90 °C for 15 min. The slurries were centrifuged at 17,636× g for 15 min at 4 °C (Avanti J-26S Centrifuge, Beckman-Coulter INC., Brea, CA, USA) to separate the chitin-rich pellet from the protein supernatant. Chitin pellets were stored at 5 °C prior to purification.

2.2. Chitin Demineralization and Deproteinization

Cricket chitin pellets were demineralized in 0.25 mol L^{-1} HCl (1:2 w/v) in a 85 °C shaking water bath for 15 min following procedures previously established [27,28]. Chitin was filtered (100 mesh) and washed with distilled water to neutrality. Demineralized cricket chitin was deproteinized with an alkali treatment of 1 mol L^{-1} NaOH (1:2 w/v) at 70 °C in a water bath for 22 h as previously described [13,14,27]. The purified chitin samples were filtered (100 mesh), washed again to neutrality, lyophilized for 4 days, ground in a blender (Waring Commercial, CT) and stored in a −20 °C freezer until further use.

The efficiency of the purification process was determined through moisture (AOAC 950.46b), ash (AOAC 920.153) and total *N(%)* (AOAC methods 984.13 (A-D)) analysis. The % protein content remaining in the chitin was calculated using Equation (1) [29]:

$$Protein\ (\%) = (N\ (\%) - 6.9\%) \times 6.25 \tag{1}$$

where *N(%)* is the chitin's total nitrogen content determined via composition analysis, 6.9% represents the nitrogen content of pure and fully acetylated chitin and 6.25 is the average nitrogen content in proteins. All proximate composition analyses are reported as percentages on a dry weight basis.

Chitin yield was determined on a dry weight basis (dwb) for each cricket species using Equation (2):

$$Chitin\ content\ (\%) = \frac{weight\ of\ chitin\ (g)}{weight\ of\ whole\ crickets\ (g)} \times 100 \tag{2}$$

2.3. Chitosan Conversion and Characterization by Fourier Transform Infrared Spectroscopy (FTIR) and Degree of Deacetylation

Purified, lyophilized cricket chitin was converted to chitosan with 67% w/v NaOH (1:20 w/v) for 2, 4, 6 and 10 h to vary the degree of deacetylation [13]. Due to employing the traditional conversion method for chitosan with concentrated alkali solution, chitosan was first filtered (100 mesh) with water until effluent ran clear. Chitosan was then washed with a minimum of 5 L of water, until effluent reached neutrality (pH 6.5), to reduce the presence of residual NaOH. Finally, chitosan was collected, lyophilized and stored at −20 °C until needed. Chitosan conversion yield (dwb) for each duration treatment was determined using Equation (3):

$$Chitosan\ yield\ (\%) = \frac{mass\ chitosan\ (g)}{mass\ chitin\ (g)} \times 100 \tag{3}$$

The eight cricket chitosan samples were evaluated by FT-IR (Thermo Scientific, Waltham, MA, USA) from 3500 to 800 cm^{-1} with a resolution of 8 cm^{-1} (n = 8) to determine the chemical structure and degree of deacetylation (DDA) [1,13,14]. Intensity of FTIR peaks with appropriate baselines at 1650 cm^{-1} (A_{1652}) and 3350 cm^{-1} (A_{3450}) were used to determine DDA Equation (4) [13,30,31].

$$DDA\ (\%) = 100 - \left(\frac{A_{1650}}{A_{3350}} \times \frac{100}{1.33}\right) \tag{4}$$

Additionally, commercial shrimp chitosan spectra were used as a reference. Cricket chitosan samples from each species, which were determined to have similar DDA values (~72%, 76% and 80%), were used in the lipid-binding capacity and antibacterial activity experiments.

2.4. Molecular Weight Determination

The average molecular weight of cricket chitosan was determined viscometrically following the method of Czechowska-Biskup, Wach [32] and Roberts and Domszy [33]. In brief, solutions (1.5–4.5 mg/mL) were prepared from ~80% DDA cricket chitosan (10 h samples) in a solvent system consisting of equal parts of 0.1 mol L^{-1} CH_3OOH and 0.2 mol L^{-1} NaCl and stirred overnight. The flow time of each chitosan solution and solvent system (5 mL) was measured in triplicate using a capillary viscometer (Cannon-Fenske, Size 75) in a 25 °C water bath. Relative viscosity, specific viscosity and reduced viscosity were calculated using measured flow times [34]. The linear relationship between reduced viscosity and concentration for each cricket chitosan was extrapolated to determine its intrinsic viscosity ($[\eta]$). Finally, using the averaged intrinsic viscosity (n = 3), the viscosity average molecular weight (M_v) for each chitosan was calculated using the Mark–Houwink Equation:

$$[\eta] = KM_v^{\alpha} \tag{5}$$

with previously determined solvent system constants: K =1.81 × 10^{-3} α =0.93 [12,33].

2.5. Anti-Obesity Effects (Lipid-Binding Capacity) of Cricket Chitosan

The in vitro lipid-binding capacity of shrimp and cricket (72, 76 and 80% DDA) chitosan samples was determined in triplicate following established procedures [35,36]. Each chitosan sample (20 mg) was dissolved in 1.25 mL 0.6 mol L^{-1} HCl and incubated for 30 min in a 37 °C water bath with constant shaking. Then, 25 g of olive oil was added to each tube, vortexed and incubated under the same conditions for 2 h. After incubation, 8 mL of phosphate buffer (pH 7.4) was added and the solution's pH was adjusted to 6.8 with 1 mol L^{-1} NaOH and incubated again for 30 min to mimic duodenal conditions. Finally, tubes were centrifuged at 697× g for 10 min and the supernatant representing unbound lipids was gravimetrically measured. The lipid-binding capacity of chitosan was calculated using Equation (6):

$$lipid\ binding\ capacity\left(\frac{g\ oil}{g\ chitosan}\right) = \frac{(25\ g\ oil - g\ unbound\ oil)}{0.02\ g\ chitosan} \tag{6}$$

2.6. Determination of Antimicrobial Activity

The antimicrobial activity of cricket and shrimp chitosan samples against two bacteria strains was determined using methods described by Aguilar-Toalá and Deering [37] with slight modifications [38]. Specifically, antibacterial activity against *Escherichia coli* ATCC 25922 and *Listeria innocua* ATCC 33090, as surrogates for foodborne pathogens *Escherichia coli* O157:H7 and *Listeria monocytogenes*, were studied [39].

2.6.1. Bacterial Stock Cultures and Sample Preparation

The lyophilized bacteria were inoculated in Brain Heart Infusion (BHI) and Mueller Hinton Broth (MHB) for *L. innocua* and *E. coli*, respectively, following ATCC guidelines. Inoculums were used to prepare freezer stocks with a final glycerol concentration of 20% (w/v), cryovials were stored at −80 °C until needed. Prior to use, 50 µL of each stock was transferred into 5 mL MHB and BHI, for *E. coli* and *L. innocua*, grown to stationary phase. As determined by plate counts, *E. coli* inoculums were 1.1 × 10^8 cfu/mL and *L. innocua* inoculums were 7.2 × 10^8 cfu/mL. Fresh cultures were prepared from freezer stocks following the same incubation procedure and used immediately for each assay.

Shrimp and cricket chitosan (72, 76 and 80% DDA) solutions (8 mg/mL) were prepared in 1% (v/v) acetic acid and stirred overnight to ensure homogenous dissolution. Solutions were then sterilized by autoclaving for 20 min at 121 °C. Additionally, 1% acetic acid was sterilized under the same conditions.

2.6.2. Antibacterial Determination

The antimicrobial activity of cricket chitosan samples was determined using 96-well sterile microplates. Chitosan samples, positive and negative control wells were prepared aseptically with a final volume of 220 µL (n = 6). For chitosan wells, 13.75 µL of each chitosan sample was added with 204.25 µL BHI, resulting in a final chitosan sample concentration of 0.5 mg/mL in each well. Positive control and negative control wells were prepared with 220 µL and 218 µL BHI, respectively. Chitosan sample and negative control wells were inoculated with 2 µL of bacterial culture (~10^6 cfu/mL) (n = 6). Finally, an additional solvent control was prepared to deconvolute its effects from chitosan on bacterial growth. Similar to the chitosan wells, 13.75 µL of 1% CH_3OOH was combined with 204.25 µL BHI and inoculated (2 µL) (n = 6).

Following preparation, microplates were incubated for 24 h at 37 °C with optical density measurements (OD = 620 nm) conducted at 0, 6, 12 and 24 h. Growth curves of samples were created by plotting optical density over the duration of the experiment. Additionally, the percent of bacterial growth inhibition at 24 h was calculated (n = 6) using Equation (7):

$$\%\,inhibition = 1 - \frac{\Delta OD_{sample,\ 24\ hr}}{OD_{control,\ 24\ hr}} \times 100 \quad (7)$$

2.7. Statistical Analysis

All experiments and analyses were conducted in triplicate, unless otherwise indicated. Statistical analysis of chitosan samples' degree of deacetylation, chitosan yield, lipid-binding capacities and percent of bacterial inhibition were performed using a one way ANOVA with 95% confidence level (Minitab 18® State College, PA, USA).

3. Results and Discussion

3.1. Cricket Chitin and Chitosan Processing

Cricket chitin and chitosan were successfully extracted, purified and converted with similarities to the commercial products. Overall, the extraction and purification processing steps produced cricket chitin with low protein and ash compositions, in comparison to their original compositions (Table 1).

The demineralization process is an important first step to remove the minerals present in the exoskeleton matrix. Traditionally, crustacean shells are demineralized with hydrochloric acid solutions, to react with the calcium carbonate intrinsically present at high concentrations (30–50%) [4,27,40]. In the case of insects, due to their environmental differences, they do not contain a large amount of minerals in their exoskeletons when compared to crustaceans [41]. Yet, the majority of insect chitin studies commonly employ the demineralization procedures optimized for crustaceans [41]. For example, two studies on grasshoppers (*Decticus verrucivorus*) and cockroach wings (*Blaberus giganteus*), employed HCl demineralization parameters that were highly concentrated (2 M), hot (75 °C) and prolonged (>2 h) [7,11]. Demineralization of cicada sloughs, silkworms, beetles and black soldier flies were performed at 100 °C for 20–30 min with 1 mol L^{-1} HCl and resulted in ash values of 3–5% [29,42,43]. Whereas silkworms, mealworms and grasshoppers were treated with the same acid concentration at lower temperatures (25–30 °C) for more than 2 h and ash contents were reported to be less than 1% [44]. In this study, *A. domesticus and G. sigillatus* chitin was demineralized with 0.25 mol L^{-1} HCl for 15 min at 85 °C with final chitin ash contents less than 0.09% (Table 1). The results of this study show that cricket chitin and likely other insect-derived chitin, do not need extensive demineralization treatment as those used for crustacean (shellfish) sources.

Table 1. Chitin composition and yield, after proteolysis and purification, with reference to whole crickets.

	Composition Analysis (%)	*A. domesticus*	*G. sigillatus*
Whole crickets, ground			
	Protein	67.4 ± 1.5 [1]	56.8 ± 0.01 [2]
	Ash	4.0 ± 0.90 [1]	18.1 ± 0.60 [2]
	Chitin [3]	5.7 ± 0.10	3.4 ± 0.10
Chitin, extracted and purified			
	Nitrogen	<6.9	7.1
	Protein (calculated) [4]	ND	1.4
	Ash	0.09	<0.01

[1] [26]; [2] [18]; [3] yield of chitin, extracted and purified, following Equation (2); [4] corrected protein concentration following Equation (1), where protein content in chitin with N% < 6.9% (due to residual inorganic materials), is determined to be "none detected" (ND) [45].

A. domesticus chitin had comparable yields (5.68%) to that extracted in other studies for other insects including other cricket species (5.1%), grasshoppers (5.7%), beetles (5.0%) and wasp pupa (6.2%) [8,13,44,46]. In contrast, the *G. sigillatus* had lower chitin yields (3.38%); however, it was still comparable to chitin extracted from silkworm chrysalis (3.1%), mealworms (2.5%), honeybees (2.5%) and wasp larvae (2.2%) [8,44,46]. This chitin yield difference between the two cricket species is likely a result of the larger body size and exoskeleton of the *A. domesticus* in comparison to *G. sigillatus*. Other insect chitin studies reported higher chitin yields (18–37%); however, these studies used molted exoskeletons or wings, which naturally have higher chitin contents, when compared to the intact insect [29,44,47].

3.2. Cricket Chitin and Chitosan Characterization

FTIR analysis of all cricket chitin and chitosan samples showed strong chemical and structural similarities to those from shrimp. Chitin from both cricket species (Figure 1) contained characteristic peaks known to chitin from both crustaceans and other insects [12,48]. The two cricket chitins did not differ from each other. Chitin has three crystalline forms, either alpha, beta or gamma, although chitin is most commonly present in its alpha form [48]. The FTIR spectra peak between 1700 and 1500 cm^{-1}, known as the Amide I band, shows that cricket chitin from both species is in the alpha crystalline form, due to the two peaks observed at ~1660 and ~1630 cm^{-1}. These two peaks represent alpha chitin's intra- and inter-molecular hydrogen bonds forming its antiparallel chain arrangement [48,49]. The hydrogen bonds between –C=O (Amide I) and –NH– (Amide II) are responsible for the peak at ~1660 cm^{-1}, while the second peak occurring at ~1630 cm^{-1} is due to hydrogen bonding between the –CH_2O side chain and –C=O [50].

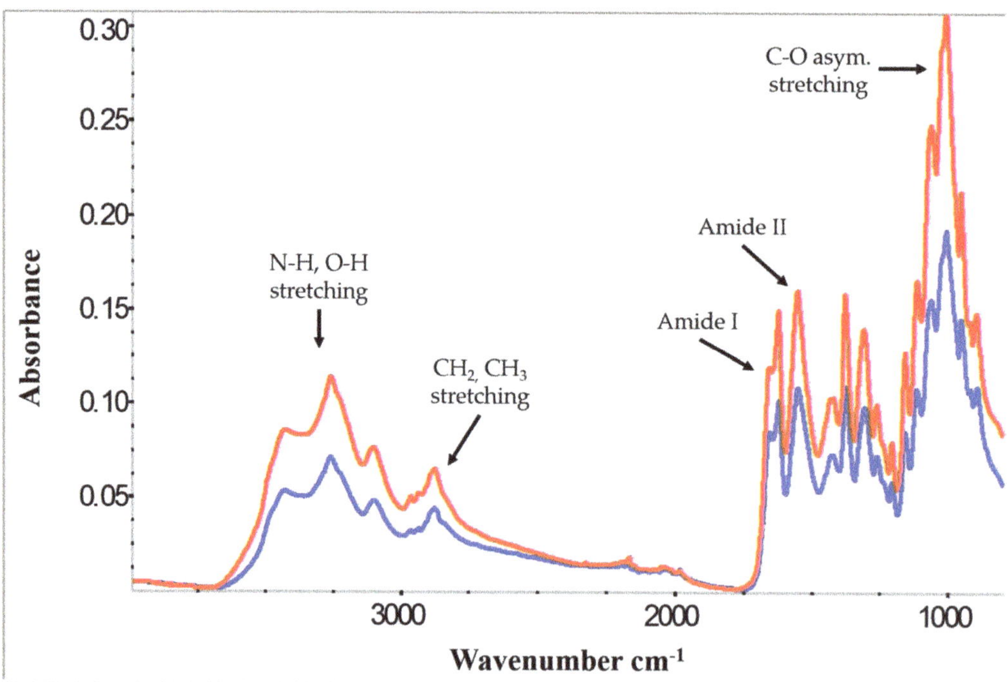

Figure 1. FTIR spectra of *A. domesticus* (**blue**) and *G. sigillatus* (**red**) purified chitin.

Spectral peaks related to chitosan's chemistry have been extensively reported. Following deacetylation of chitin, chitosan samples produced from both cricket species were in good alignment with commercially purchased shrimp chitosan (Figure 2). Two spectral peaks, the OH peak (~3450 cm^{-1}) and the –CO–NH (1650 cm^{-1}) of the Amide I band, were used to calculate the DDA of the cricket and shrimp chitosans [13,30,31] (Table 2). Deacetylating both the *A. domesticus* and *G. sigillatus* chitin for 2 h resulted in chitosan with a DDA of ~73%, while the commercial shrimp chitosan was ~70%. Colorado potato beetle (*Leptinotarsa decemlineata*) and larvae chitin processed under similar conditions for 3 h produced chitosan with 71% and 64% DDA, respectively [12]. *A. domesticus* chitin reached 76% DDA after 4 h, whereas *G. sigillatus* chitin required 6 h to reach a similar DDA. After 10 h of deacetylation, cricket chitosan reached a DDA of ~ 80%, similar to that reported by Chae, Shin [13], which produced 84% DDA chitosan from the African field cricket (*Gryllus bimaculatus*) chitin after 9 h of deacetylation under similar conditions. In Figure 2, as the deacetylation time of cricket chitin increased, acetyl groups were continuously replaced with amine groups leading to a higher DDA [30]. Shrimp chitosan had a lower DDA (70%) and therefore a greater peak intensity at ~1650 cm^{-1} compared to the two cricket chitosans deacetylated for 10 h (~80% DDA).

The chemical deacetylation of cricket chitin to chitosan had similar yields (Table 2) comparable to other insects and crustaceans [45]. Approximately 70–80% of the *A. domesticus* chitin mass was recovered after deacetylation, whereas of *G. sigillatus* cricket chitin yielded about 60–65% chitosan. Additionally, deacetylation duration had little effect on the overall conversion yields. *G. sigillatus* chitosan conversion yields were not affected by duration time and *A. domesticus* chitosan yield was the highest after 6 h and the lowest after 10 h. However, the 6 and 10 h deacetylation had chitosan yields that were not significantly different after 2 and 4 h of deacetylation.

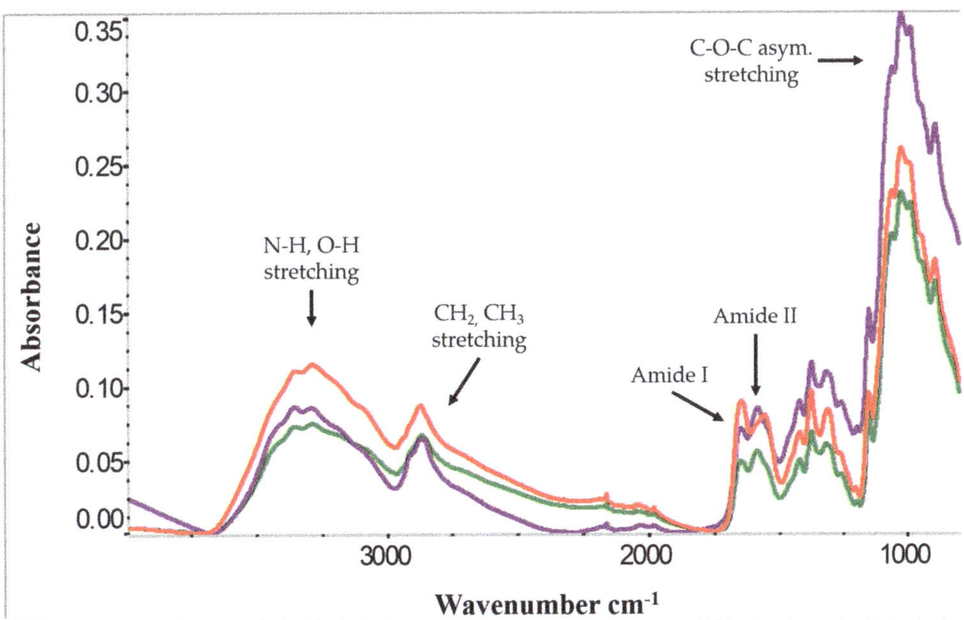

Figure 2. FTIR Characterization of *A. domesticus* (**green**) and *G. sigillatus* (**purple**) chitosan samples after 10 h of deacetylation, compared to commercial shrimp chitosan (**red**).

Table 2. Yield and degree of deacetylation (DDA) of cricket chitin converted to chitosan.

	Conversion Time (Hours)	*A. domesticus* [3]	*G. sigillatus* [3]
DDA (%) [1]			
	2	72.5 ± 1.0 [a]	73.5 ± 1.4 [a]
	4	76.3 ± 1.3 [b]	74.9 ± 1.3 [a]
	6	79.1 ± 1.9 [c]	77.2 ± 1.8 [b]
	10	79.4 ± 1.3 [c]	81.3 ± 1.1 [c]
Chitosan yield (%) [2]			
	2	76.0 ± 6.7 [ab]	65.0 ±1.6 [a]
	4	77.3 ± 1.9 [ab]	63.7 ± 1.2 [a]
	6	80.5 ± 2.1 [b]	60.3 ± 3.3 [a]
	10	69.0 ± 2.2 [a]	62.3 ± 0.9 [a]
Chitosan molecular weight [4]	-	344 kDa	524 kDa

[1] DDA (%) for each cricket species, average of triplicate determinations ± standard deviation, was calculated using Equation (4); [2] chitosan yield (%), average of triplicate determinations ± standard deviation, was calculated using Equation (3); [3] values that do not share the same letter (a–c) within a column, for DDA% or chitosan yield, are statistically different ($p < 0.05$); [4] intrinsic viscosity, measured in triplicate, was used to calculate molecular weight average using Equation (5).

3.3. Molecular Weight Determination

The average molecular weight was determined to be 344 kDa for *A. domesticus* chitosan, while *G. sigillatus* chitosan had a larger molecular weight of 524 kDa (Table 2). While the *A. domesticus* cricket is physically larger than *G. sigillatus*, its molecular weight was determined to be smaller. This suggests that *A. domesticus* may have more chitin fibers in its exoskeleton of a smaller weight, while *G. sigillatus cricket's* exoskeleton may have larger, but fewer, chitin fibers in its exoskeleton. The size of chitosan from these two cricket species is in good agreement with another cricket chitosan study. Kim, Song [22] studied the molecular weight of *G. bimaculatus* crickets using dynamic light scattering and found its molecular weight to be ~308 kDa. As a result, *A. domesticus* and *G. bimaculatus* crickets have similar molecular weights, but are smaller than *G. sigillatus* crickets. Additionally, cricket chitosan is relatively similar in size to other insect chitosan previously studied, such as that sourced from blowfly (501 kDa) and housefly larvae (426 kDa) [51,52]. However, compared to other species of insect chitosan, *A. domesticus* and *G. sigillatus* cricket chitosan are much larger in size. Cicada sloughs, grasshoppers, mealworms, silkworm chrysalis, black soldier flies and beetles ranged in molecular weights between 2.7 and 15 kDa [12,44,53]. As a result, insect chitosan has repeatedly been stated to be of low molecular weight compared to crustacean chitin and chitosan, which can range from 50 to 2000 kDa [54,55]. The results of this study, as well as others, confirm that insect chitosan may also be characterized as high molecular weight.

3.4. Lipid-Binding Capacity

Commercial shrimp chitosan and its water-soluble derivatives, has been reported as an anti-obesity or hyperlipidemic treatment in a variety of in vitro and in vivo studies [24]. Chitosan exhibits both hypocholesteremic and hypolipidemic properties through its ability to regulate lipid metabolism, achieved by electrostatic interaction with or physical entrapment of, targeted molecules [24]. The focus of this study was to evaluate the lipid-binding capacity of chitosan through interaction and entrapment of dietary lipids in vitro, one of the reported mechanisms of chitosan to reduce lipid metabolism in a high lipid diet. The consumption of lipids in the presence of chitosan, is thought to lead to the glucosamine in chitosan to be positively charged in the stomach due to its low pH, followed by its binding to negatively charged lipid molecules such as triglycerides and fatty acids [56,57]; this emulsion formation has been demonstrated previously with fluorescent microscopy [35]. When the lipid–chitosan emulsion is transferred from the acidic stomach conditions to the alkaline conditions of the duodenum, the chitosan precipitates and physically entraps the emulsified lipid droplets through gel formation [35,36,56]. Ultimately, this entrapment prevents lipids to be accessed and digested by the body and results in their excretion through feces.

Chitosan from both cricket species showed high lipid-binding capacity, between ~160 and 220 g of oil per gram of chitosan (Table 3). To the best of our knowledge, this is the first study evaluating the lipid-binding capacity of chitosan from edible insects. There were no significant differences on lipid-binding capacity between the different samples; therefore, no correlation between physicochemical properties such as molecular weight or DDA can be made from this study. However, these results suggest that edible cricket chitosan could be as effective as commercial shrimp chitosan in binding lipids under gastric conditions. Panith, Wichaphon [35] showed that chitosan from shrimp shells, produced with a low molecular weight (~30 kDa), had inferior lipid-binding capacity compared to medium and high molecular weight chitosan samples (890 kDa and 8900 kDa). In contrast, low molecular weight chitosan (~25 kDa), from an unspecified marine species, had superior-binding capacity compared to larger molecular weight chitosan (408 kDa) [56]. In this study, the molecular weights of cricket and shrimp chitosans were determined to be high molecular weight. Therefore, it is possible the molecular weights of the samples of this study were too similar to result in significant changes between their lipid-binding capacities. Further research on this topic is required since many insect chitosan studies report low molecular weights [12,44,58] and it is unknown how cricket chitosans' lipid-binding capacities would compare to other insect-based chitosans due to opposing literature conclusions.

Table 3. Cricket and shrimp chitosan lipid-binding capacity.

	Lipid-Binding Capacity (g Oil per g Chitosan) [1]		
DDA (%)	***A. domesticus***	***G. sigillatus***	***Shrimp***
~72	210.8 ± 21.1	163.5 ± 17.8	168.5 ± 36.8
~76	221.8 ± 25.4	159.0 ± 15.9	-
~80	168.7 ± 10.2	180.5 ± 21.6	-

[1] Average of triplicate determinations ± standard deviation, calculated with Equation (3). DDA= degree of deacetylation. Values were not statistical different ($p > 0.05$).

Overall, the lipid-binding capacities of edible cricket and shrimp chitosan in this study were found to be different than those reported for crustacean chitosan. Panith, Wichaphon [35] showed the highest-binding capacity to be at ~28 g oil/g chitosan, which is much lower than the results of this study. These differences are likely due to the different in vitro methods applied. For example, our study gravimetrically measured unbound oil, whereas Panith et al. (2016) solvent-extracted entrapped lipids and then gravimetrically measured bound oil. These differences in methods could lead to some over-estimation in our study. Additionally, Panith et al. (2016) centrifuged at 4000× *g*, whereas in this study, our method called for 696× *g* to separate bound lipids from unbound lipids. It is possible the larger centrifugal force used in other studies destabilized the chitosan-lipid gel and physically removed chitosan entrapped lipids; this could result in an under-estimation of lipid-binding capacity reported by Panith et al. (2016). In contrast, Zhou and Xia [36] reported a much larger lipid-binding capacity value of ~1200 g of oil bound per gram of chitosan. However, the methods for determining the-binding capacity did not differ, making it difficult to understand the differences in values reported.

3.5. Antimicrobial Activity

The effect of edible cricket chitosan on inhibiting growth of *Listeria innocua* ATCC 33090 and *Escherichia coli* ATCC 25922 over 24 h is shown in Figure 3. For each bacteria strain, the effect of shrimp chitosan on growth was also evaluated as a reference for current commercial (shrimp) chitosan. Additionally, the solvent used, 1% acetic acid, is an antimicrobial agent itself, and thus its effect on measured antibacterial properties was evaluated (Table 4) [59,60]. At a concentration of 0.50 mg/mL, all chitosan samples were able to effectively inhibit growth of *L. innocua* over 24 h, with ~100% inhibition at hour 24 (Table 4), as shown by no increase in the optical density of the samples during the duration of the experiment (Figure 3). Acetic acid, at a final concentration of 0.06%, resulted in *L. innocua* growth to be inhibited by 38% (Table 4). Therefore, *L. innocua* inhibition by chitosan samples was due to both the presence of chitosan, as well as acetic acid.

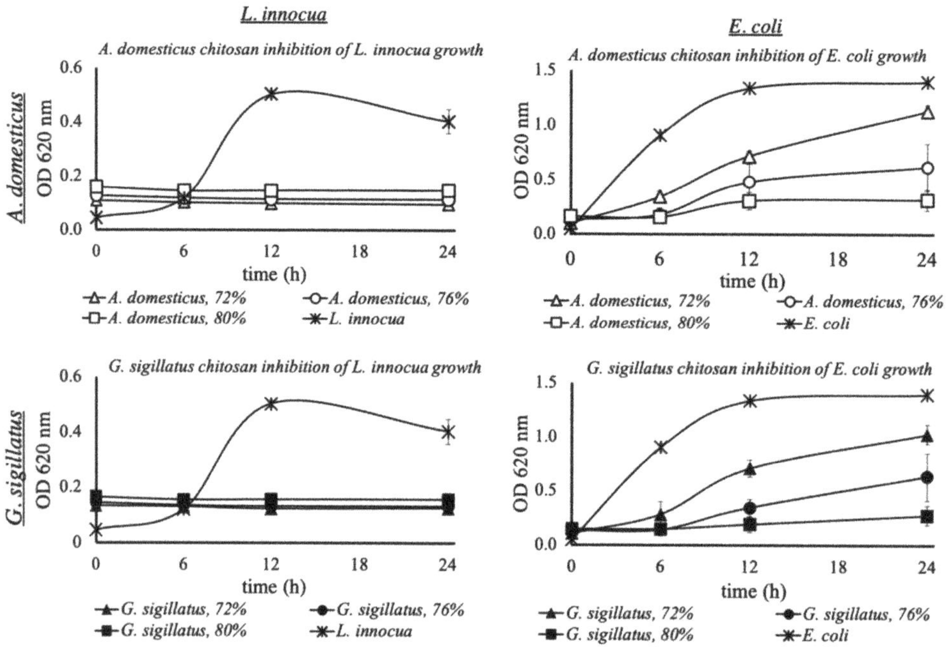

Figure 3. Antimicrobial effect of chitosan from edible *A. domesticus* (empty shapes) and *G. sigillatus* (filled shapes) crickets, with varying degrees of deacetylation including 72% (triangles), 76% (circles) and 80% (squares), on the growth of *Listeria innocua* (left column) and *Escherichia coli* (right column) over 24 h of incubation (37 °C).

Table 4. Effect of edible cricket chitosan (0.50 mg/mL) with varying degrees of deacetylation on bacteria growth inhibition (%) after 24 h incubation.

Inhibitors	Bacteria Growth Inhibition (%) [1]			
		Degree of Deacetylation (%)		
		72	76	80
***Escherichia coli* ATCC 25922**				
A. domesticus chitosan	-	26.6 ± 0.7 [a]	65.6 ± 2.5 [b]	89 ± 1.2 [c]
G. sigillatus chitosan	-	33.8 ± 1 [a]	64.2 ± 2.7 [b]	94.2 ± 0.9 [c]
Commercial shrimp chitosan	21.4 ± 2.4	-	-	-
Acetic acid [3]	5.4 ± 0.2	-	-	-
***Listeria innocua* ATCC 33090**				
A. domesticus chitosan	-	100 [2]	100	100
G. sigillatus chitosan	-	100	100	100
Commercial shrimp chitosan	100	-	-	-
Acetic acid [3]	37.7 ± 0.7	-	-	-

[1] Inhibition (%) values represent mean of six replicates ± standard error, determined using Equation (7), where different letters (a, b, c) indicate significant difference ($p < 0.05$) between degree of deacetylation for each cricket chitosan (rows); [2] 100% represents complete inhibition in chitosan samples inoculated with *L. innocua*, where no change in optical density at 620 nm was observed (Figure 3, left); [3] acetic acid (0.06% v/v final concentration) quantifies the contribution of solvent to chitosan's antimicrobial activity.

One generally accepted mechanism of chitosan's antibacterial activity is a result of the positive charged $+NH_3$ glucosamines interacting with negatively charged surface components of bacterial cells, which ultimately interrupts or prevents vital cell functions [54]. Therefore, it is thought that an increase of DDA increases the number of positively charged glucosamines that are able to bind with cell components, leading to an increase in antimicrobial activity. This was seen when the DDA of *G. sigillatus* cricket chitosan increased from 72 to 76 and 80%, which resulted in a sequential increase of growth inhibition of *E. coli*; showing approximately 90% growth inhibition at 80% DDA (Table 4). However, DDA did not have an apparent effect on the inhibition of *L. innocua* growth at the chitosan concentration chosen (Figure 3). Decreasing the chitosan sample concentrations sequentially may allow for differentiation between chitosan samples with different degrees of deacetylation on the efficacy of inhibiting *L. innocua* growth, while minimizing the contribution due to acetic acid. For both bacteria, growth inhibition achieved by 72% DDA cricket chitosan was similar to that of commercial shrimp chitosan, suggesting that cricket chitosan can be as effective as current market options.

The results of this study are similar to other insect chitosan antimicrobial assays; however, direct comparisons are difficult to make due to differing microorganisms, chitosan concentrations employed and assaying methods (i.e., MIC, MBC, zone of inhibition, etc.). For example, chitosan derived from larvae and adult Colorado potato beetles were found to have minimal bactericidal concentrations (MBC) of 0.32 and 1.25 mg/mL against *L. monocytogenes*, where MBC values represent the lowest concentration of an antimicrobial agent required to kill bacteria in an inoculated sample. Chitosan derived from two different grasshopper species were found to have a MBC of 0.32 and 0.63 mg/mL against *L. monocytogenes*. Another study quantified the effect of mealworm (*Tenbrio molitor*) and crustacean-derived chitosan on cell counts of foodborne pathogens *E. coli* O157:H7, *L. monocytogenes* and *Salmonella enterica* serovar Typhimurium, over 48 h [61]. The study found in general that crustacean and insect chitosan antibacterial activity at 1.5 mg/mL led to unchanged or reduced following 24 h after inoculation for all three bacteria, with recovery of bacterial counts detected between 24 and 48 h (when inoculated with 10^6 cfu/mL, pH 6.2). At 24 h, the antibacterial activity of crustacean chitosan was slightly greater for *E. coli* and *S.* Typhimurium, whereas mealworm chitosan was more effective against *L. monocytogenes*. However, the study did not characterize or report the specifications of the chitosan samples, such as molecular weight or % deacetylation, so it is difficult to understand the basis of these differences. Another study evaluated the antimicrobial efficacy of two different grasshopper species against a variety of organisms, including that of *L. monocytogenes*. The MBC of one grasshopper chitosan (*Calliptamus barbarous*) was 0.32 mg/mL, whereas the MBC of chitosan from the other grasshopper species (*Oedaleus decorus*) was 0.63 mg/mL [38]. The chitosan from the two species were reported to have a DDA between 70 and 75%, although the specific DDA of each chitosan was not reported. It is possible the differences in MBC are due to slight differences in DDA. As shown in this current study (Figure 3), a 4% change in the cricket chitosan DDA (i.e., from 72 to 76 and 80%) significantly improved *E. coli* inhibition (Table 4). No difference in chitosan antimicrobial efficacy was seen between the two cricket species, likely a result of standardizing DDA prior to antimicrobial testing.

The results of this study, in conjunction with others, elucidate the importance to evaluate each newly derived chitosan against specific bacterial strains to determine its anticipated antimicrobial activity. Overall, cricket chitosan derived from *A. domesticus* and *G. sigillatus* were as effective at inhibiting microbial growth of surrogate foodborne pathogens, as the traditional and commercial crustacean (shrimp) chitosan. A concentration of 0.50 mg/mL was an effective concentration for all chitosan samples against *L. innocua*; however, higher concentrations of the lower DDA chitosan samples may be required to enhance the antimicrobial effect against *E. coli*. Further studies on the effect of edible cricket chitosan on bacterial cell wall material are required to understand the mechanisms of inhibition by these chitosan samples.

4. Conclusions

Chitin from two edible cricket species commonly reared in the United States were successfully collected as a by-product of cricket protein hydrolysis and adequately isolated through demineralization and deproteinization processing steps. *A. domesticus* crickets were found to yield slightly higher amounts of purified chitin, compared to *G. sigillatus* crickets. FTIR results confirmed that cricket chitin was in its alpha-crystalline form, similar to that isolated from most crustacean and insect species. Cricket chitin was successfully converted to chitosan with approximately 72%, 76% and 80% degree of deacetylation, achieved by varying deacetylation times using concentrated alkaline treatments. Structural analysis of cricket chitosan samples was chemically similar to that of commercial shrimp chitosan. Finally, the average molecular weight of chitosan derived from *A. domesticus* crickets was determined to be 344 kDa, while that from *G. sigillatus* had a larger molecular weight of 524 kDa. In contrast to the findings of other insect-derived chitosan studies, the results of this study suggest that cricket chitosan can be recognized as a high molecular weight polymer, similar to commercially available chitosan with potential bioactive properties.

Lipid-binding capacity of all chitosan samples were determined to be between 160 g and 220 g oil per 1 g chitosan, although physicochemical properties did not have any significant effect on lipid binding. Additionally, the edible cricket chitosan samples effectively inhibited *E. coli* and *L. innocua*. The degree of deacetylation did not have an effect on the antibacterial properties of cricket chitosan against *L. innocua*, whereas the antimicrobial activity of *A. domesticus* and *G. sigillatus* chitosan was more effective against *E. coli* at higher DDA values, compared to shrimp chitosan. Thus, edible cricket chitosan has the potential to inhibit the growth of foodborne pathogens, as a naturally derived antimicrobial agent.

As the consumption of insects and insect-containing food products is rapidly growing, an increase of insect protein production and thereof chitin by-products will become more available. This study concludes that chitosan derived from U.S. reared edible crickets have physicochemical and bioactive properties similar to commercial crustacean (e.g., shrimp) chitosan. As a result, there is potential for the mass production of cricket-based chitosan as the consumer acceptability for arthropods widens outside the traditional source (crustaceans).

Author Contributions: Conceptualization, M.M. and A.M.L.; methodology, M.M. and A.M.L.; investigation, M.M.; data curation, M.M. and A.M.L.; writing—original draft preparation, M.M.; writing—review and editing, M.M. and A.M.L.; supervision, A.M.L.; funding acquisition, A.M.L. All authors have read and agreed to the published version of the manuscript.

Funding: This work is supported by the USDA National Institute of Food and Agriculture, Hatch Act formula funds project 1019794.

Institutional Review Board Statement: Not Applicable.

Informed Consent Statement: Not Applicable.

Data Availability Statement: All findings discussed are based on the data contained within this current study.

Acknowledgments: The authors acknowledge Aguilar-Toala, Deering and Hansel for their support with the antimicrobial assays performed.

Conflicts of Interest: The authors declare no conflict of interest.

References

1. Sivashankari, P.R.; Prabaharan, M. 5-Deacetylation modification techniques of chitin and chitosan. In *Chitosan Based Biomaterials*; Jennings, J.A., Bumgardner, J.D., Eds.; Woodhead Publishing: Cambridge, UK, 2017; Volume 1, pp. 117–133.
2. Kumar, M.N.V.R. A review of chitin and chitosan applications. *React. Funct. Polym.* **2000**, *46*, 1–27. [CrossRef]
3. Hamed, I.; Özogul, F.; Regenstein, J.M. Industrial applications of crustacean by-products (chitin, chitosan, and chitooligosaccharides): A review. *Trends Food Sci. Technol.* **2016**, *48*, 40–50. [CrossRef]
4. Zargar, V.; Asghari, M.; Dashti, A. A Review on Chitin and Chitosan Polymers: Structure, Chemistry, Solubility, Derivatives, and Applications. *ChemBioEng Rev.* **2015**, *2*, 204–226. [CrossRef]

5. No, H.K.; Meyers, S.P. Preparation and Characterization of Chitin and Chitosan—A Review. *J. Aquat. Food Prod. Technol.* **1995**, *4*, 27–52. [CrossRef]
6. Majtán, J.; Bíliková, K.; Markovič, O.; Gróf, J.; Kogan, G.; Šimúth, J. Isolation and characterization of chitin from bumblebee (*Bombus terrestris*). *Int. J. Biol. Macromol.* **2007**, *40*, 237–241. [CrossRef]
7. Kaya, M.; Lelešius, E.; Nagrockaitė, R.; Sargin, I.; Arslan, G.; Mol, A.; Baran, T.; Can, E.; Bitim, B. Differentiations of Chitin Content and Surface Morphologies of Chitins Extracted from Male and Female Grasshopper Species. *PLoS ONE* **2015**, *10*, e0115531. [CrossRef]
8. Kaya, M.; Sofi, K.; Sargin, I.; Mujtaba, M. Changes in physicochemical properties of chitin at developmental stages (larvae, pupa and adult) of Vespa crabro (wasp). *Carbohydr. Polym.* **2016**, *145*, 64–70. [CrossRef] [PubMed]
9. Zelencova, L.; Erdoğan, S.; Baran, T.; Kaya, M.; Erdogan, S. Chitin extraction and chitosan production from Chilopoda (*Scolopendra cingulata*) with identification of physicochemical properties. *Polym. Sci. Ser. A* **2015**, *57*, 437–444. [CrossRef]
10. Greven, H.; Kaya, M.; Sargin, I.; Baran, T.; Kristensen, R.M.; Sørensen, M.V. Characterisation of chitin in the cuticle of a velvet worm (*Onychophora*). *Turk. J. Zoöl.* **2019**, *43*, 416–424. [CrossRef]
11. Kaya, M.; Sargin, I.; Sabeckis, I.; Noreikaite, D.; Erdonmez, D.; Salaberria, A.M.; Labidi, J.; Baublys, V.; Tubelytė, V. Biological, mechanical, optical and physicochemical properties of natural chitin films obtained from the dorsal pronotum and the wing of cockroach. *Carbohydr. Polym.* **2017**, *163*, 162–169. [CrossRef]
12. Kaya, M.; Baran, T.; Erdoğan, S.; Menteş, A.; Özüsağlam, M.A.; Çakmak, Y.S. Physicochemical comparison of chitin and chitosan obtained from larvae and adult Colorado potato beetle (*Leptinotarsa decemlineata*). *Mater. Sci. Eng. C* **2014**, *45*, 72–81. [CrossRef] [PubMed]
13. Chae, K.-S.; Shin, C.-S.; Shin, W.-S. Characteristics of cricket (*Gryllus bimaculatus*) chitosan and chitosan-based nanoparticles. *Food Sci. Biotechnol.* **2018**, *27*, 631–639. [CrossRef] [PubMed]
14. Ibitoye, B.E.; Lokman, I.H.; Hezmee, M.N.M.; Goh, Y.M.; Zuki, A.B.Z.; Jimoh, A.A. Extraction and physicochemical characterization of chitin and chitosan isolated from house cricket. *Biomed. Mater.* **2017**, *13*, 025009. [CrossRef]
15. Liceaga, A.M. Approaches for Utilizing Insect Protein for Human Consumption: Effect of Enzymatic Hydrolysis on Protein Quality and Functionality. *Ann. Entomol. Soc. Am.* **2019**, *112*, 529–532. [CrossRef]
16. Hall, F.; Johnson, P.E.; Liceaga, A. Effect of enzymatic hydrolysis on bioactive properties and allergenicity of cricket (*Gryllodes sigillatus*) protein. *Food Chem.* **2018**, *262*, 39–47. [CrossRef] [PubMed]
17. Hall, F.; Liceaga, A. Effect of microwave-assisted enzymatic hydrolysis of cricket (*Gryllodes sigillatus*) protein on ACE and DPP-IV inhibition and tropomyosin-IgG binding. *J. Funct. Foods* **2020**, *64*, 103634. [CrossRef]
18. Hall, F.G.; Jones, O.G.; O'Haire, M.E.; Liceaga, A.M. Functional properties of tropical banded cricket (*Gryllodes sigillatus*) protein hydrolysates. *Food Chem.* **2017**, *224*, 414–422. [CrossRef]
19. Berezina, N.; Hubert, A.; Berro, F.; Levon, J.-G.; Le Roux, K.; Socolsky, C. Chitin, Hydrolysate and Production of at Least One Desired Product from Insects by Means of Enzymatic Hydrolysis, Comprising a Combination of Steps Performed Prior to the Enzymatic Hydrolysis. U.S. Patent Application No. 15/541,162, 2019.
20. Berezina, N.; Hubert, A.; Berro, F.; Levon, J.-G.; Le Roux, K.; Socolsky, C. Chitin, Hydrolysate and Method for the Production of One or More Desired Products by Means of Enzymatic Hydrolysis, Including Pre-Treatment with an Oxidising Agent. U.S. Patent Application No. 15/541,174, 2018.
21. Berezina, N.; Hubert, A.; Berro, F.; Levon, J.-G.; Le Roux, K.; Socolsky, C. Chitin, Hydrolysate and Method for the Production of One or More Desired Products from Insects by Means of Enzymatic Hydrolysis. U.S. Patent Application No. 15/541,186, 2018.
22. Kim, M.-W.; Song, Y.-S.; Han, Y.S.; Jo, Y.H.; Choi, M.H.; Park, Y.-K.; Kang, S.H.; Kim, S.-A.; Choi, C.; Jung, W.-J. Production of chitin and chitosan from the exoskeleton of adult two-spotted field crickets (*Gryllus bimaculatus*). *Entomol. Res.* **2017**, *47*, 279–285. [CrossRef]
23. Jin, Q.; Yu, H.; Wang, X.; Li, K.; Li, P. Effect of the molecular weight of water-soluble chitosan on its fat-/cholesterol-binding capacities and inhibitory activities to pancreatic lipase. *PeerJ* **2017**, *5*, e3279. [CrossRef] [PubMed]
24. Xia, W.; Liu, P.; Zhang, J.; Chen, J. Biological activities of chitosan and chitooligosaccharides. *Food Hydrocoll.* **2011**, *25*, 170–179. [CrossRef]
25. Rabea, E.I.; Badawy, M.E.-T.; Stevens, C.V.; Smagghe, G.; Steurbaut, W. Chitosan as Antimicrobial Agent: Applications and Mode of Action. *Biomacromolecules* **2003**, *4*, 1457–1465. [CrossRef]
26. Luna, G.C.; Martin-Gonzalez, F.S.; Mauer, L.; Liceaga, A. Cricket (*Acheta domesticus*) protein hydrolysates' impact on the physicochemical, structural and sensory properties of tortillas and tortilla chips. *J. Insects Food Feed.* **2021**, *7*, 109–120. [CrossRef]
27. Percot, A.; Viton, A.C.; Domard, A. Optimization of Chitin Extraction from Shrimp Shells. *Biomacromolecules* **2003**, *4*, 12–18. [CrossRef] [PubMed]
28. Elshaarawy, R.F.M.; Mustafa, F.H.A.; Herbst, A.; Farag, A.E.M.; Janiak, C. Surface functionalization of chitosan isolated from shrimp shells, using salicylaldehyde ionic liquids in exploration for novel economic and ecofriendly antibiofoulants. *RSC Adv.* **2016**, *6*, 20901–20915. [CrossRef]
29. Sajomsang, W.; Gonil, P. Preparation and characterization of α-chitin from cicada sloughs. *Mater. Sci. Eng. C* **2010**, *30*, 357–363. [CrossRef]
30. Czechowska-Biskup, R.; Jarosińska, D.; Rokita, B.; Ulański, P.; Rosiak, J.M. Determination of degree of deacetylation of chitosan-comparision of methods. *Prog. Chem. Appl. Chitin Deriv.* **2012**, *17*, 5–20.

31. Domard, A.; Rinaudo, M. Preparation and characterization of fully deacetylated chitosan. *Int. J. Biol. Macromol.* **1983**, *5*, 49–52. [CrossRef]
32. Czechowska-Biskup, R.; Wach, R.A.; Rosiak, J.M.; Ulański, P. Procedure for determination of the molecular weight of chitosan by viscometry. *Prog. Chem. Appl. Chitin its Deriv.* **2018**, *XXIII*, 45–54. [CrossRef]
33. Roberts, G.A.; Domszy, J.G. Determination of the viscometric constants for chitosan. *Int. J. Biol. Macromol.* **1982**, *4*, 374–377. [CrossRef]
34. Chen, R.H.; Tsaih, M.L. Effect of temperature on the intrinsic viscosity and conformation of chitosans in dilute HCl solution. *Int. J. Biol. Macromol.* **1998**, *23*, 135–141. [CrossRef]
35. Panith, N.; Wichaphon, J.; Lertsiri, S.; Niamsiri, N. Effect of physical and physicochemical characteristics of chitosan on fat-binding capacities under in vitro gastrointestinal conditions. *LWT* **2016**, *71*, 25–32. [CrossRef]
36. Zhou, K.; Xia, W.; Zhang, C.; Yu, L. (Lucy) In vitro binding of bile acids and triglycerides by selected chitosan preparations and their physico-chemical properties. *LWT* **2006**, *39*, 1087–1092. [CrossRef]
37. Aguilar-Toalá, J.E.; Deering, A.J.; Liceaga, A.M. New Insights into the Antimicrobial Properties of Hydrolysates and Peptide Fractions Derived from Chia Seed (*Salvia hispanica* L.). *Probiotics Antimicrob. Proteins* **2020**, *12*, 1571–1581. [CrossRef]
38. Kaya, M.; Baran, T.; Asan-Ozusaglam, M.; Cakmak, Y.S.; Tozak, K.O.; Mol, A.; Mentes, A.; Sezen, G. Extraction and characterization of chitin and chitosan with antimicrobial and antioxidant activities from cosmopolitan Orthoptera species (Insecta). *Biotechnol. Bioprocess Eng.* **2015**, *20*, 168–179. [CrossRef]
39. Hu, M.; Gurtler, J.B. Selection of Surrogate Bacteria for Use in Food Safety Challenge Studies: A Review. *J. Food Prot.* **2017**, *80*, 1506–1536. [CrossRef]
40. Kurita, K. Chitin and Chitosan: Functional Biopolymers from Marine Crustaceans. *Mar. Biotechnol.* **2006**, *8*, 203–226. [CrossRef] [PubMed]
41. Berezina, N.; Hubert, A. Marketing and Regulations of Chitin and Chitosan from Insects. *Chitin Chitosan* **2019**, *2019*, 477–489. [CrossRef]
42. Zhang, M.; Haga, A.; Sekiguchi, H.; Hirano, S. Structure of insect chitin isolated from beetle larva cuticle and silkworm (*Bombyx mori*) pupa exuvia. *Int. J. Biol. Macromol.* **2000**, *27*, 99–105. [CrossRef]
43. Purkayastha, D.; Sarkar, S. Physicochemical Structure Analysis of Chitin Extracted from Pupa Exuviae and Dead Imago of Wild Black Soldier Fly (*Hermetia illucens*). *J. Polym. Environ.* **2019**, *28*, 445–457. [CrossRef]
44. Luo, Q.; Wang, Y.; Han, Q.; Ji, L.; Zhang, H.; Fei, Z.; Wang, Y. Comparison of the physicochemical, rheological, and morphologic properties of chitosan from four insects. *Carbohydr. Polym.* **2019**, *209*, 266–275. [CrossRef]
45. Mohan, K.; Ganesan, A.R.; Muralisankar, T.; Jayakumar, R.; Sathishkumar, P.; Uthayakumar, V.; Chandirasekar, R.; Revathi, N. Recent insights into the extraction, characterization, and bioactivities of chitin and chitosan from insects. *Trends Food Sci. Technol.* **2020**, *105*, 17–42. [CrossRef] [PubMed]
46. Marei, N.H.; El-Samie, E.A.; Salah, T.; Saad, G.R.; Elwahy, A.H. Isolation and characterization of chitosan from different local insects in Egypt. *Int. J. Biol. Macromol.* **2016**, *82*, 871–877. [CrossRef] [PubMed]
47. Kaya, M.; Baran, T. Description of a new surface morphology for chitin extracted from wings of cockroach (*Periplaneta americana*). *Int. J. Biol. Macromol.* **2015**, *75*, 7–12. [CrossRef] [PubMed]
48. Kumirska, J.; Czerwicka, M.; Kaczyński, Z.; Bychowska, A.; Brzozowski, K.; Thöming, J.; Stepnowski, P. Application of Spectroscopic Methods for Structural Analysis of Chitin and Chitosan. *Mar. Drugs* **2010**, *8*, 1567–1636. [CrossRef] [PubMed]
49. Liu, S.; Sun, J.; Yu, L.; Zhang, C.; Bi, J.; Zhu, F.; Qu, M.; Jiang, C.; Yang, Q. Extraction and Characterization of Chitin from the Beetle Holotrichia parallela Motschulsky. *Molecules* **2012**, *17*, 4604–4611. [CrossRef] [PubMed]
50. Jang, M.-K.; Kong, B.-G.; Jeong, Y.-I.; Lee, C.H.; Nah, J.-W. Physicochemical characterization of α-chitin, β-chitin, and γ-chitin separated from natural resources. *J. Polym. Sci. Part A Polym. Chem.* **2004**, *42*, 3423–3432. [CrossRef]
51. Ai, H.; Wang, F.; Yang, Q.; Zhu, F.; Lei, C. Preparation and biological activities of chitosan from the larvae of housefly, Musca domestica. *Carbohydr. Polym.* **2008**, *72*, 419–423. [CrossRef]
52. Song, C.; Yu, H.; Zhang, M.; Yang, Y.; Zhang, G. Physicochemical properties and antioxidant activity of chitosan from the blowfly Chrysomya megacephala larvae. *Int. J. Biol. Macromol.* **2013**, *60*, 347–354. [CrossRef]
53. Khayrova, A.; Lopatin, S.; Varlamov, V. Black Soldier Fly Hermetia illucens as a Novel Source of Chitin and Chitosan. *Int. J. Sci.* **2019**, *8*, 81–86. [CrossRef]
54. Raafat, D.; Sahl, H.G. Chitosan and its antimicrobial potential—A critical literature survey. *Microb. Biotechnol.* **2009**, *2*, 186–201. [CrossRef]
55. Varun, T.K.; Senani, S.; Jayapal, N.; Chikkerur, J.; Roy, S.; Tekulapally, V.B.; Gautam, M.; Kumar, N. Extraction of chitosan and its oligomers from shrimp shell waste, their characterization and antimicrobial effect. *Veter. World* **2017**, *10*, 170–175. [CrossRef]
56. Czechowska-Biskup, R.; Rokita, B.; Ulanski, P.; Rosiak, J. Radiation-induced and sonochemical degradation of chitosan as a way to increase its fat-binding capacity. *Nucl. Instrum. Methods Phys. Res. Sect. B Beam Interact. Mater. Atoms* **2005**, *236*, 383–390. [CrossRef]
57. Dimzon, I.K.D.; Ebert, J.; Knepper, T.P. The interaction of chitosan and olive oil: Effects of degree of deacetylation and degree of polymerization. *Carbohydr. Polym.* **2013**, *92*, 564–570. [CrossRef]

58. Kaya, M.; Sargin, I.; Mulerčikas, P.; Labidi, J.; Salaberria, A.M.; Cakmak, Y.S.; Kazlauskaitė, S.; Erdonmez, D.; Baublys, V. Conversion of Waste Parasitic Insect (Hylobius abietis L.) into Antioxidative, Antimicrobial and Biodegradable Films. *J. Renew. Mater.* **2019**, *7*, 215–226. [CrossRef]
59. Fraise, A.; Wilkinson, M.; Bradley, C.; Oppenheim, B.; Moiemen, N. The antibacterial activity and stability of acetic acid. *J. Hosp. Infect.* **2013**, *84*, 329–331. [CrossRef] [PubMed]
60. Halstead, F.D.; Rauf, M.; Moiemen, N.S.; Bamford, A.; Wearn, C.M.; Fraise, A.P.; Lund, P.A.; Oppenheim, B.A.; Webber, M.A. The Antibacterial Activity of Acetic Acid against Biofilm-Producing Pathogens of Relevance to Burns Patients. *PLoS ONE* **2015**, *10*, e0136190. [CrossRef] [PubMed]
61. Ibañez-Peinado, D.; Ubeda-Manzanaro, M.; Martínez, A.; Rodrigo, D. Antimicrobial effect of insect chitosan on Salmonella Typhimurium, Escherichia coli O157:H7 and Listeria monocytogenes survival. *PLoS ONE* **2020**, *15*, e0244153. [CrossRef]

polysaccharides

Article

Chitosan and Collagen-Based Materials Enriched with Curcumin (*Curcuma longa*): Rheological and Morphological Characterization

Eduardo P. Milan [1], Mirella Romanelli V. Bertolo [2,*], Virginia C. A. Martins [2], Stanislau Bogusz Junior [2] and Ana Maria G. Plepis [1,2]

1 Interunits Graduate Program in Bioengineering (EESC/FMRP/IQSC), University of Sao Paulo (USP), Sao Carlos 13560-970, Brazil; epmilan@usp.br (E.P.M.); amplepis@iqsc.usp.br (A.M.G.P.)
2 Sao Carlos Institute of Chemistry, University of Sao Paulo (USP), Sao Carlos 13560-970, Brazil; virginia@iqsc.usp.br (V.C.A.M.); stanislau@iqsc.usp.br (S.B.J.)
* Correspondence: mirella.bertolo@usp.br

Abstract: In this study, chitosan and collagen (Ch: Col)-based materials containing curcumin (Cur) as a bioactive compound were developed for wound-healing purposes. The effects of incorporating curcumin and increasing its concentration on both the rheological properties of the formed solutions and the morphological and thermal properties of the three-dimensional scaffolds obtained from them were evaluated. Rheology showed that the presence of curcumin resulted in solutions with a solid-like behavior (G′ > G″), higher collagen denaturation temperatures, and higher viscosities, favoring their use as biomaterials for wound healing. A greater cross-linking effect was observed at higher curcumin concentrations, possibly between the amino groups from both polymers and the hydroxyl and keto groups from the polyphenol. Such cross-linking was responsible for the delay in the onset of degradation of the scaffolds by 5 °C, as revealed by thermogravimetric analysis. Moreover, the pore diameter distribution profile of the scaffolds changed with increasing curcumin concentration; a greater number of pores with diameters between 40 and 60 μm was observed for the scaffold with the highest curcumin content (50 mg), which would be the most suitable for the proposed application. Thus, the materials developed in this study are presented as promising biomaterials for their biological evaluation in tissue regeneration.

Keywords: chitosan; collagen; curcumin; *Curcuma longa*; scaffolds; rheology

Citation: Milan, E.P.; Bertolo, M.R.V.; Martins, V.C.A.; Bogusz Junior, S.; Plepis, A.M.G. Chitosan and Collagen-Based Materials Enriched with Curcumin (*Curcuma longa*): Rheological and Morphological Characterization. *Polysaccharides* **2022**, 3, 236–249. https://doi.org/10.3390/polysaccharides3010013

Academic Editor: Azizur Rahman

Received: 22 January 2022
Accepted: 15 February 2022
Published: 17 February 2022

1. Introduction

Wounds are discontinuities created in healthy tissue that can be caused by chemical, physical, and immunological processes, among others. The wound-healing process comprises several steps, ranging from healing through inflammation to the induction of cell proliferation [1,2]. One of the main factors that can negatively affect the wound-healing process is the attack of free radicals, which are highly unstable species capable of oxidizing the new cells that have been proliferating around the wound, thus damaging them and delaying the healing process [3].

The use of biomaterials stands out as one of the best alternatives for the wound-healing process due to their biocompatibility, non-toxicity, capacity to induce cell interactions and responses, and their capacity for tissue regeneration [4,5]. In this sense, chitosan and collagen are two options of polymers that are extensively used as biomaterials, individually or in combination [6–9]; chitosan, a polysaccharide derived from chitin, has advantages due its antimicrobial activity and its positive charge in solution [10–12]; collagen, the most abundant protein found in bone and tendons, acts as an extracellular matrix in the control of the function, structure, and shape of tissues [13].

One of the ways to improve both the physicochemical and biological properties of chitosan- and collagen-based biomaterials involves the addition of nutraceutical compounds that also act as cross-linkers; curcumin, a yellowish pigment found in turmeric (*Curcuma longa*), is one of these compounds [14]. Its keto and hydroxyl groups can promote the cross-linkage of polymers, thus increasing their structural stability, and its numerous biological properties (such as anti-inflammatory, anti-fungal, anti-cancer, and anti-invasive properties) also improve the performance of the formulated biomaterials [4,15,16]. Moreover, the proven antioxidant activity of curcumin makes its application in preventing oxidation and scavenging free radicals useful in the wound-healing process.

The interaction between curcumin with both chitosan and collagen has already been elucidated in several studies in the literature, many of which propose the use of these biomaterials for application in wound healing. Dharunya et al. (2016) [4] developed an aerogel based on cross-linked collagen with curcumin, which presented controlled anti-proteolytic activity, making it a suitable 3D scaffold for biomedical applications. In another study, Rezaei et al. (2018) [17] developed chitosan and collagen scaffolds containing curcumin nanoparticles; the scaffolds with the best physicochemical properties were those containing the polymers in a 2:1 ratio (chitosan: collagen), with and without curcumin nanoparticles incorporated. However, the scaffolds without the nanoparticles did not cause any up-regulation in TGF-β1 or Smad7 mRNA expression; therefore, they were unable to accelerate the wound-healing process, while the scaffolds containing curcumin were effective in this acceleration [18]. In a recent study, a combination of chitosan, collagen, and curcumin was tested in the form of electrospun nanofibers containing polyethylene oxide, which also targeted the wound-healing process [19]. Curcumin was successfully released from the nanofibers for up to 3 days and did not cause any significant cytotoxicity to human dermal fibroblasts, which also makes its use and incorporation in other types of biomaterials, such as three-dimensional scaffolds, promising.

As far as we know, the aforementioned studies with the combination of chitosan, collagen, and curcumin did not conduct any rheological characterizations of these materials, which, once well detailed (with oscillation, temperature, and flow tests, for example), can be combined with the thermal, morphological, and structural properties observed for the scaffolds. Moreover, rheology can unravel how the incorporation of curcumin may affect the stability of the polymeric matrix against deformation, shear, and temperature at the molecular level. Thus, the aim of this study was to develop chitosan- and collagen-based materials with different concentrations of curcumin and to analyze their rheological profiles (viscoelastic and steady shear), as well as the structural and thermal characteristics of the scaffolds formed. The materials developed were suitable for the proposed application with their solid-like behavior, high collagen denaturation temperatures, and high viscosities.

2. Materials and Methods

2.1. Materials

All solvents and reagents were of analytical grade and were used as such. The squid pens used as a source of β-chitin for chitosan preparation were obtained at Miami Comércio e Exportação de Pescados Ltda (Cananéa, SP, Brazil). The bovine tendon used for collagen extraction was obtained at Casa de Carnes Santa Paula(São Carlos, SP, Brazil). Curcumin was purchased from a local manipulation drugstore (São Carlos, SP, Brazil).

2.2. Methods

2.2.1. Chitosan Obtention

Chitosan was obtained from squid pens (*Doryteuthis* spp.) through a demineralization, deproteinization, and deacetylation process adapted from Horn, Martins, Plepis (2009) [20]. Briefly, the procedure involved the removal of impurities with a 0.55 mol L^{-1} HCl solution, followed by the extraction of β-chitin in an alkaline medium (0.3 mol L^{-1} NaOH, 1 h, 80 °C); finally, a new alkaline treatment (NaOH 40% *w/w*, 80 °C, 3 h, N_2 flow) led to the deacetylation of β-chitin and obtention of chitosan, which was washed to neutrality and dried. The reaction yield was 26.44%, and the chitosan acetylation degree and molecular weight were 5.68 ± 0.05% and 205 kDa, respectively, as determined according to the procedures developed by Lavertu et al. (2003) [21] and Rinaudo (2006) [22].

2.2.2. Collagen Obtention

The extraction of collagen from the bovine tendon followed the procedure described by Horn, Martins, and Plepis (2009) [20] and started with the removal of unwanted organic materials by washing in a 0.9% (*w/w*) saline solution (NaCl) and in deionized water until completely clean. Then, the tendon was submitted to hydrolysis in an alkaline solution (pH > 12) containing chlorides and sulfates of Na^+, K^+, and Ca^{2+} for 96 h. After this period, the tendon was removed from alkaline conditions and placed in another solution containing sulfates and chlorides of Na^+, K^+, and Ca^{2+} ions for 6 h. The excess salts were removed by washing in a 3% (*w/w*) boric acid solution and deionized water (3 h each), followed by washing in a 0.3% (*w/w*) EDTA solution and deionized water (3 h each). Finally, type I collagen was extracted and maintained in a pH 3.5 acetic acid (HAc) solution. Its concentration of 1% (*w/w*) was determined by weighing after lyophilization in a Freeze Dryer Modulyo model (Edwards High Vacuum International, West Sussex, UK).

2.2.3. Curcumin Purification

The purification process of commercial curcumin was carried out by dissolving the powder in a 1% HAc/ethanol solution (20:1 ratio) at room temperature and under protection from light until complete solubilization. The solution was then filtered, the ethanol was removed by evaporation for 24 h, and the purified curcumin solution was lyophilized for 16 h to obtain a thin powder.

2.2.4. Preparation of Chitosan/Collagen/Curcumin Scaffolds

To prepare the solutions containing chitosan, collagen, and curcumin, both polymers were solubilized separately in 1% (*w/w*) HAc at a concentration of 1% (*w/w*), leading to the Ch and Col solutions. These solutions were mixed in a 1:1 ratio at room temperature, leading to Ch: Col (without curcumin) solution. Next, curcumin (Cur) was dissolved in 1 mL of HAc 1% at concentrations of 10, 20, and 50 mg; 1 mL of Cur solution was mixed with 20 g of the polymeric solutions, resulting in the Ch: Col: Cur10, Ch: Col: Cur20, and Ch: Col: Cur50 solutions. For the preparation of scaffolds, the air was removed from the solutions, which were placed in Teflon® molds, frozen, and lyophilized.

2.2.5. Fourier Transform Infrared Spectroscopy (FTIR)

FTIR analysis was conducted to identify the main functional groups of chitosan, collagen, and curcumin, as well as to verify the interactions between them in the Ch: Col: Cur20 sample. For that, chitosan and curcumin (previously dried) were mixed in $KBr_{(s)}$; collagen, in turn, was diluted in HAc with a pH 3.5 (final concentration of 0.2%), placed in Teflon® molds, and dried under air flow to form a film through the casting method. The equipment used was an FTIR Shimadzu IR Affinity–1 (Shimadzu, Kyoto, Japan), and the spectra were obtained in the region of 4000–400 cm^{-1} with 32 scans and a resolution of 4 cm^{-1}.

2.2.6. Rheological Assays

The rheological study was performed with the Ch: Col, Ch: Col: Cur10, Ch: Col: Cur20, and Ch: Col: Cur50 solutions in an AR-1000 N stress-controlled rheometer (TA Instruments, New Castle, DE, USA) with a cone/stainless steel plate geometry of 60 mm in diameter, angle of $0° 30'$, and a fixed gap of 15 μm. The temperature was controlled with a Peltier system. Prior to measurements, the prepared solutions were stored under refrigeration and protected from light. The rheological assays started with the determination of the linear viscoelastic region (LVR) of the solutions in the strain sweep measurements, from 0.01 to 100 Pa, at a frequency of 1.0 Hz and at 25 °C; then, the behavior of the viscous (G") and elastic (G′) moduli of the solutions were evaluated as a function of temperature from 25 to 75 °C with a heating ratio of 5 °C min^{-1} at a fixed strain and frequency of 10% and 1 Hz, respectively. Finally, flow tests determined the viscosity of the solutions while varying the shear rate from 1 to 1000 s^{-1} at 25 °C.

2.2.7. Thermogravimetric Analysis (TGA)

The thermogravimetric profile of the chitosan/collagen/curcumin scaffolds was assessed with TGA-Q50 equipment (TA Instruments, New Castle, DE, USA). A sample of 5–6 mg was used, with a temperature range from 25 to 800 °C at 10 °C min^{-1} under a synthetic air atmosphere (60 mL min^{-1}).

2.2.8. Scanning Electron Microscopy (SEM)

The morphology of the scaffolds—both the surface and cross-sectional surface—was observed using a ZEISS LEO 440 instrument (Zeiss, Cambridge, UK) with an OXFORD detector (model 7060) and an electron beam of 20 kV. Before the analysis, the scaffolds were covered with a 6 nm gold layer. The diameters of the scaffolds pores and channels were calculated using the UTHSCSA Image Tool software. For each scaffold, 20 measurements were performed in the surface and cross-sectional surface photomicrographs at 500× and 1000× magnification, respectively.

2.2.9. Statistical Analysis

The Shapiro–Wilk test was used to verify the data distribution. The results for the pore and channel size were examined using analysis of variance (ANOVA), followed by Tukey's test, with a significance level of 5%.

3. Results and Discussion

3.1. Fourier Transform Infrared Spectroscopy (FTIR)

The FTIR spectra of chitosan, collagen, curcumin, and the Ch: Col: Cur20 scaffold are shown in Figure 1. The spectra allow the identification of bands referring to the main functional groups of both polymers and curcumin; this was also a structural identification of this compound, which was of commercial origin. In Figure 1a, the characteristic bands of chitosan are observed, with an emphasis on the broad band in the region of 3700–3100 cm^{-1}, which is related to the axial deformations of the O-H and N-H bonds present in the polysaccharide structure (Table 1). Such deformations are also responsible for the band of lower intensity observed in the region of 3300 cm^{-1} for collagen in the spectrum in Figure 1b. Regarding curcumin, the main bonds that lead to the broad band observed between 3360 and 3450 cm^{-1} are the O-H bonds of the phenolic rings present in the curcuminoid structure (Figure 1c).

Other characteristic bands can be observed along the spectra, such as those present in the region of 1660–1653 cm^{-1}; in the case of the polymers, these bands arise due to the stretching of the C=O bonds of the carbonyls that comprise amide I [23,24]. In relation to curcumin, the band at 1660 cm^{-1} refers to the C=O bonds of the beta-diketone found in the central region of the molecule, which can lead to an intermolecular hydrogen atom transfer and to a keto-enol tautomerism [25].

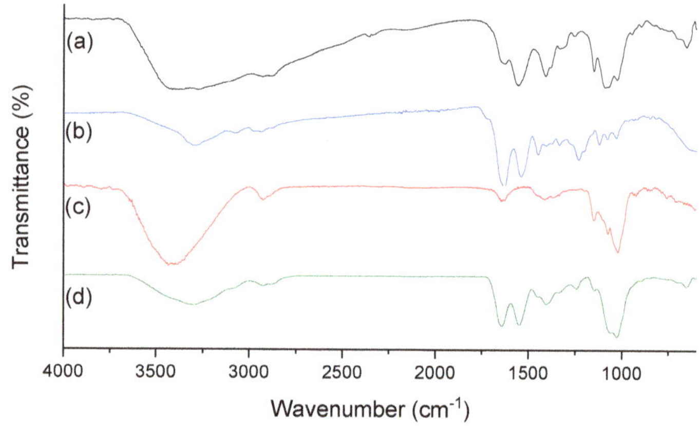

Figure 1. FTIR spectra of chitosan (**a**), collagen (**b**), curcumin (**c**), and Ch: Col: Cur20 scaffold (**d**).

Table 1. Characteristic FTIR bands of chitosan, collagen, and curcumin.

Wavenumber (cm^{-1})	Chitosan	Collagen	Curcumin
3700–3100	O-H and N-H deformation		
3300		O-H and N-H deformation	
3450–3360			O-H phenol deformation
2930–2880	C-H axial deformation		C-H axial deformation
1660			C=O carbonyl and keto-enol tautomerism
1655 (sh) *	Amide I		
1653 (sh) *		Amide I	
1640	O-H water bond		
1560	Amide II		
1558		Amide II	
1456		C-H pyrrolidine rings deformation	
1413	CH_2 axial deformation		
1385	CH_2 out-of-plane deformation		
1400-1350			C-O alcohol and phenol deformation
1240	O-H axial deformation		
1238		Amide III	
1190–960	C-O axial deformation		
1000	O-H and N-H deformation		C-H alkene deformation

* (sh) = stretching bands.

Other prominent bands are the ones in 1624 and 1524 cm^{-1}, which are related to the protonated amino groups of chitosan (NH_3^+), the ones in 1558 and 1238 cm^{-1}, which are characteristic of collagen amides II and III, respectively, and the band at 1000 cm^{-1}, which is related to the deformation of C-H bonds of the alkenes that are present in the structure of curcumin. The presence of these characteristic bands of the three components in the spectrum of Ch: Col: Cur20 (Figure 1d) shows the effectiveness of the incorporation of curcumin in the polymeric matrix, which will be attested by the rheological studies below.

3.2. *Rheological Assays*

3.2.1. Strain Sweep Measurements

The rheological study of chitosan and collagen solutions without curcumin and with curcumin at different concentrations started with the sweeping of their elastic and viscous moduli as a function of %strain; one of the main parameters for the characterization of the viscoelastic behavior of polymeric solutions is their linear viscoelastic region (LVR), that is, the region in which the G′ and G″ moduli remain constant, regardless of the deformation to which the solution is subjected. The LVR is a direct indicator of the resistance of the solution to the applied deformation, and from its extension, it is possible to evaluate the structural strength of the polymeric network formed [26].

Figure 2a shows the graphs of G′ and G″ as a function of %strain; in all cases, the elastic modulus exceeded the viscous one (G′ > G″), which is characteristic of a solid-like behavior [26,27]. This behavior indicates that the solutions of chitosan and collagen had a gel-like structure and that their gel-like response was not affected by the incorporation of curcumin into the system. From the curves, it was possible to determine the parameters shown in Table 2: the first one, γ_{LVR}, is the critical deformation of the solutions, that is, the maximum deformation that can be applied before the LVR ends and the moduli decrease; the greater γ_{LVR} is, the more resistant to the applied deformation the solution will be. Likewise, the G'_{LVR} parameter is the value of the elastic modulus at the limit of the LVR, and it starts to decrease as soon as the LVR ends. For the solutions in this study, the incorporation of curcumin and the increase in its concentration in the chitosan and collagen matrix did not affect γ_{LVR} value, but interfered in the elastic behavior, leading to G'_{LVR} values up to 30 Pa lower in the Ch: Col: Cur10 solution when compared to the solution without curcumin. The increase in curcumin concentration, in turn, led to a gradual increase in the elastic moduli of the solutions, going from 43.96 to 57.46 Pa in Ch: Col: Cur50.

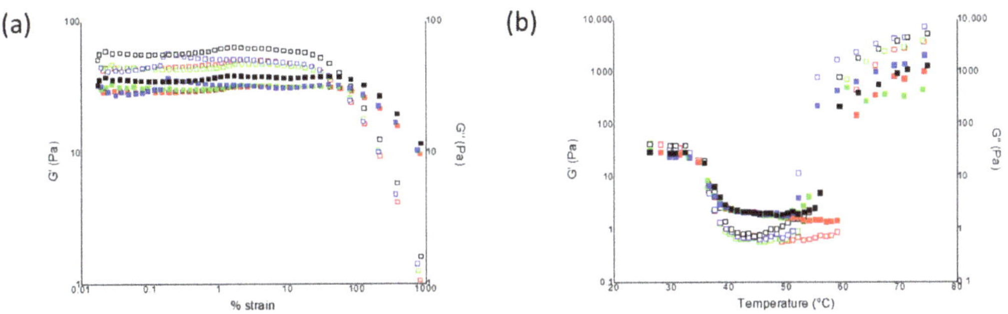

Figure 2. G′ and G″ moduli as a function of %strain (at 1 Hz and 25 °C) (**a**); G′ and G″ moduli as a function of temperature (25 to 75 °C, 5 °C min^{-1}, at 10% strain and 1 Hz) (**b**). G′ (empty symbols) and G″ (full symbols) for Ch: Col (▪), Ch: Col: Cur10 (▪), Ch: Col: Cur20 (▪), and Ch: Col: Cur50 (▪).

Table 2. Rheological parameters obtained from strain sweep, temperature sweep, and flow measurements: γ_{LVR}, critical deformation at the limit of the LVR; G'_{LVR}, elastic modulus value at the limit of the LVR; tanδ, G″/G′ ratio; G′–G″ difference at 1% strain; T and $G'_{crossover}$; η_0, viscosity at zero shear rate, obtained through the adjustment of the curves with the Cross model.

Parameter	Ch: Col	Ch: Col: Cur10	Ch: Col: Cur20	Ch: Col: Cur50
γ_{LVR} (%)	17.9	18.6	17.8	18.5
G'_{LVR} (Pa)	77.6	44.0	46.9	57.5
tanδ	0.5	0.7	0.7	0.6
G′-G″	35.7	13.7	18.5	24.0
$T_{crossover1}$ (°C)	33.8	32.9	34.7	35.3
$G'_{croosver1}$ (Pa)	14.5	15.5	13.8	13.2
$T_{crossover2}$ (°C)	59.8	53.9	51.6	55
$G'_{croosver2}$ (Pa)	7.4	4.1	2.7	3.5
η_0 (Pa s)	35.3	42.8	57.5	61.3

This effect of curcumin incorporation on the decrease in the elastic character of the solutions can also be observed in the tanδ parameter; as this parameter is the ratio between the viscous and elastic moduli, tanδ values < 1 indicate the predominance of G′ [28]. As shown in Table 2, tanδ increased with the incorporation of curcumin into the polymeric matrix (which again points to a decrease in the elastic character of the solutions) and slightly decreased at higher curcumin concentrations.

Finally, the last parameter obtained from the strain sweep curves was the difference between the moduli at 1% strain (inside the LVR); G′–G″ indicates whether the moduli are approaching or retreating from each other and reinforces the effect that the addition of the polyphenols brings to the system. As expected, the difference between the moduli decreased with the incorporation of curcumin, since the elastic modulus decreased. In general, this behavior can be explained by the greater number of interactions that were formed between the polyphenols of curcumin and the polymeric network of chitosan and collagen; phenolic compounds tend to accentuate the viscous moduli of polymeric solutions and, consequently, decrease their elastic character due to the new interactions formed [29]. However, the opposite effects were observed for the strain parameters in the solutions with higher concentrations of curcumin (i.e., a further increase in G'_{LVR} and in G′–G″, as well as a further decrease in tanδ for Ch: Col: Cur20 and Ch: Col: Cur50), which may be an indication of a probable cross-linking of the polymeric network with higher concentrations of polyphenols [30].

3.2.2. Temperature Sweep Measurements

Figure 2b shows the curves of the G′ and G″ moduli as a function of temperature; in general, all solutions presented curves with similar profiles, starting with the predominance of G′ at 25 °C. From 25 to 35 °C, a small plateau was observed in which both the elastic and viscous moduli remained constant, thus reflecting the stability of the polymeric matrix.

Right after this region, the first crossover between the moduli occurred, in which G′ = G″ and G″ soon overcame the elastic modulus; this effect was associated with the thermal denaturation of collagen, whose triple helix collapsed into a random coil [31]. The temperature at which such an effect happens can also be called the collagen denaturation temperature (T_d) and is easily obtained at the point where G′ = G″ or where tanδ reaches its maximum [30]. In this study, the addition of curcumin to the polymeric matrix delayed the denaturation of collagen by about 1.5 °C (from 33.8 °C in Ch: Col to 35.3 °C in Ch: Col: Cur50) (Table 2), which is a new indicator of the possible cross-linking induced by this polyphenol.

From 35 to 50 °C, the elastic modulus of the solutions started to rise again, and in a more accentuated way than the predominant viscous modulus; the second crossover occurred above 50 °C for all solutions, which was an indicator of the gelling process in which G′ became greater than G″ again; this is also representative of a liquid–solid transition. This gelling point is associated with the elimination of energized water molecules at high temperatures, which rearranges the polymeric chains and leads to an increase in the elastic behavior of the solutions [27,29]. The addition of curcumin to the polymeric matrix led to a gelling point that was about 8 °C and 6 Pa lower for the Ch: Col: Cur20 solution when compared to Ch: Col, which was probably related to the initially smaller number of interactions between the polymers and water molecules due to the polyphenols' presence.

3.2.3. Flow Measurements

Figure 3 shows the viscosity curves as a function of the shear rate (s^{-1}) for the solutions; in the adopted shear rate range (from 1 to 1000 s^{-1}), all chitosan and collagen solutions—containing or not containing curcumin—presented a Newtonian behavior, that is, a decrease in viscosity with the application of shear. This behavior was associated with the ordering of polymeric chains, which were initially randomly distributed, when in the presence of shear [31].

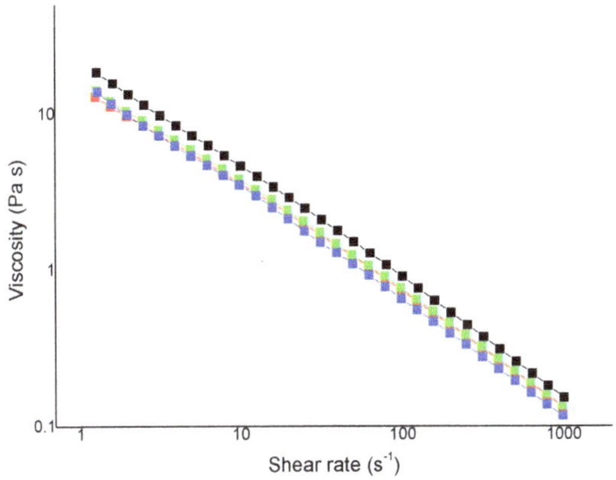

Figure 3. Viscosity (in Pa s) as a function of shear rate (s^{-1}) for Ch: Col (▪), Ch: Col: Cur10 (▪), Ch: Col: Cur20 (▪), and Ch: Col: Cur50 (▪).

From the adjustment of the curves with the Cross Equation (Equation (1), where η_0 is the zero-shear viscosity (Pa s), η_∞ is the viscosity limit at infinite shear (Pa s), γ is the shear rate (s^{-1}), k is the consistency index (s), and n is the rate index), it was possible to estimate the values of η_0, that is, the viscosity at zero shear (Table 2).

$$\frac{\eta - \eta_\infty}{\eta_0 - \eta_\infty} = \frac{1}{(1 + (k\gamma)^n)} \quad (1)$$

The incorporation of curcumin and the increase in its concentration promoted a linear effect of increasing the η_0 value of the solutions from 35.32 Pa s in Ch: Col to 61.27 Pa s in Ch: Col: Cur50. This effect of increasing the viscosity of polymeric solutions with the incorporation of polyphenols was already observed by Almeida et al. (2018) [32], who attributed it to the greater stiffness of a polymeric system containing curcumin.

Therefore, the results of the rheological assays (both oscillatory and steady shear) pointed to a proven interaction of curcumin with the polymeric network of chitosan and collagen, probably by cross-linking the amino groups of the polymeric matrix of collagen and chitosan with the OH and the keto groups of the curcumin structure. Moreover, a clear effect of curcumin concentration on the cross-linking of the polymeric system was observed. This cross-linking led to solutions with a more accentuated solid-like behavior, higher collagen denaturation temperatures, and higher viscosities, thus favoring their use as biomaterials for wound healing and tissue regeneration [30].

3.3. Thermogravimetric Analysis (TGA)

Once the rheological behavior of the chitosan and collagen solutions containing curcumin at different concentrations was elucidated and the effects that the polyphenol addition had on the polymeric matrix were evaluated, the solutions were frozen and lyophilized for the obtention of scaffolds. The thermal stability of the scaffolds was investigated for the scaffolds without curcumin (Ch: Col), as well as for the scaffolds with the lowest and highest amounts of polyphenol (that is, Ch: Col: Cur10 and Ch: Col: Cur50, respectively).

The thermogravimetric curves obtained, as well as their derivatives, are shown in Figure 4. All of the scaffolds with and without curcumin presented four stages of weight loss, whose percentages are displayed in Table 3. The first stage of weight loss, from 25 to 100 °C, represents the loss of absorbed water, ranging from 16.96 to 18.51%; next, the loss of structural water bound to the polymeric molecules in the scaffolds took place from 100 to 200 °C. The incorporation of curcumin led to percentages of bound water that were about 3.6% lower in relation to those of the Ch: Col scaffold (9.08%), which reinforces what the previous rheological results had pointed out: The presence of the curcuminoid decreased the interaction sites in the polymeric network available for water contact, which also led to lower gelling temperatures.

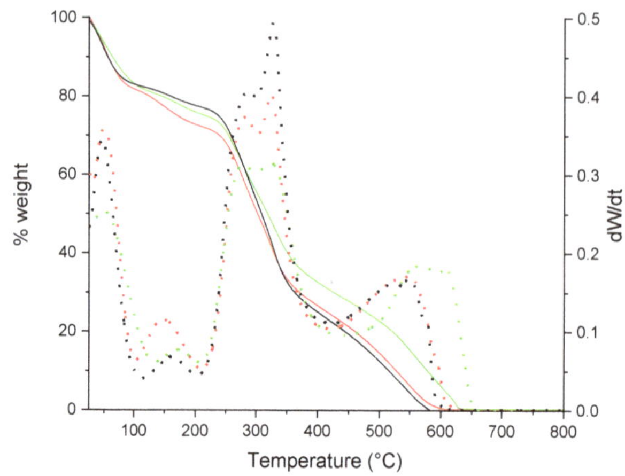

Figure 4. Thermogravimetric curves (solid lines) and their derivatives (dotted lines) for: (–) Ch: Col, (–) Ch: Col: Cur10, and (–) Ch: Col: Cur50.

Table 3. Weight loss and T_{onset} of the scaffolds determined through thermogravimetric analysis.

Scaffold	Weight Loss (%)				T_{onset} (°C)
	25–100 °C	100–200 °C	200–410 °C	410–750 °C	
Ch: Col	18.09	9.08	47.37	25.18	254.38
Ch: Col: Cur10	18.51	5.49	46.8	28.8	259.53
Ch: Col: Cur50	16.96	5.39	53.86	24.19	259.89

The third stage of weight loss of the thermogravimetric curves, from 200 to 410 °C, is related to the degradation of the collagen structure, and it presented the highest weight loss values, as well as the highest derivatives. From the peak of the derivatives in the third stage, it was possible to calculate T_{onset}, that is, the temperature at which the polymeric network started to degrade for each scaffold; again, the results of curcumin incorporation were promising, leading to T_{onset} values that were about 5 °C higher than in the case of the Ch: Col scaffold (Table 3). This result may be related to the cross-linking effect promoted by polyphenol, as mentioned in the rheological results. Finally, the last stage of weight loss, from 410 to 750 °C, refers to the decomposition and carbonization of the degradation products from the previous stage [11].

3.4. Scanning Electron Microscopy (SEM)

3.4.1. Surface and Cross-Sectional Surface Images

Figure 5 shows the appearances of the scaffolds obtained, as well as their surface and cross-sectional surface morphologies. The incorporation of curcumin into the polymeric matrix of chitosan and collagen led to changes in the coloration of the scaffolds, which changed from white (in the case of Ch: Col) to yellowish with the highest concentration of polyphenols (Ch: Col: Cur50). Regarding the surface morphology of the scaffolds (Figure 5a–c), all of them presented interconnected pores in the polymeric network, which is advantageous and desirable for the proposed application, since scaffolds must allow the transport of nutrients, absorb fluids and moisture, and allow cell migration and proliferation [33]. Moreover, there was a clear increase in the number of pores with the inclusion of curcumin compared to the compact matrix observed for Ch: Col. No agglomerates or precipitates of curcumin were observed along the surface of the scaffolds, which indicated the homogeneity of the mixture and the complete miscibility of the polyphenol in the polymeric matrix.

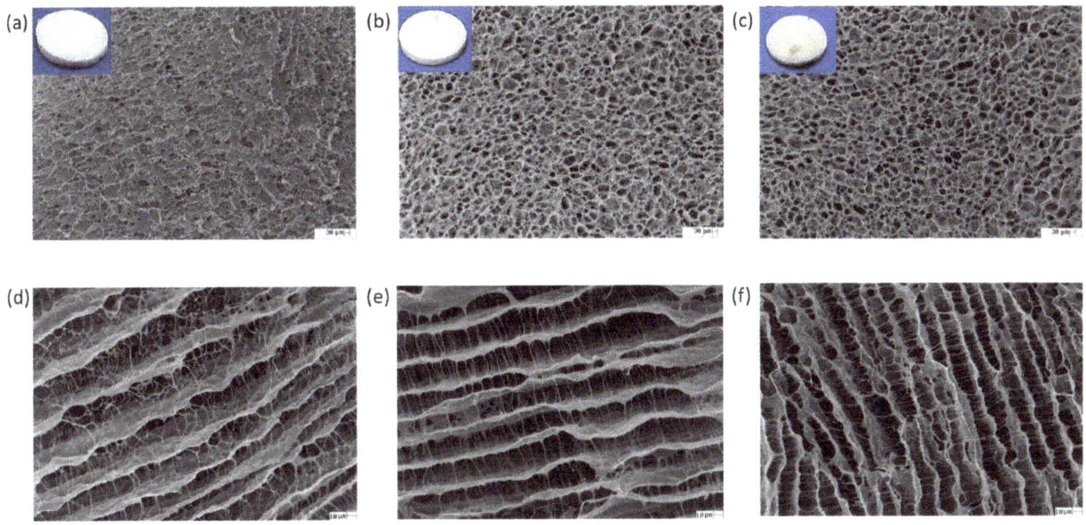

Figure 5. SEM images of the surface (200× magnification) and cross-sectional surface (1000× magnification), respectively, of Ch: Col (**a**,**d**), Ch: Col: Cur 10 (**b**,**e**), and Ch: Col: Cur 50 (**c**,**f**).

Regarding the cross-sectional surface micrographs of the scaffolds (Figure 5d–f), the polymeric fibers were channeled-distributed, and the addition of curcumin in its highest concentration led to a greater compaction of these channels, which was probably due to the cross-linking effect caused by the polyphenol. No visible agglomerates or precipitates of curcumin were observed in these cases either.

3.4.2. Pore and Channel Size

Once the surface and cross-sectional surface morphologies of the chitosan, collagen, and curcumin scaffolds were analyzed, the pore diameters of their surfaces were measured. The pore size of scaffolds intended to be applied as biomaterials in wound healings is one of the factors that most influences their performance and efficiency in the application; ideally, a scaffold must have pores with an adequate size to absorb moderate amounts of fluids (such as exudates that are released from the wound or injury) and to allow the maintenance of a moist environment in the wound bed at the same time [34].

Table 4 shows the average pore diameter values obtained for each sample. Although no statistically significant difference was observed between the samples, a clear tendency towards an increase in the pore diameter was noted for the scaffold with the highest concentration of curcumin, Ch: Col: Cur50, with an average diameter that was about 10 μm larger than the average diameter of the scaffolds without curcumin (Ch: Col) and the scaffolds with curcumin in a lower concentration (Ch: Col: Cur10). This effect of increasing the pore diameter of scaffolds based on biopolymers with the inclusion of bioactive compounds, such as phenols and polyphenols, was already observed by Bertolo et al. (2020) [35]; the inclusion of pomegranate peel extract (rich in phenolic compounds) in chitosan and gelatin scaffolds led to an increase in the diameter of the surface pores.

Table 4. Average pore size and channel size values for the scaffolds.

Scaffold	Pore Size (μm)	Channel Size (μm)
Ch: Col	35.15 ± 10.86 [a]	36.47 ± 7.30 [a]
Ch: Col: Cur10	35.14 ± 12.64 [a]	28.62 ± 2.36 [b]
Ch: Col: Cur50	45.31 ± 8.75 [a]	18.48 ± 2.30 [c]

[a, b, c] In the same column, values with different superscript letters indicate statistically different samples according to ANOVA and the Tukey test ($p \leq 0.05$).

Regarding the diameter of the channels observed in the cross-sectional images of the scaffolds, the effect of curcumin incorporation was opposite to the effect on the surface pores; Table 4 shows the average values obtained for the channels in each scaffold; in general, polyphenol promoted a greater compaction of the scaffolds' structure, leading to channels that were less spaced from each other. The difference in the size of the channels was statistically significant among the three scaffolds, and in the case of Ch: Col: Cur50, the channels (18.48 ± 2.30 μm) presented half the average diameter observed for Ch: Col (36.47 ± 7.30 μm).

Figure 6 shows the distribution of the pore diameter values measured for each of the scaffolds; the sample without curcumin (Figure 6a) presented a higher number of pores with diameters between 20 and 40 μm; with the addition of curcumin at its lowest concentration (Figure 6b), there was a shift of this higher number of pores to the region of 40–50 μm. When the amount of curcumin was increased (Figure 6c), the pore diameter distribution profile of the scaffolds changed completely, and a greater number of pores with diameters between 40 and 60 μm was observed.

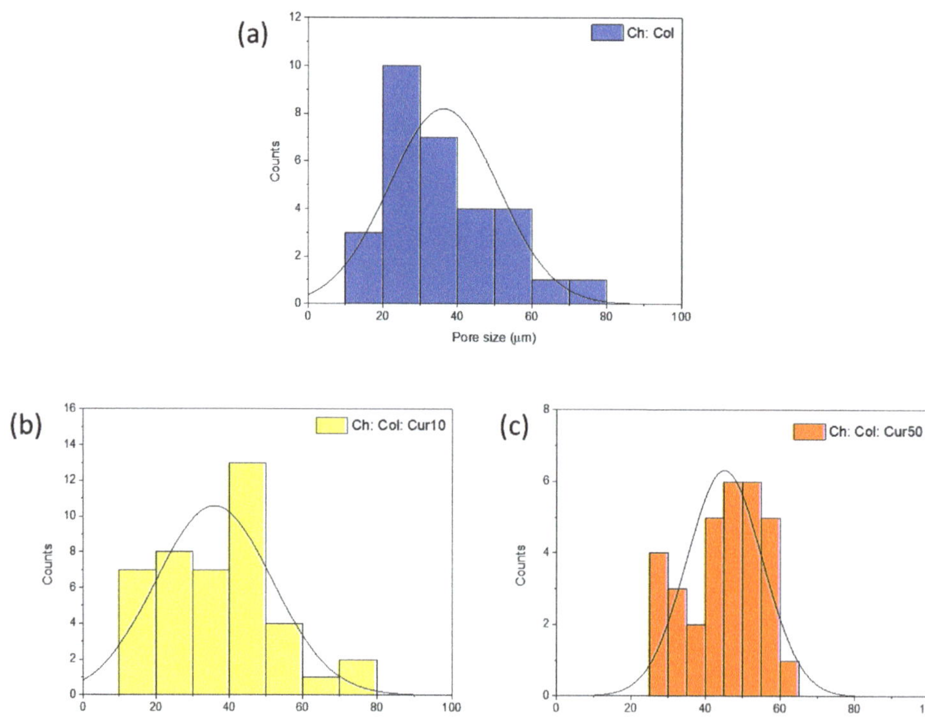

Figure 6. Pore size distribution for Ch: Col (**a**), Ch: Col: Cur10 (**b**), and Ch: Col: Cur50 (**c**).

Many studies in the literature have dealt with the optimal pore size for scaffolds that will be applied for tissue regeneration; the pore size is affected by many factors, including the choice of biopolymers and the presence of bioactive compounds (such as curcumin in this study). The type of tissue to be regenerated and that will be in contact with the scaffold also affects the optimal range desired. Ideally, pores with a range of 90 to 130 μm in diameter allow cell migration and proliferation, but in vitro results of cell growth in pores with diameters smaller than 100 μm have already been reported [36]. Furthermore, reports in the literature have already noted that microvascular epithelial cells generally require even smaller pores of about 40 μm for proliferation and differentiation [37]. In this case, the Ch: Col: Cur50 scaffold would be the most suitable for application in wound-healing applications, since it has a greater number of pores with diameters of around 40 μm and a structure with a greater number of channels that are interconnected with each other by smaller distances.

4. Conclusions

Curcumin was evaluated as a valuable bioactive compound in the development of chitosan- and collagen-based materials for wound-healing purposes due to its antioxidant, anti-inflammatory, and antiseptic properties. The effects of curcumin incorporation and concentration were evaluated for the rheological properties of Ch: Col solutions; the polyphenols decreased the elastic behavior of the polymers at low concentrations, but the opposite tendency of further increasing G′ was observed with higher curcumin concentrations (20 and 50 mg). Likewise, both the elastic and viscous moduli were affected by temperature, and curcumin incorporation delayed the denaturation temperature of collagen (T_d) of the scaffolds by about 1.5 °C. The TGA results also showed promising effects of curcumin on delaying the temperature at which scaffolds start to degrade by about 5 °C. The cross-linking brought by the polyphenols was also reflected in the greater viscosity and stiffness of the solutions, as well as in the greater compaction of the polymer channels in the cross-sectional surface images obtained with SEM. The diameters of the surface pores increased according to the increase in curcumin concentration, ranging mainly from 40 to 60 μm for Ch: Col: Cur50, the most suitable scaffold for the continuation of the biological and in vitro studies that are necessary for the proposed application.

Author Contributions: Conceptualization, E.P.M., V.C.A.M. and A.M.G.P.; methodology, E.P.M. and V.C.A.M.; writing—original draft preparation, E.P.M. and M.R.V.B.; writing—review and editing, V.C.A.M., S.B.J. and A.M.G.P.; supervision, A.M.G.P.; project administration, A.M.G.P. All authors have read and agreed to the published version of the manuscript.

Funding: This study was supported by the Coordenação de Aperfeiçoamento de Pessoal de Nível Superior—Brasil (CAPES)—Finance Code 001.

Institutional Review Board Statement: Not applicable.

Informed Consent Statement: Not applicable.

Acknowledgments: The authors would like to thank the Center of Analytical Chemical Analysis of IQSC/USP for the infrastructure that was made available for the FTIR and SEM analyses. We also thank Eder Tadeu Gomes Cavalheiro and Ana Paula Garcia Ferreira (IQSC/USP) for the access to the TGA-Q50 instrument. Finally, we thank the Fundação de Amparo à Pesquisa do Estado de São Paulo (FAPESP) for the grant to M.R.V.B. (2019/18748-8).

Conflicts of Interest: The authors declare no conflict of interest.

References

1. Pattanayak, S.P.; Sunita, P. Wound healing, anti-microbial and antioxidant potential of *Dendrophthoe falcata* (L.f) Ettingsh. *J. Ethnopharmacol.* **2008**, *120*, 241–247. [CrossRef] [PubMed]
2. Yu, R.; Zhang, H.; Guo, B. Conductive biomaterials as bioactive wound dressing for wound healing and skin tissue engineering. *Nano-Micro Lett.* **2022**, *14*, 751. [CrossRef] [PubMed]

3. Wu, M.; Lu, Z.; Wu, K.; Nam, C.; Zhang, L.; Guo, J. Recent advances in the development of nitric oxide-releasing biomaterials and their application potentials in chronic wound healing. *J. Mater. Chem. B* **2021**, *9*, 7063–7075. [CrossRef] [PubMed]
4. Dharunya, G.; Duraipandy, N.; Lakra, R.; Korapatti, P.S.; Jayavel, R.; Kiran, M.S. Curcumin cross-linked collagen aerogels with controlled anti-proteolytic and pro-angiogenic efficacy. *Biomed. Mater.* **2016**, *11*, 045011. [CrossRef]
5. Cheng, A.; Schwartz, Z.; Kahn, A.; Li, X.; Shao, Z.; Sun, M.; Ao, Y.; Boyan, B.D.; Chen, H. Advances in porous scaffold design for bone and cartilage tissue engineering and regeneration. *Tissue Eng. Part B Rev.* **2019**, *25*, 14–29. [CrossRef]
6. Amiri, N.; Ajami, S.; Shahroodi, A.; Jannatabadi, N.; Amiri Darban, S.; Fazly Bazzaz, B.S.; Pishavar, E.; Kalalinia, F.; Movaffagh, J. Teicoplanin-loaded chitosan-PEO nanofibers for local antibiotic delivery and wound healing. *Int. J. Biol. Macromol.* **2020**, *162*, 645. [CrossRef]
7. Bayat, S.; Amiri, N.; Pishavar, E.; Kalalinia, F.; Movaffagh, J.; Hashemi, M. Bromelain-loaded chitosan nanofibers prepared by electrospinning method for burn wound healing in animal models. *Life Sci.* **2019**, *229*, 57–66. [CrossRef]
8. Chattopadhyay, S.; Raines, R.T. Collagen-based biomaterials for wound healing. *Biopolymers* **2014**, *101*, 821–833. [CrossRef]
9. Chen, Y.; Lee, K.; Yang, Y.; Kawazoe, N.; Chen, G. PLGAcollagen-ECM hybrid meshes mimicking stepwise osteogenesis and their influence on the osteogenic differentiation of hMSCs. *Biofabrication* **2020**, *12*, 025027. [CrossRef]
10. Elhendawi, H.; Felfel, R.M.; El-Hady, A.; Bothaina, M.; Reicha, F.M. Effect of synthesis temperature on the crystallization and growth of in situ prepared nanohydroxyapatite in chitosan matrix. *Biomaterials* **2014**, *2014*, 897468. [CrossRef]
11. Bertolo, M.R.V.; Martins, V.C.A.; de Guzzi Plepis, A.M. Effects of calcium phosphates incorporation on structural, thermal and drug-delivery properties of collagen:chitosan scaffolds. *Int. J. Adv. Med. Biotechnol.-IJAMB* **2020**, *2*, 25–35. [CrossRef]
12. Sultankulov, B.; Berillo, D.; Sultankulova, K.; Tokay, T.; Saparov, A. Progress in the development of chitosan-based biomaterials for tissue engineering and regenerative medicine. *Biomolecules* **2019**, *9*, 470. [CrossRef] [PubMed]
13. Lin, K.; Zhang, D.; Macedo, M.H.; Cui, W.; Sarmento, B.; Shen, G. Advanced collagen-based biomaterials for regenerative biomedicine. *Adv. Funct. Mater.* **2019**, *29*, 1804943. [CrossRef]
14. Hatcher, H.; Planalp, R.; Cho, J.; Torti, F.M.; Torti, S.V. Curcumin: From ancient medicine to current clinical trials. *Cell. Mol. Life Sci.* **2008**, *65*, 1631–1652. [CrossRef]
15. Abbas, M.; Hussain, T.; Arshad, M.; Ansari, A.R.; Irshad, A.; Nisar, J.; Hussain, F.; Masood, N.; Nazir, A.; Iqbal, M. Wound healing potential of curcumin cross-linked chitosan/polyvinyl alcohol. *Int. J. Biol. Macromol.* **2019**, *140*, 871–876. [CrossRef] [PubMed]
16. Fathima, N.N.; Devi, R.S.; Rekha, K.B.; Dhathathreyan, A. Collagen–curcumin interaction—A physico-chemical study. *J. Chem. Sci.* **2009**, *121*, 509–514. [CrossRef]
17. Rezaei, M.; Oryan, S.; Nourani, M.R.; Mofid, M.; Mozafari, M. Curcumin nanoparticle-incorporated collagen/chitosan scaffolds for enhanced wound healing. *Bioinspired Biomim. Nanobiomater.* **2018**, *7*, 159–166. [CrossRef]
18. Rezaei, M.; Oryan, S.; Javeri, A. Curcumin nanoparticles incorporated collagen-chitosan scaffold promotes cutaneous wound healing through regulation of TGF-β1/Smad7 gene expression. *Mater. Sci. Eng. C* **2019**, *98*, 347–357. [CrossRef]
19. Jirofti, N.; Golandi, M.; Movaffagh, J.; Ahmadi, F.S.; Kalalinia, F. Improvement of the wound-healing process by curcumin-loaded chitosan/collagen blend electrospun nanofibers: In vitro and in vivo studies. *ACS Biomater. Sci. Eng.* **2021**, *7*, 3886–3897. [CrossRef]
20. Horn, M.M.; Martins, V.C.A.; Plepis, A.M.G. Interaction of anionic collagen with chitosan: Effect on thermal and morphological characteristics. *Carbohydr. Polym.* **2009**, *77*, 239–243. [CrossRef]
21. Lavertu, M.; Xia, Z.; Serreqi, A.N.; Berrada, M.; Rodrigues, A.; Wang, D.; Buschmann, M.D.; Gupta, A. A validated ^{1}H NMR method for the determination of the degree of deacetylation of chitosan. *J. Pharm. Biomed. Anal.* **2003**, *32*, 1149–1158. [CrossRef]
22. Rinaudo, M. Chitin and chitosan: Properties and applications. *Prog. Polym. Sci.* **2006**, *31*, 603–632. [CrossRef]
23. Cai, L.; Shi, H.; Cao, A.; Jia, J. Characterization of gelatin/chitosan ploymer films integrated with docosahexaenoic acids fabricated by different methods. *Sci. Rep.* **2019**, *9*, 8375. [CrossRef] [PubMed]
24. Batista, T.M.; Martins, V.C.A.; Plepis, A.M.G. Thermal behavior of in vitro mineralized anionic collagen matrices. *J. Therm. Anal. Calorim.* **2009**, *95*, 945–949. [CrossRef]
25. Chen, X.; Zou, L.-Q.; Niu, J.; Liu, W.; Peng, S.-F.; Liu, C.-M. The stability, sustained release and cellular antioxidant activity of curcumin nanoliposomes. *Molecules* **2015**, *20*, 14293–14311. [CrossRef]
26. Steffe, J.F. *Rheological Methods in Food Process Engineering*; Freeman Press: East Lansing, MI, USA, 1996.
27. Silva-Weiss, A.; Bifani, V.; Ihl, M.; Sobral, P.; Gómez-Guillén, M.C. Structural properties of films and rheology of film-forming solutions based on chitosan and chitosan-starch blend enriched with murta leaf extract. *Food Hydrocoll.* **2013**, *31*, 458–466. [CrossRef]
28. Razmkhah, S.; Mohammad, S.; Razavi, A. Dilute solution, flow behavior, thixotropy and viscoelastic characterization of cress seed (*Lepidium sativum*) gum fractions. *Food Hydrocoll.* **2017**, *63*, 404–413. [CrossRef]
29. Romanelli Vicente Bertolo, M.; Leme, R.; da Conceição Amaro Martins, V.; de Guzzi Plepis, A.M.; Bogusz Junior, S. Rheological characterization of the influence of pomegranate peel extract addition and concentration in chitosan and gelatin coatings. *Polysaccharides* **2021**, *2*, 648–660. [CrossRef]
30. Kaczmarek, B.; Mazur, O. Collagen-based materials modified by phenolic acids—A review. *Materials* **2020**, *13*, 3641. [CrossRef]
31. Mandala, I.; Savvas, T.; Kostaropoulos, A. Xanthan and locust bean gum influence on the rheology and structure of a white model-sauce. *J. Food Eng.* **2004**, *64*, 335–342. [CrossRef]

32. Almeida, C.M.R.; Magalhães, J.M.C.S.; Souza, H.K.S.; Gonçalves, M.P. The role of choline chloride-based deep eutectic solvent and curcumin on chitosan films properties. *Food Hydrocoll.* **2018**, *81*, 456–466. [CrossRef]
33. Leong, K.F.; Chua, C.K.; Sudarmadji, N.; Yeong, W.Y. Engineering functionally graded tissue engineering scaffolds. *J. Mech. Behav. Biomed. Mater.* **2008**, *1*, 140–152. [CrossRef] [PubMed]
34. Negut, I.; Dorcioman, G.; Grumezescu, V. Scaffolds for wound healing applications. *Polymers* **2020**, *12*, 2010. [CrossRef] [PubMed]
35. Bertolo, M.R.V.; Martins, V.C.A.; Horn, M.M.; Brenelli, L.B.B.; Plepis, A.M.G. Rheological and antioxidant properties of chitosan/gelatin-based materials functionalized by pomegranate peel extract. *Carbohydr. Polym.* **2020**, *228*, 115386. [CrossRef]
36. Chong, E.J.; Phan, T.T.; Lim, I.J.; Zhang, Y.Z.; Bay, B.H.; Ramakrishna, S.; Lim, C.T. Evaluation of electrospun PCL/gelatin nanofibrous scaffold for wound healing and layered dermal reconstitution. *Acta Biomater.* **2007**, *3*, 321–330. [CrossRef]
37. Bruzauskaite, I.; Bironaite, D.; Bagdonas, E.; Bernotiene, E. Scaffolds and cells for tissue regeneration: Different scaffold pore sizes—different cell effects. *Cytotechnology* **2016**, *68*, 355–369. [CrossRef]

polysaccharides

MDPI

Article

Reversible Metal Ion/Complex Binding to Chitin Controlled by Ligand, Redox, and Photochemical Reactions and Active Movement of Chitin on Aquatic Arthropods

Stefan Fränzle * and Felix Blind

IHI (International Institute) Zittau, TU Dresden, Markt 23, D-02763 Zittau, Germany; felix.blind@mailbox.tu-dresden.de

* Correspondence: stefan.fraenzle@tu-dresden.de

Abstract: There is strong adsorption of metal ions and their complexes to chitin, which depends on both the oxidation and complexation states of many of the said elements (whereas others display chemical reactions detectable via electrochemical methods while being retained by chitin); thus, ad- and desorption at ambient water concentrations (often in the nMol/L range) are controlled by the presence and photochemical properties (concerning Eu and probably U and Ag) of mainly biogenic organic matter (both DOC and POC, and DON). With chitin forming the outer hull of mobile organisms (animals), this biopolymer is expected to take part in metal distribution in aquatic (limnetic and riverine) ecosystems. Having studied the attachment of many different elements to both crayfish and grafted (marine shrimp) chitin, with the highest accumulations observed in Bi, V, Ni, and LREEs, one should consider secondary biochemical transformations which take place at different water and sediment levels. After chitin had been embedded into sediment, methanogenesis (which requires Ni), Bi, and Sb biomethylations and photodesorption in the illuminated water column will occur if there are appropriate organics, causing the vertical separation of Eu from other REEs, at least during the daytime. Eutrophication will enhance both the production and especially the photooxidation rates of organics in water because phosphorylated sugars and lipids are formed quantitatively within min P, which enter water and undergo Eu-mediated photooxidation much more readily. Another biopolymer, gelatin, acts as an inert matrix-enhancing organic photooxidation product via Eu, producing chemical waves, indicating autocatalysis upon light impact. From the redox-related photodesorption of metal analytes from chitin, both sensors and devices for (light-assisted) electrochemical energy conversion are being developed by our workgroup. The electrochemical determination of adsorption thermodynamics on chitin is thus directly linked to its applications in environmental monitoring and technology.

Keywords: chitin–metal ion interactions; redox; photoredox desorption of M; daily vertical migration of aquatic arthropods; mineral (catalyst) formation from decaying chitin; Eu photochemistry; devices modeling chitin-related M dynamics on arthropods and their electrochemical readout

check for updates

Citation: Fränzle, S.; Blind, F. Reversible Metal Ion/Complex Binding to Chitin Controlled by Ligand, Redox, and Photochemical Reactions and Active Movement of Chitin on Aquatic Arthropods. *Polysaccharides* **2022**, *3*, 515–543. https://doi.org/10.3390/polysaccharides3030031

Academic Editor: Azizur Rahman

Received: 12 April 2022
Accepted: 14 July 2022
Published: 28 July 2022

1. Introduction

1.1. Introduction

Chitin was previously employed to purify wastewater from HM ions [1,2] and fractionate radionuclides obtained in nuclear technology (produced via the fission and neutron irradiation of $^{235;238}U$ and HREE nuclides) [3]. Chitin was also demonstrated to act as the site/basis of photoinduced and possibly thermoelectric electric effects [4], concerning wing interspace and egg cocoons of the hornet *Vespa orientalis*. Here, however, photoactivity is not caused by the content/adsorption of metal ions or their complexes, but by that of xanthopterin.

Infrared spectroscopy is used to detect deformation/bond pair bending and vibration along the bond axes of some di- or polyatomic dissymmetrical (otherwise, e.g., in N_2 or CO_2, transitions are "forbidden" and very weak) molecules or ions, including polymers. Certain structural motifs correspond to (that is, can be identified by) certain IR absorption bands which generally are located between some 3 and 40 μm wavelengths. Hence, chemical reactions that modify such a molecule are likely to cause changes in IR spectra, too, and thus a comparison of IR spectra taken before and after some process can be used to constrain reactions that do or do not take place during interaction with certain reagents. For IR spectra of both chitin and chitosan, see [5]. Possible reagents which might modify chitin in this framework include:

- Hot water, possibly causing the hydrolysis of chitin acetamide groups to yield chitosan or even (N-acetyl-)glucosamine;
- Adsorbed and photoexcited Eu(III);
- Free radicals made from co-adsorbed organic matter attacked by Eu(III)*.

Furthermore, REE ions are known to catalyze the ring-opening polymerization of lactones (with gluconolactone being a product of glucose photooxidation; the ring-opening polymerization method was employed in the preparation of a PLA "biogenic" polymer [6]) and sometimes lactams. In turn, the processes discussed in this paper are unlikely to go on "forever", yet the reactions will likely go on during timescales of zooplankton between molting episodes, given our data on the continuation of photochemical transformations via Eu(III) retained by chitin. Concerning Eu and REEs $\neq$ Eu partitions on and in arthropods, it is likely that different REEs follow different pathways of uptake and release in ambient (illuminated) conditions. The binding capacity of metal ions including Eu(III) on chitin is rather constant [7], with a few exceptions [8].

However, there has been little theoretical or even practical concern for the ecological impacts of this adsorption, e.g., when animals feed on arthropods, lichens covered by chitin and inhabiting areas with high HM (e.g., Pb, Cd, Cu, Zn, Bi, V, and REEs) or toxic LM (Al, Ba, and Be) levels, or the role of chitin in ecosystem-scale element transport. Nor did people consider effects other than pH that might influence adsorption equilibria, such as the presence of ligands or the oxidation state of metal ions. We did so in recent years [9,10]. Now, we are:

(1) Constructing model devices to study the photochemical effects of chitin's retention of certain ions, especially Eu, with an emphasis on:
(2) Photo-assisted electrochemical energy conversion [11,12];
(3) Ecological consequences involving vertical aquatic transport and fractionation by influences of light, O_2, and waterborne ligands [10,13] on metal uptake, release, and sediment deposition by chitin shrouds in mobile aq. microarthropods.

In addition, water flows might bring oxidation or reduction products no longer binding to chitin to remote sites (caves, etc.). The former, especially chromate, is toxic and genotoxic [14], and the latter reductions possibly permit the production of food organics without light or chemical energy sources other than those made available by the movement of chitin or adsorbed items. Different speciation forms of elements binding to chitin differ in binding stability, including complexes [10,15], (pnictogen) biomethylation products [16], and different oxidation states. The interconversion of speciation forms might occur through oxidants, reductants, or even photochemical processes; the latter include:

- Photoredox—(Eu and U, possibly Fe and Cu);
- The cleavage of M-CH_3- (pnictogens, Hg, Sn, and Pb);
- M′-CO bonds, with $[M(CO)_6]$ and $[Fe(CO)_5]$ formed in and vented from landfills [17].

Both the exposition of arthropods, lichens, and animals feeding on either and element cycles in which aq. arthropod mobility is included would change the distribution/partition of elements among chitin, water, and sediment. Concerning REEs, previous data on the abundances of Y, La, Sm, Gd, and Yb in a model tank accommodating daphnia and mussels partly covered by chitin [18] in addition to goldfish, water plants, and sediment can be linked to our previous work on ligand's effects on REE adsorption on chitin surfaces [9,10]. There are photochemical effects of Eu (not studied by Yang) which can be exploited in the conversion of light and chemical energy [11] and surely will influence the distribution of Eu in such (tank or open-water) ecosystems once there are mobile carriers of chitin, i.e., animals, especially arthropods. The argument will extend to other elements involved in photochemical processes, while chemoclines inducing changes in the redox states of elements should be less important in arthropods as these animals are mostly obligately aerobic. An exception is the dynamics of Mn, which does change oxidation state near the O_2/water line in the Pourbaix diagram [19], with MnO_2 then catalyzing a multitude of oxidations. Test cells were described to study such effects with an emphasis on Eu, since the latter was previously often used as a redox tracer in geochemistry, often neglecting possible influences of photochemistry. We could show that Eu can be readily photoreduced by quite different organic compounds [11,12], with the Eu^{2+} product then losing the adsorptive grip to chitin. Accordingly, Eu should undergo fractionation from other REEs and trivalent elements in illuminated parts of the biosphere and hydrosphere. Unless there are high ambient concentrations and local enrichments in mine tailing waters, it is hard to study photochemical effects on Eu distribution and Eu dynamics in open fields. In addition, there is just a single site on Earth where Eu-rich particles (in fact, radioactive particles with prevailing ^{152}Eu content among radionuclides) have been released into open surface waters: particles (with some 200 in total being isolated) in the River Yenisei near the Zheleznogorsk military nuclear fuel processing plant (formerly called "Krasnoyarsk-26") in which Eu was left over from neutron flow control rods used in Pu production reactors [20]. Here, fortunately, Eu was kept in partly sintered particles in a way that meant leaching and liquid-phase dynamics with or without the involvement of the said photochemistry could not occur and be studied. Additionally, Eu is more mobile than other REEs co-leached from bared lignite when facing illumination [12,13].

Data from test cells filled with chitin, which emulate flows of water and organics over Eu(III)-loaded chitin, illuminated in various aspects and manners, can add important information to our understanding of element cycles in lakes and ponds. This includes exposition estimates for benthic, periodically migrating organisms, and the range of common biogeochemical redox tracers such as Eu, Ce, and Cr can be considerably extended. Here, photochemical contributions and those of oxidations catalyzed by MnO_2 on the chitin adsorption/desorption of different elements needed to be considered. MnO_2 in turn might form [21] when Mn^{2+} is leached from chitin in oxygenated (a precondition to sustain life in most arthropods) and slightly alkalinic conditions, and then oxidizes quite a sizable number of other elements such as Cr, Ce, and Cl.

1.2. Hypothesis

The metal fractionation process promoted by the muscle activity and chitin cover of arthropods is depicted in Figure 1.

Eu(III) adsorption to chitin is known to surpass that of other elements including REEs $\neq$ Ce [8] and depends on the oxidation state; the only ligand that is prone to photooxidation by Eu(III) (11] yet is capable of increasing the adsorption of this ion is glycinate [10]. Formation constants of REE(III) (REE = La, Ce, Pm, and Eu) glycinatocomplexes in water are extremely small ($\log \beta < 1$) [22], whereas the electrostatic effects of protonated glycine at a negatively charged chitin interface prevail in the enhanced adsorption of La or Eu [10]. Photochemistry mediated by Eu(III) and product patterns for simple carboxylates and the corresponding acids (pH range 1.5–6) were studied by [23], and Mn-mediated oxidation pathways were studied by [24], whereas Eu organic ligand complexes should be of minor importance at the ligand concentrations observed in open waters [25,26].

The release of Eu only occurs if:

(a) Chitin-covered plankton dwells close to the surface during daylight hours;
(b) There is sufficient and appropriate $C;N_{org}$ in water.

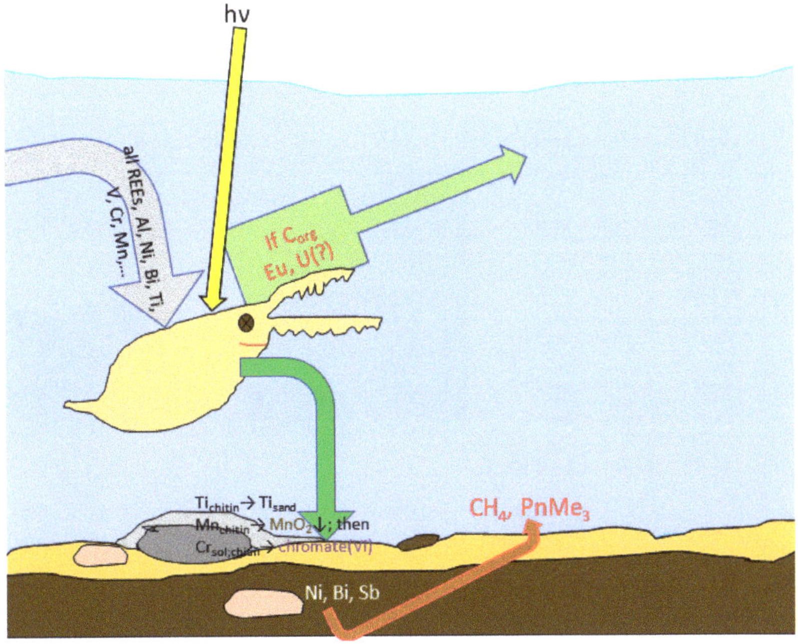

Figure 1. Daphnias bind (gray arrow) and transport (dark green arrow) metal ions by their diurnal migration patterns. If there is photoactive or especially phosphorylated organic matter, Eu (and U) will be released next to surface. Nighttime contact with sediment possibly transfers some components to the surface of arthropods (bottom). Actual quantitative extent of the depicted process is discussed in this paper.

However, the levels of nitrate/other nutrients control the abundance and growth rates of pico- and phytoplankton; the latter produces and gives away organic matter suitable for photooxidation by Eu (humic acids and certain free and fatty oligosaccharides). The nocturnal presence of chitin-covered animals near the bottom of a water body (Figure 2) allows elements to be taken up directly or indirectly from sediment parts, which can only be released when the above conditions are met.

In turn, the C/P (dissolved) ratio is known to determine the biodiversity and productivity (in terms of organics released from microorganisms such as all POC, DOC, and DON) of some water volumes. It is thus commonly used as a simple measure of the extent of eutrophication [27]. Previous work showed that even sugars and other organic compounds stable towards Eu(III)-mediated photooxidation, such as ribose or mannose, can be readily processed after phosphorylation [12]. It is unlikely that this effect is related to the complexation of phosphate ester groups because other compounds such as toluene, naphthalene, or chlorobenzene [14] do react with excited Eu(III), although they are not coordinated by Eu(III) or other REEs. It rather is an activation very similar to that seen in biology. The coupled chitin-bound Eu/P/organics cycle inferred from these statements and experiments is shown in the following Figure 2.

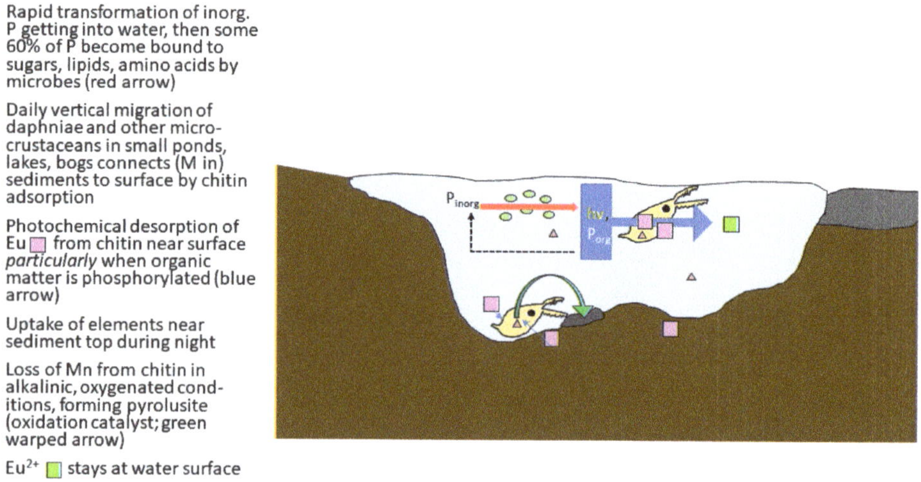

Figure 2. Phosphorus (P) cycling in limnetic biotopes interferes with transport of element by Eu because P-ated dissolved organic matter is much more susceptible to photooxidation by Eu(III) (light pink squares on top (insolated) water flea and in solution)) than others. Phosphorylation is accomplished by (cyano-)bacteria within minutes [28] and partly reversed by the said photooxidation and extracellular phosphatases (broken arrow) (cp. [29]), while Eu^{2+} is lost (desorbed, eluted) from Daphnia surface (green square). During night (bottom Daphnia), some part of Mn (rose-colored triangles) is lost in oxidizing conditions to form pyrolusite deposits (black) which in turn oxidize trivalent Cr, Ce, phenols, Cl^-, etc. Eu is re-adsorbed from both water (the local levels suffice) and near sediment during night and transported to the surface (top Daphnia in morning (at dawn)).

This process is thought to dominate at least in amictic water pools but is also relevant in monomictic water pools. This recapture of Eu on the chitin of some animals is unlikely to happen in small, fast-swimming/moving arthropods such as shrimps, water beetles, and gammarids, while it can and must be considered in big, slow ones (crayfish) and, of course, in cases of water slowly running over extended stationary chitin fields such as the surface of lichens [30]. The performance of the test cell under different illumination conditions permits an estimate of the photo-assisted transport and fractionation of metals in shallow (mono- or amictic) lakes and ponds with a high abundance of microcrustaceans, testing the ideas summarized in Figures 1 and 2.

Dissolved P is rapidly transformed into organic matter by bacteria and cyanobacteria, reducing free HPO_4^{2-} to just picomol/L amounts (moreover, such polysaccharides are known to suspend Eu, Tb, and other REEs [31]). This excludes the precipitation of REE phosphates, while phosphorylated sugars undergo ready photooxidation by Eu(III). Accordingly, once there is eutrophication, Eu released next to a water surface will increase, producing the sizable net upward transport of europium in a- or monomictic waters. Additionally, the ratio of elements such as Eu but also Mn, Cr, and Ce towards a background of strongly accumulating trivalent ions such as Al, La, and Bi [32] should vary with time of day and available $C;N_{org}$.

2. Materials and Methods

The reported work was carried out using purified chitin from shrimp shells (Sigma-Aldrich; St.Louis, MO, USA) (see Table 1). Grafted chitin was used as supplied (prepurified). The average acetylation extent of natural chitin was some 80–85% ([1,2]), with the remainder being "free" amino groups like in chitosan. The calibration line on partition equilibria [12] was required to determine whether the partition of adsorbed metal ions on chitin introduced into sediment and another sample immersed in water just a few cm distant from each other [9,32] was "normal" or biased by some subsurface biogeochemical process, as shown in the following graph (Figure 3):

Table 1. Devices and substances used in this study.

Equipment Item/Substance	Precise Denomination	Supplier	Remarks
Multimeter	Peaktech 3430	Peaktech Ahrensburg, Germany	Recording photocurrents and voltages
Cyclic voltammetry gear	Radiometer Solartron 402	Tacussel Lyon, France	
FTIR spectrometer	Tensor 27	Bruker Karlsruhe, Germany	
Electrode materials	Cu sheet	unspecified	Electrolytic Cu quality, purchased from local retailer
	Ta wire	American Elements, Los Angeles, CA, USA	We used Ta wire or sheet electrodes to avoid corrosion observed with Cu wire grids in chitin wraps with a couple of substrates and Eu(III)
Chitin	Chitin washed	Sigma-Aldrich St.Louis, MO, USA	From marine shrimp *Pandalus borealis*
Eu salts	Eu(III) trifluoromethanesulfonate	Sigma-Aldrich St.Louis, MO, USA	Nitrate or perchlorate also worked
Solvent	Dimethylformamide	VWR Darmstadt, Germany	
Microbes producing substrates	Baker´s yeast	Local retailer	Does convert mannose and others into compounds more reactive towards Eu(III)*, e.g., ethanol
Organic substrates of photooxidation	Glucose		
	Ethanolamine	Sigma-Aldrich St.Louis, MO, USA	
	Malonic acid	Sigma-Aldrich	Deuterated isotopomers produce far larger photocurrents
	Ethanediol	St.Louis, MO, USA	Deuterated isotopomers produce far larger photocurrents
	Glycerol	Sigma-Aldrich	Deuterated isotopomers produce far larger photocurrents
	Glycine	Laborchemie Apolda, Apolda, Germany Merck, Darmstadt, Germany	Partial photodecomposition (disproportionation) into ethanolamine, CO_2, and CN^-
	N-methylglycine	Sigma-Aldrich St.Louis, MO, USA	= sarcosine; does photooxidize only when kept in matrix
	Mannose	Sigma-Aldrich	Little phototransformation except when in matrix
	Mannose-1,5-diphosphate	St.Louis, MO, USA	
	Ribose-5-phosphate	Sigma-Aldrich	Unsubstituted ribose is photoinert with Eu(III)

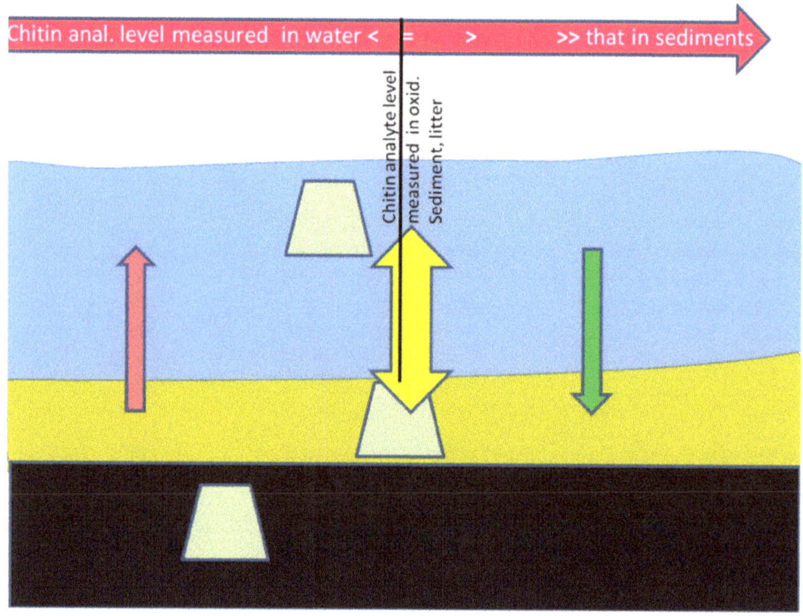

Figure 3. Partition of element ion sorbates on chitin immersed in water (top) and in wet sediment (bottom). Chitin flakes are located so close to each other that equilibrium (yellow double arrow) can be expected (not implying levels of a given element on flakes in water and in sediment are equal) unless there are subsurface processes absorbing metals into or releasing them from the sediment. Black: anoxic sediment layer where there is sulfate reduction mobilizing Ba, Pb (direction of pink arrow), and Cp. [12].

This bias does report chemical processes either involving this ion or its counter-ion which occur in sediment, including biochemical transformations or photochemical processes in the water column (Figures 1 and 2) keeping metals from sediment or making them dissolve. The substances were used as supplied, without additional purification measures such as distillation or recrystallization.

FTIR spectroscopy was carried out using a standard spectrometer (Bruker Tensor 27); the analyte was placed as a powder on some planar support (diamond or Ge) and pressed to it. Fourier-transform FT IR spectroscopy draws upon the fact that every curve—including spectra—can be displayed as the sum of several sine functions. Nowadays, it is the standard method of IR spectroscopy. Cyclic voltammetry is used to understand the electrochemical behavior of metal ions and complexes adsorbed and reacting on chitin. Chitin itself is completely silent in CV [12,30] although it contains traces of redox-active elements Fe, Ti, and Cu.

The transport and fractionation processes which take place at an illuminated chitin interface could be simulated using the following simple device in which *photochemical* H_2 production from acidic Eu^{2+} [33] on the second electrode was deliberately precluded by wrapping the flask with Al foil (Figure 4).

The position of either oxidation state of Eu (+II or +III) and thus sites of photoreactions and of possible reoxidations (air, hν [33], or at the counter-electrode) can both be surveyed using fluorescence caused by a colloquial blue LED. The oxidation state of Eu in either the supernatant or at the bottom electrode was checked using its typical (color of) fluorescence (Figure 5).

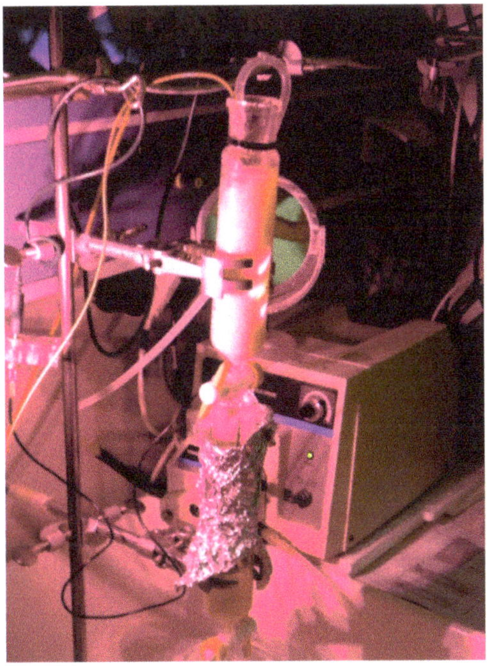

Figure 4. Illumination of the test system: Eu^{2+} formed via photooxidation of glucose was passed down to the lower cavity with dripping water to give away the "extra" electrons to make (a) H_2 and (b) a negative photocurrent. Illuminating the bottom cavity instead of the upper one produced no photocurrent whatsoever. A Becquerel effect was not observed with metal electrodes but only with illuminated semiconductive electrodes such as Se.

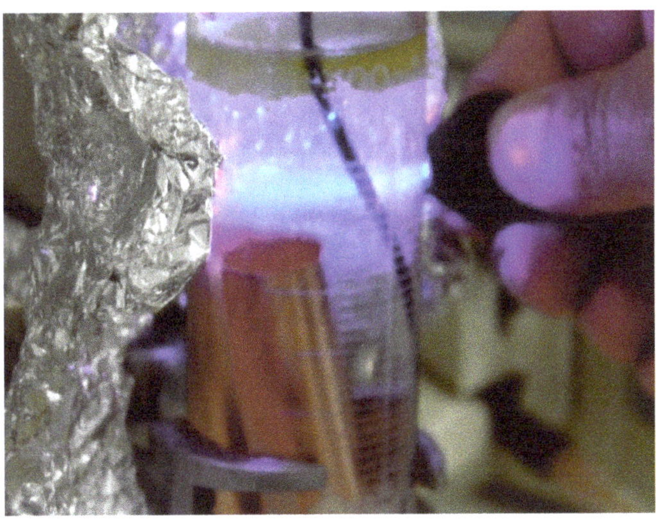

Figure 5. Turquoise fluorescence indicates presence of Eu^{2+} washed down from illuminated upper cavity containing chitin and organic substrate. Eu(III) would appear orange to pinkish if emitted at all in this chemical setting.

Experiments on the metal adsorption on chitin were conductd in a 50/50 by volume mixture of water and DMF [10]. Because adsorption to chitin depends on pH and chitin adsorption does enhance the acidity of protons at the saccharide ring [10,11], we had to find a buffer system which caused neither precipitation nor the excessive complexation of REEs (there is precedence in the published literature concerning the erroneous determination of log β of REE hydroxocomplexes due to using phosphate or carbonate buffers [34]). Inappropriately chosen buffers would either compromise adsorption studies (cp. [10]), or the buffers themselves would experience photooxidation by Eu(III), like acetate [23], or even do both (oxalate/Hox^-).

The chitin suspension was buffered by adding an equimolar mixture made from aniline and $PhNH_3^+$ trifluoroacetate ([10], quintuplicate setups). Ce(IV) is quickly reduced in pure DMFl however (observable by the bleaching of an originally yellow solution), cyclic voltammetry still can be carried out in this medium (Ce(III/IV) $\approx$ 1.62 V vs. SCE, sweep rate $\pm$ 100 mV/s [10,11,35]). Electrochemical data obtained in this manner can then be used to estimate complex stability constants log β by (downward) potential shift. Practical experiments were carried out with arthropods (crayfish *Orconectes limosus*) and compared to results obtained from chitin flakes placed at the same sites [32]. Animals (invertebrates only) were treated according to the published legal regulations. When applying chitin adsorption for environmental analysis, one would take eight times the amount of DMF/1.6 M Li^+. This does remove the topmost some 15–20 μm of chitin in one step to obtain all the matter adsorbed representing some time range from the last few minutes to several months ago [30,35], taking (small) background into account (for optimization parameters, also see [9,10,12]).

Electrochemical studies to understand photochemical effects were carried out using (a) a cyclic voltammograph Radiometer Solartron 402 (Tacussel, France) and (b) homebuilt cells fitted to commercial multimeters (Peaktech, Ahrensburg, Germany). Both reported data to a Dell PC, using either solutions or Petri dishes filled with gelatin containing sets of electrode pairs. For the simulation of biofilms, gelatin (commercial sources (local supermarket retailer)) was doped with Eu(III) (1 mM/L of water used to induce swelling and solidification, with 10% total gelatin content) and organics to be processed by Eu-mediated photochemistry [36]. The matrix effects on Eu-mediated photooxidation were studied by embedding potential substrates of photooxidation which are known to react not at all (aniline and N-methyl glycines) or negligibly fast (mannose, glycine, and cyclohexanol) with excited Eu(III) in gelatin (setup Figure 6). In some experiments, additionally, doped gelatin was incubated with baker´s yeast (*Saccharomyces cerevisiae*) to study the consequences of metabolism, possibly producing (more) photoreactive metabolites, e.g., ethanol from mannose. *Saccharomyces cerevisiae* (and other fungi such as molds) were shown to be much more robust towards LREEs including Eu [36,37] than any bacteria investigated previously or than archaea (the biodiversity of which is reduced to a remainder of aerobic Thaumarchaeota only in REE mining fields [38]), that is, methanogenesis and anything other than aerobic NH_3 oxidation are suppressed by the addition of REE^{3+} or their solid oxides [38].

In electrochemical experiments, Pt wire was generally replaced with Ta (wire or sheet) as an electrode material to exclude contributions from the heterogeneous catalytic cleavage of CH bonds promoted by PGMs (for a comparison of results, cp. [11]). In addition, unlike Cu or other electrode materials, Ta withstands attack by all the components studied:

a. Dark, organic matter only;
b. The addition of Eu(III);
c. The start of illumination, studying events/changes taking place just after light was admitted (within a few s);
d. Measuring changes in prolonged irradiation, extending over several days until there was no photoinduced voltage or current left.

Figure 6. Two pairs of electrodes were placed in gelatin in a Petri dish. Gelatin contained 0.9 mM/kg Eu and 5 mM/kg organic substrate. Electrode pairs were used to determine and compare voltages at different sites of lines in Petri dish. The device was isolated from air by a plexiglass™ cover plate through which electrode wires were passed, with glue closing the drilled holes. Upon illumination, slow chemical waves formed, indicated by reversal of sign of voltage at different times at the two pairs of electrodes, e.g., when consuming mannose, sarcosine, or aniline. Period length was about 24 h, and propagation took place at about 1 mm/h. Distance among electrode pairs was 2.8 cm.

The maximum current densities were much higher in this Ta system than when using Cu, too.

The setup for immobilized Eu and substrates is displayed in Figure 6 ([36]; photograph by S.F.):

3. Results and Discussion

3.1. Results in Lab Systems and Using Isolated Chitin

None of the three ions indigenous to chitin in sizable amounts, which undergo redox transitions within the range of biological redox potentials (Fe, Ti, and Cu [7,10]), gave electrochemical signals when dissolved in DMF containing Li^+ [12]. Moreover, the electrochemical study of adsorbed metal ions was readily feasible in both DMF/Li^+ solution [10] and solid state (with either causing reduced Eu to leach from chitin surfaces). Though analogous work concerning "real" biofilms has not yet been carried out, information concerning electrochemistry and Eu-mediated photochemistry in gelatin are available. Data on water-exposed parts of crayfish (the antennae and "neck") and on chitin flakes in same media strongly correlate (cp. [12]); so do data obtained from leg tips and sediment-buried chitin flakes.

During our experiments on photo-assisted fuel cells using chitin, we noted that $[Eu(OH)]^+$ ions formed by the photooxidation of aq. glucose produce:

(a) H_2;
(b) Variable electrochemical signals at a Cu sheet electrode (photograph, Figure 5; screenshot of electrochemical signal over time, Figure 7).

In the lower of the two cavities, while the top electrode is covered/surrounded by chitin flakes. Meanwhile, glucose is rapidly photoconsumed, and lipids (e.g., olive oil) undergo hydrolysis (saponification) catalyzed by REE^{3+} at 25 °C already (own observations); there are thus two different processes occurring at the downstream Cu electrode, that is, H_2 evolution and the formation of an initially negative photocurrent which, however, becomes positive after some time and shows slight additional oscillations (Figure 7).

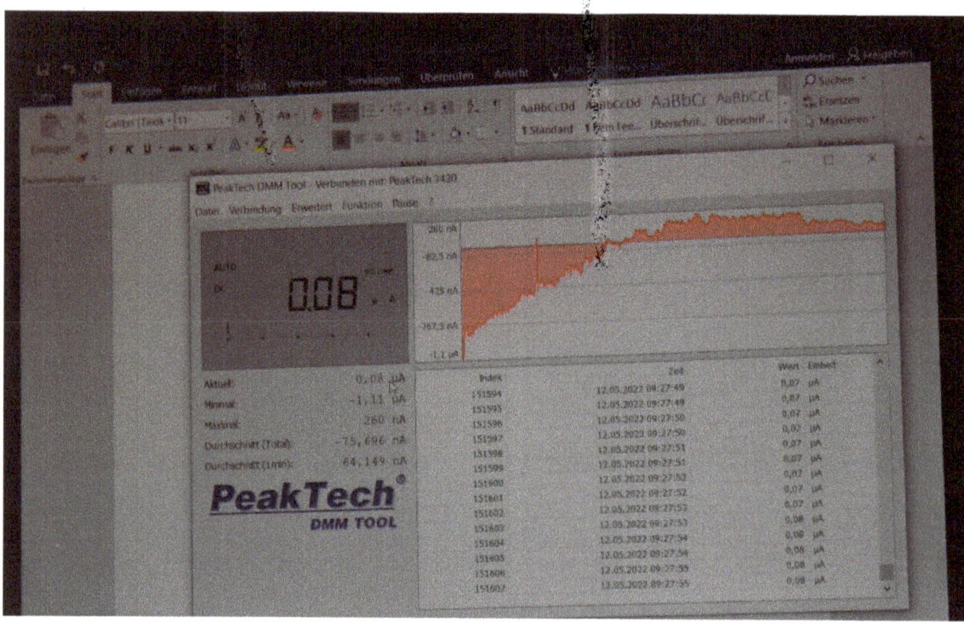

Figure 7. Development of photocurrent in Figure 4 setup over time.

Thus, some part of the H atoms taken from the photooxidation of glucose (cp. [11]) can be used for both H_2 formation and the reduction of CO_2, depending on respective overvoltages. Current oscillations are less pronounced with Ta electrodes.

Typical evidence of chemical waves [36] mentioned above included multiple reversals of voltage signs, much like those observed at a given pair of electrodes, while there was some time lag before the same happened at a neighboring pair of electrodes. During ongoing photolysis, sometimes, strange smells and color changes of solution or gelatin (the solvents water, DMF, and gelatin are all inert towards direct photooxidation by Eu) were noticed, usually during the early phases, while some gas bubbles and turbidity were noticed in both solvents and in gelatin systems. Apparently, both the liberation of amines and mineralization down to CO_2 occurred.

The co-adsorption of glycine in addition to REEs to chitin (pzzp = 6.13; H_0 = 11.0 in neat DMF [39]) (for the meaning of H_0, see the list of abbreviations at the end of this paper) was shown to enhance that of LREE, except Sm. A multitude of ligands such as caffeic acid, taken to represent humic acids, malonate, or hydroxamates did so only with Dy and Yb [10]. The calibration line describing the partition of adsorbed metal ions M^{x+} (constant x) on chitin introduced into sediment and another sample immersed in water just a few cm distant from each other [12] was defined as "normal", which was determined like in previous work [12]; a deviation from this expectation indicated chemical processes in sediment or biochemical or photochemical transformation either involving this ion or its counter-ion.

The relative stability of the substrate complexes could be estimated from data on the dissociation of gluconatocomplexes (Pr(III) [40]) and redox potentials for $Eu^{2+/3+}$ and

Eu^{2+}/metal in trimethyl phosphate vs. data in water or DMF, CH_3CN [41]. The rather high value of the former potential in TMP [41] indicated the rather weak solvation of Eu(III) in phosphate esters, suggesting that this weak solvation with phosphorylated sugars, alcohols, or ethanolamines does facilitate the attack of Eu(III)* on ester alkyl groups. The said gluconatocomplex and its galactonate, gulonate analogs, all have log $\beta_1 \approx 2.7$ [40], quite comparable to those of glycollato- or lactatocomplexes [25]. Accordingly, the mode of binding should be identical to that of simple 2-hydroxycarboxylate complexes (cp. [25]), omitting the binding of Eu(III) by/at additional hydroxy groups. Whereas glycollate is photochemically inert with Eu(III) and lactate displays little reactivity, glyconitrile, lactonitrile, malate, citrate, and gluconate are all rapidly photooxidized [11]. This emphasizes that there is no relationship whatsoever between complex stability and photochemical lability in this system; accordingly, non-ligating compounds/anions can also be photodehydrogenated by Eu(III)*. Generally speaking, solvents in which this $Eu^{2+/3+}$ photochemistry takes place are distinguished by low potentials (data from [41,42]), that is, there is the pronounced solvation of the trivalent ion (there is no reaction in CH_3CN, for example).

After exploiting its use as a photocatalyst support in the photooxidation of glucose, the following spectra were obtained, now showing the almost irreversible retention of Eu on biopolymers (Figure 8a,b).

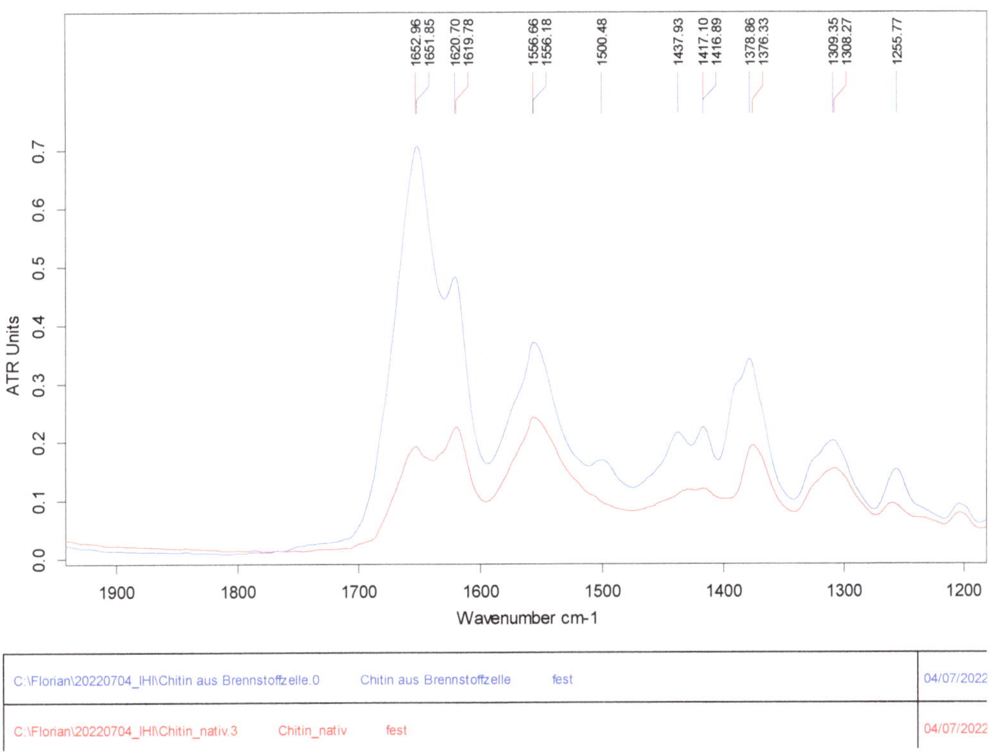

(**a**)

Figure 8. *Cont.*

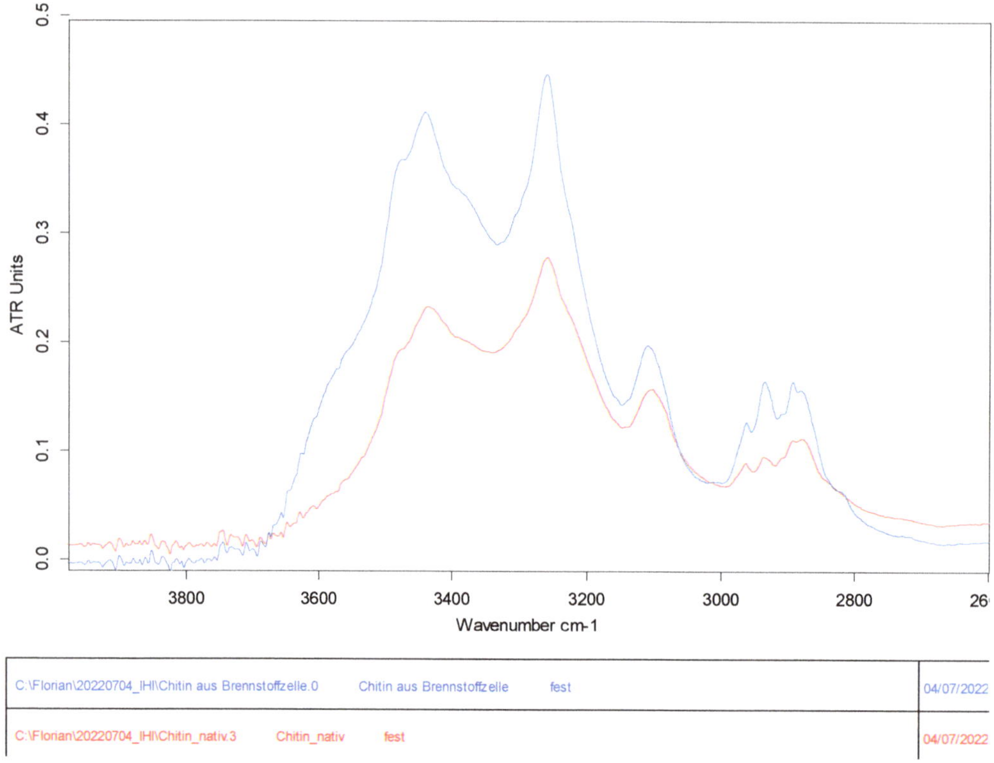

(**b**)

Figure 8. (**a**): Native chitin (red) and its degradation after photooxidation mediated by Eu(III); spectral region 1950–1150 cm^{-1} (≈5.1–8.7 μm). (**b**): Native chitin from *Pandalus borealis* (Sigma-Aldrich) and its photooxidation product; spectral 2600–4000 cm^{-1}. The intermediate range around 2200 cm^{-1} could not be measured because there were absorption bands of the supporting diamond crystal. Spectra were taken by F.Prasse (Hochschule Zittau-Görlitz, Saxony). "Brennstoffzelle" means fuel cell.

For Fe complexes modeling the transition from slightly brackish water to fungal injections to chitin adsorption, chitin adsorption for energies for corresponding complexes were estimated using data published previously in [12] (Figure 9).

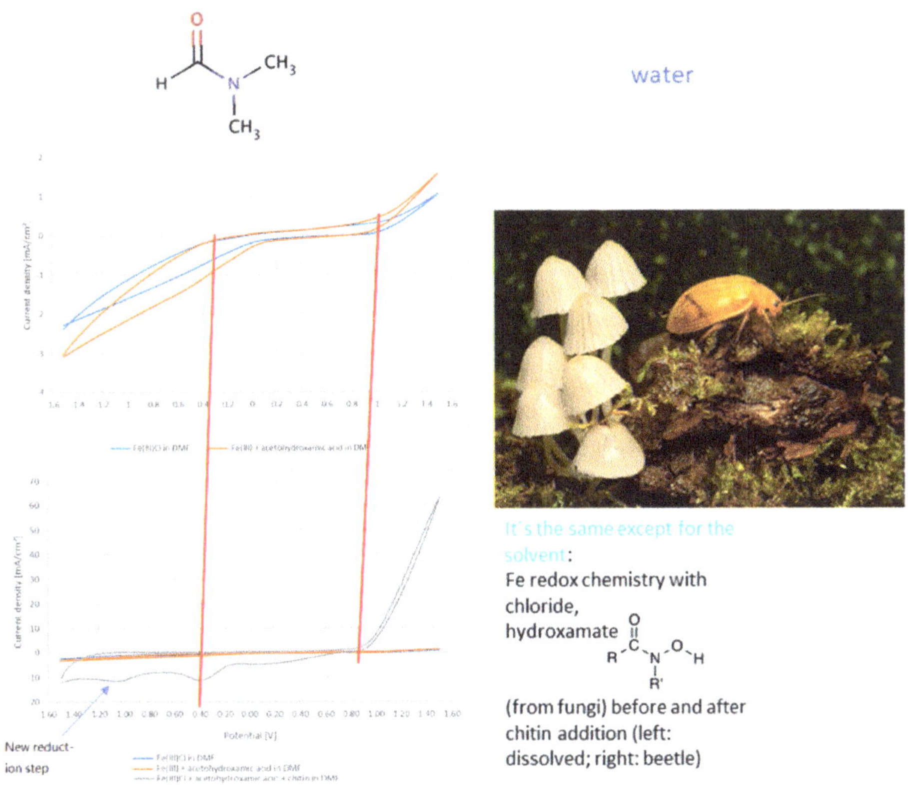

Figure 9. Derivation of chitin adsorption free energies of metal complexes which are likely to occur in soils from electrochemical data. Note that a new reductive peak turns up upon chitin adsorption of (probably Cl^- -containing Fe hydroxamatocomplex) at −1.1 V vs. SCE. The lines are inclined to the left, indicating an (uneven) depression of redox potential (determined by CV) upon adsorption to chitin, meaning that the Fe(III) complex (which is very stable in the case of Fe(III) [43]) more strongly adsorbs to chitin than its Fe(II) counterpart. For the CV diagrams, cp. [12]. Picture taken from a conference contribution by S.F.

Even though the solvation energies of trivalent ions in water exceed those measured in DMF (data from [41,44]), about three times more metal ions are transferred from metalliferous slurries in water than to chitin from equally concentrated solutions of the same material in DMF [13].

3.2. *Chitin in Organisms, Multilayer Films on Them and Pathways of Metals, Signal Compounds in Eusocial Arthropods: Evidence from Ants*

(Some kinds of) ants can live, build nests, and reproduce in areas covered by bared, weathered lignite. In these areas, one might anticipate high environmental levels of such compounds which ants use for intraspecific communication [13], that is, certain long-chain hydrocarbons. The organics released from local lignite (from abandoned and partly flooded Glückauf pit, Zittau/Olbersdorf, SE Saxony, FRG) include benzene, toluene, and valero- or methylated butyrolactone, whereas C_{22} alkane, 1-eicosene n-$C_{18}H_{37}CH{=}CH_2$, and a C_{13}-amine [13] were isolated from the waxy cuticle overlying chitin in these small animals. Accordingly, they do not interfere with ant communication (Figure 10 (left side)):

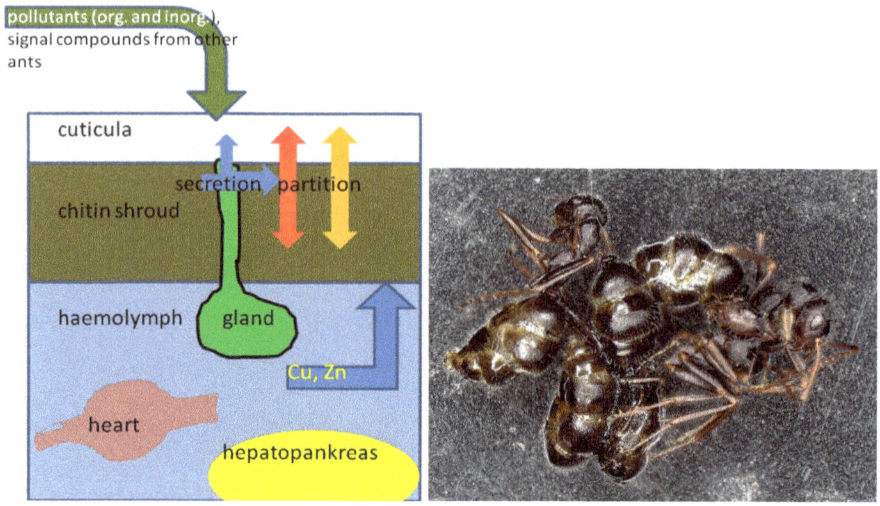

Figure 10. Left: cross-section of an ant showing the waxy cuticle absorbing non-polar organics and the underlying chitin layer with arrows indicating matter exchange equilibria. Inner organs of an ant were exposed to hemolymph liquid percolating the body freely, without blood vessels, while there was a heart, glands, and other organs (sketch by S.F.). Cu levels are very high due to hemocyanin transporting dioxygen in most arthropods, while Zn is very common in enzymes. Actual levels on chitin were 1–2 µg/g DM each. Ants shown in right photo (S.F.; [13,45]) were about 4 mm long.

In ants which inhabit degrading lignite, there is evidence of effects which might even enhance fractionation against Eu via photochemistry: La, Ce, Sm, or Dy enter ants mainly via the mouth, whereas a sizable percentage of total Eu is observed to be absorbed in/at the surface chitin layer [13]. When organic matter is available from the environment, Eu^{2+} formed by its photoreduction will be transported on top of the chitin layer. There is an analogous separation between Zn and Cd [13], which are taken up in different ways. It is an open issue as to whether the difference among (Eu vs. other REEs) pathways is related to Eu-mediated photochemistry, with Eu^{2+} possibly following Ca^{2+} ion channels in the chitin layer (Figure 10 left). If the effect seen in lignite-dwelling ants also holds for limnetic zooplankton, the larger contribution at the outer chitin surface (which in ants even is covered by an additionally waxy cuticle almost completely free of any metal ions) should enhance the extent of chitin-related transport and fractionation in upper, euphotic water layers, too (cp. Figures 2 and 11).

Budelmann [32] studied element concentrations in water and on different parts of crayfish *Orconectes limosus* which were or were not (leg tips and telson) exclusively exposed to water (this animal does not dig into sediment) rather than sediment. This changed the adsorption on grafted chitin dislocated in water and in sediment, both dislocated next to the traps the crayfish were caught in. The effects in a real outdoor system—representing several water bodies, including ponds, a creek, and a flooded quarry in upper Lusatia studied in August, 2020—are summarized in the following Table 2. The upper limits for mixing ratios of Eu in water could only be given owing to the lesser sensitivity of ICP-MS-based analytical determination and probably lower contents than for La and even-Z REE Sm. Animals were caught during the night, thus excluding the said photochemistry associated with Eu. Hence, the ratios observed on chitin should represent the upper limits reached over a day. The pH of all four water bodies was similar, about 7.5. Nevertheless, the Eu/La, Eu/Sm, and Eu/Bi ratios in chitin were rather high (black number are ratios by weight, and red ones are by stoichiometry) (Table 2):

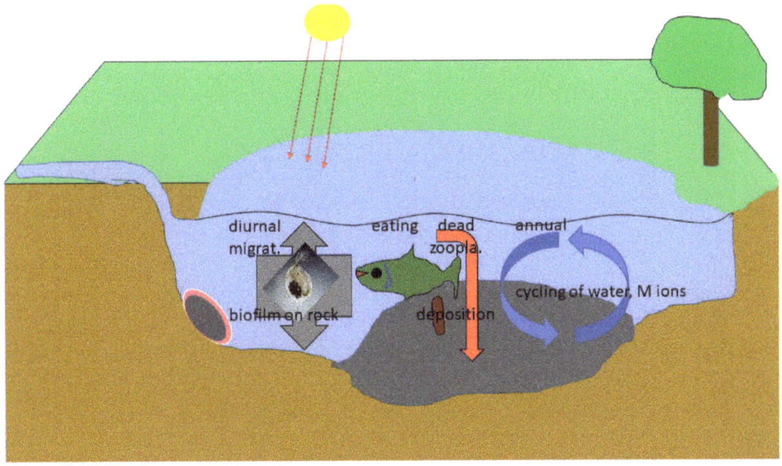

Figure 11. Different polysaccharides are involved in retention of M ions in limnetic environments, including biofilms (mainly glucan, left) and chitin actively (diurnal migration of copepods and water fleas) and passively (deposition of dead arthropods), switching among water depth levels. Upon illumination along appropriate organic matter, Eu—hitherto firmly adsorbed to chitin to an exceptional extent—photooxidizes the latter by CH bond photocleavage. Transport by animal movement (cp. Figures 1 and 2) competes well with that caused by annual circulation.

Some elements thus deposited after downward transport on chitin might be critical in sediment-located biochemical pathways, such as Ni and Co in methanogenesis, which require certain minimum levels of either metal. The same holds for La and other LREEs in methanol oxidation. In turbid waters or those rich in humic acids, Eu^{2+} will only be released close to the surface. Chlorophyll (Mg porphyrine) undergoes rapid M exchange with REEs including Eu, and Eu porphyrines are thus formed as much superior photocatalysts compared to simple Eu(III) salts or complexes in terms of photocurrent [45]. The same process should occur in phytoplankton including phototrophic cyanobacteria. Apparently, Eu is lost from photosynthetic organs due to photoreduction, as shown by its depletion from pine needles or birch or whortleberry leaves relative to the BCF levels of other LREEs at these light-exposed sites (calculated from data in [46,47]).

If this holds for phytoplankton, too, the oral uptake of Eu by zooplankton in turn should also be reduced, whereas the chitin shells of, e.g., water fleas are partly transparent. The behavior of Mn^{2+} on chitin (although at fixed (buffered) pH (aniline/$PhNH_3^+$ in water/DMF)) alone and in combination with REEs as a result of ligand addition was studied in our workgroup by M.Erler [10]. This concerned ligands known or likely to exist in groundwaters derived from all plant roots, fungi, soil, such as hydroxypolycarboxylates, glycine, and hydroxamates, plus humic matter, models thereof (caffeic acid), and SCN^- (for comparison). He found the rather high retention of Mn^{2+} to chitin in all samples in the mixture, indicating that the loss of Mn from chitin to aerated alkaline waters as MnO_2 will not occur readily. In turn, the MnO_2-catalyzed transformations of both metals and ligands [19,21,24] on chitin and in free water will only occur with already existing pyrolusite deposits, rather than forming such deposits to any significant extent.

Table 2. Relative metal adsorption by chitin (grafted or in different parts of *O. limosus*) in several water bodies in Upper Lusatia (Saxony, FR Germany). Blanks denote missing data (too little for analytical determination), and the upper limits in water are due to all Eu levels in water itself being below determination limit (90 ng/L or ≈ 600 pMol/L, that is, much higher than for other REEs or Bi or Cr).

Site, Element Ratios	Water	Antennae	Carapace	Mandi-bles	Leg Tips (Usually Immersed into Sediment)	Telson (Lower Part)	Branching Ratio on (Grafted) Chitin for Eu, M′ (Sedim.-/Water-Exposed Chitin)
Irmerteich ("Irmer´s pond"), former lignite open pit, 6.3 m deep							Eu: no data
Eu/La	<2.9 <2.65	5.19 4.74	0.78 0.71	5.51 5.04	4.2 3.84	12.7 11.6	La 0.39
Eu/Sm	<3.46 <3.42	16.2 16.0	5.8 5.74				Sm: no data
Eu/Bi	<0.75 <1.03	6.3 8.66	11.6 15.95	20.3 27.9	12.2 16.8	1.03 1.41	Bi 1.65
Hartauer Lache ("shallow pond located at Zi-Hartau"); shallow pool rich in SO_4^{2-}; bottom covered by clay							Eu: no data
Eu/La	<0.8 <0.7		2.56 2.34				La 10.9
Eu/Sm	<1.64 <1.62		4.82 4.77				Sm
Eu/Bi	<3.4 <4.7		3.76 5.17				Bi 1.9
Landwasser creek							Eu: data for sedim. Only
Eu/La	<0.94 <0.86	59.7 54.6	10.4 9.51	25.5 23.3	11.8 10.8		La 10
Eu/Sm	<1.35 <1.34		39.1 38.7				Sm: data for sedim. Only
Eu/Bi	<90 <124	1.42 1.95	6.14 8.44	7.4 10.2	6.20 8.53		Bi 2.2
Ostritz flooded quarry; chitin was treated with boiling water prior to analysis							Eu
Eu/La	<2.5 <2.3	14.5 13.25		4.61 4.21		14.2 13.0	La 4.5
Eu/Sm	<4.8 <4.75	80.0 79.2		9.6 9.5		36.5 36.1	Sm 9.2
Eu/Bi	<22.5 <30.9	11.7 16.1		7.1 9.8		9.6 13.2	Bi 1.13

The release of other metals rather abundant in chitin such as Fe, Ti, Al (10–20 µg/g each), and Zn [7,10] by the bacterial degradation of "dead" chitin in sediment [48] and subsequent reactions by inference are unlikely to produce significant amounts of photo-semiconductors such as TiO_2, $CaTiO_3$ (aerated top-zone of sediment), ZnS, $CuFeS_2$ just (about 0.5 mm [28]) below, or of Fe oxides such as magnetite. This estimate renders photo-electrochemistry related to the chitin-based shuttling of metals unlikely. An example would be the Dhar-Mukherjee (1934) preparation of amino acids from polyalcohols, sugars, and nitrate mediated by TiO_2 (unspecified crystal structure) [49] occurring at the bottom/shore of a water body.

3.3. Discussion

Chitin adsorption kinetics and the method to investigate them were described in previous papers [9,11,31]; adsorption is essentially completed within 10 min [7,9,10,34], with few ligands slowing down the pathway to equilibrium, such as citrate with Fe(III) [10]. Understanding element (Pb, Cu, Zn, Cd, and As) fractionations by binding to chitin the way it was depicted in Figures 1 and 2 calls for electrochemical studies on isolated (suspended and dissolved) polysaccharide strands which can be amended with arbitrary amounts and kinds of metal ions or complexes. The progression of metal(-oid) diffusion through thick chitin layers [35] can be measured through the repeated dissolution of plane layers 2–3 μm thick using DMF/Li^+ and analyzing the fractions one by one [10,35]. However, as mentioned before, there are also irreversible effects which require consideration: Figure 8a,b show substantial changes in either part of the FTIR spectrum (blue ("spent" chitin retaining Eu) vs. red (native state)]). The final spectrum differs considerably from that of chitosan (Figure 1 in introduction) and of acetamide (not shown here), especially in the 1400–1500 cm^{-1} region, arguing against simple photoinduced acetamide hydrolysis in chitin. However, the corresponding spectral features introduced by Eu-mediated photochemistry are also present in 2-oxopropanol (hydroxyacetone) (Figure 12).

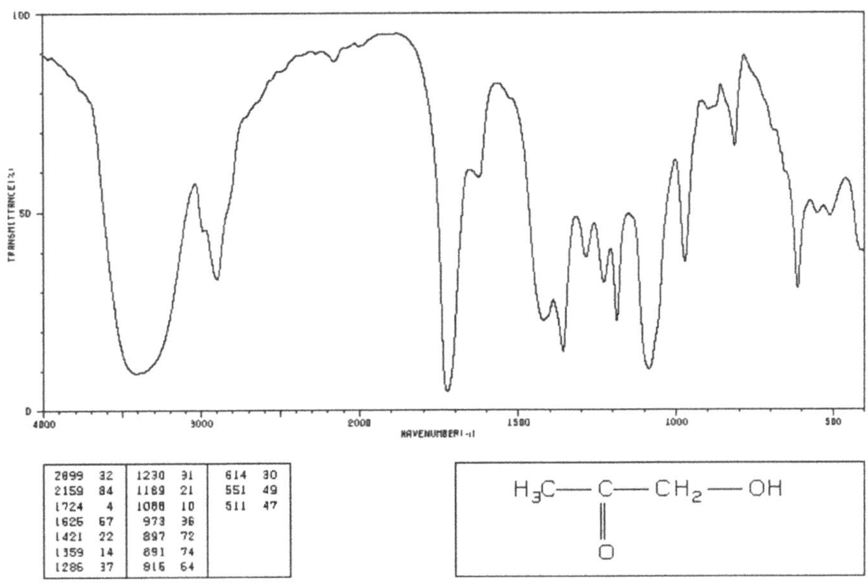

2899	32	1230	31	614	30
2159	84	1189	21	551	49
1724	4	1088	10	511	47
1626	67	973	36		
1421	22	897	72		
1359	14	891	74		
1286	37	816	64		

Figure 12. FTIR spectrum of hydroxyacetone [50].

Hence, it is more likely that the photocatalytic passivation ("poisoning") of Eu(III) bound to chitin is related to the oxidation of the neighbor -CH(OH)- structural motif than to hydrolysis. Given the results with amines, the deprotection of the glucosamine moiety would also explain the decrease in photoactivity, but this is not compatible with the shown spectra. The fact that both glucuronic acid and hyaluronic acid readily undergo photooxidation in this system [11], producing sizable photocurrents, argues against the alternative assumption that the exocyclic -CH_2OH group is involved in the process. The oxidation of a ring similar to oxidizing propylene glycol into hydroxyacetone thus appears to outcompete hydrolysis; it is not yet possible to distinguish between the two pathways of photooxidation of the polysaccharide by Eu(III), i.e., either direct or indirect using free radicals.

The numeric method was described previously, too [7,10]. It is used to calculate element fractionation on chitin exposed to (a) different ambient ligands and (b) a water–

sediment interface indicating active transport through this interface to pass the chitin flakes (or benthic animals partly penetrating the interface, while other parts covered by chitin are kept clear to be exposed to water only) In open waters, chitin and biofilms will have some negative charges due to pzzp values (3.5; [51]) and thus preferentially adsorb cations (both Na^+ and the like and organic ones derived from amines and amino acids, e.g., H_3N^+-CH_2COOH). Unlike Li^+ or Na^+, organic cations provide **bridges** and (charged) anchor groups at chitinous interfaces to enhance the adsorption of ions (cp. ion exchange and complexation chromatography). However, this must be studied for every single kind of ion because even different REEs differ considerably with respect to causing either increases or decreases in adsorption [10]. The sugar complex stabilities of REEs are unknown, with a few exceptions for certain hexose uronic acids (log $\beta \approx 2.7$; [40]), that is, they are quite similar to simple lactate and glycolate complexes [25,26,34,44]. Published examples of carbohydrate photochemical transformations [52] include photooxidations of glucose, mannose, and galactose by $TiCl_4$ in methanol to yield pentodialdoses [53,54]), or those of glucose, mannose, and arabinose by $FeCl_3$ (causing C_1/C_2 photocleavage) or C_2/C_3 photocleavage in fructose in water [53]. Accordingly, biofilms should be inert towards photochemical attack by transition metal ions except when there are very high levels of Fe(III) or of Eu(III) after enrichment in biofilm. We found that lignite and chitin are inert towards illuminated $FeCl_3$ [13], while Eu(III)* particularly attacks lignite [13].

Biofilms leak phosphatases to open water [27–29], possibly controlling photochemical turnover rates by Eu(III) in shallow waters according to Figure 2. Accordingly, one needs more pieces of information on kinds of organic matter in photic stagnant waters to estimate photoreactivity before and after phosphorylation, because groups of organic compounds differ considerably with respect to rates of photooxidation by Eu(III). The photooxidation sensitivity of various substrates towards Eu(III) varies with the content of other elements, especially that of N, P, and S, which are most important in biogenic organic compounds. The typical abundance ranges of photoreactive and other organic and heteroorganic compounds in freshwater biotopes can be found in [28]. The transformation of alcohols or sugars into phosphate esters generally increases the rates and yields of photooxidation, while the introduction of S blocks this process [9,11,12], and that of N causes quite different changes. These effects can be both significant activation (ethanolamine vs. ethanediol, glycine vs. glycolic acid, and $HOCH_2CN$ vs. ethanol) to partial (glycinate vs. propionate) or even total inactivation (aniline vs. toluene); amines usually do not react, unlike the respective quaternary ammonium salts [12]. The actual effect seems to be related to the IP of the said organics [45].

P contents in biomass (that is, (C/P)) are directly correlated with the rates of reproduction/cell budding and other metabolic activity, given the compositions of ATP, nucleic acids, and of CO_2 acceptors in both photo- and chemolithoautotrophs [55]. These processes in turn afford matter that can undergo photooxidation by Eu(III), with phosphorylated organics being much more reactive. When phyto- or zooplankton does grow, CO_2 + NO_3^-, phosphates, or N-poor foods, respectively, are transformed into organics parts which are described by the above empirical rules on photooxidation rates, and the latter in turn are related to the **ecological stoichiometry** of local organisms [11,12]. Among the sugars mainly producing plant tissues and wood, glucose, xylose, and oligomers are photooxidized by Eu(III), although without the permanent lowering of pH because there is CH homolytic bond cleavage by Eu(III)* rather than electron transfer from organic ligands like in LMCT photochemistry [53], whereas, unlike the corresponding phosphorylated compounds, mannose hardly reacts and ribose does not react at all. Phyto- and zooplankton bacterial cells are destroyed by surf to produce peculiar kinds of foam near shores, while dead matter sinks down as detritus. Phytoplankton (and higher plants, macroalgae) and cyano- and thiophotobacteria (which photosynthesize using H_2S as a hydrogen/electron source) all contain relatively little N and P, yet both elements are essential for making more reactive (towards Eu(III)*) sugars, carboxylates, lipids, and certain amino acids by biological processes. Photosynthetic net productivity will control production-, wave- and surf-related mechanical activity, which causes the release of organics which then undergo photooxidation. Of course, the organics previously contained in the cells or plankton organisms then are exposed to photooxidation, with the net outcome likely controlled by ecological stoichiometry [55]. There are two principal kinds of biomass, namely heterotrophs (animals, fungi (Opisthoconta), and most bacteria) rather rich in N and P and autotrophs such as plants and phytoplankton distinguished by much lower (and more variable) N- and P-contents. Chitin only exists in the former *Opisthoconta*, except for in lichen symbiosis (for a list, see [56]). Since (most) animals are mobile, aquatic animals can contribute to the vertical transport of chemical elements in pools and lakes, and Eu desorption by photochemical processes on chitin will become a dominant process during the daytime, changing the distribution of water columns. The next drawing (Figure 13) shows the likely results, including our own works [32] insofar as the measured levels that suffice for analysis (cp. Table 2).

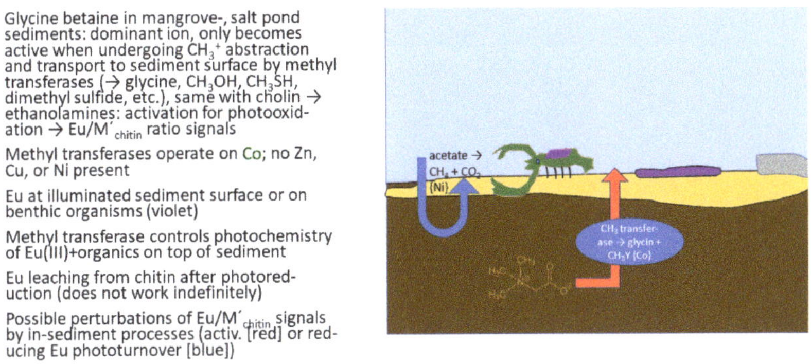

Figure 13. Activities of methyl transferases; CH_4-making Ni proteins control conversion of methyl compounds CH_3Z including acetate by Eu(III)* (violet) on top of sediment (adsorption of Eu, e.g., to chitin litter, Ti-rich sand, and benthic organisms).

Phosphorylation by bacteria and cyanobacteria is rapid, but extracellular phosphatases might return parts of the material into P_{inorg} ($H_2PO_4^-/HPO_4^{2-}$, depending on local pH and pyrophosphate) again [27,29]. Organic P amounts to some 60% of total phosphorus in lakes [57], suggesting the phosphorylated species will control the extent of C;N^{org} photooxidation by Eu(III) and thus that of Eu^{2+} formation next to the surface. Amino acids and oligosaccharides are reported to be main components of DOM at levels of 5–50 μMol C/l total DOM in Lake Constance (Überlingen "fjord") top waters.

H_2 evolution in this process is an interesting phenomenon given the sizable overpotential for this process observed on a Cu electrode (some 0.5 V, [58]). In fact, $Eu^{2+/3+}$ transition in water occurs at a potential which does not permit H_2 production at nearly neutral pH (between pH 6 and 9). Meanwhile, the Pourbaix diagram of Eu [19] shows that there will be no H_2 evolution at an 0.3 V overpotential-electrode between pH 1 and 10.5. However, the formation of carbonato- or sulfatocomplexes of Eu(III) reduces the potential in ocean or limnetic waters significantly, in favor of H_2 evolution. Both the photodetachment of Eu^{2+} during the photooxidation of organics [32,35] and the "stripping"/partition of adsorbed ions to biofilms or ion-exchanging minerals and biopolymers (such as rotting wood) in top sediment layers will decrease concentrations of M^{x+}, M'^{y+}, ... adfilms on chitin, inevitably producing even more negative potential. The difference may be in the order of 100 mV when changing from 0.2 M NaCl to freshwater (see Figure 1 in [59]). Formula (1) was derived for the estimation of complex stabilities [11]:

$$\text{Log } \beta = a \times \Sigma\sigma + b \quad (1)$$

and now can be applied to calculate the relative absorption strengths (thermodynamic, not optical) of metal ions in biofilms (consisting mainly of polysaccharides [37], using $\sigma_{OH} = -0.37$ [60]) or other polymers given that the binding functional groups are known; $\sigma \approx -0.44$ holds for chitin in pure water. In slightly alkalinic freshwaters containing other complexing ions such as sulfate [32,34], polysaccharides [28], humic-, or amino acids [12], effective σ values are somewhat lower than -0.44. The retention of traces of dissolved ions from water or soil liquids or water films on ground minerals [7,9,15] thus depend on the polymeric nature of the ligand, with

$$\log \beta' \approx \log \beta_{\text{monomer ligand}} + \log n \tag{2}$$

where n is the extent of polymerization/polycondensation (while there will be significant complexation at monomer only if $[M] \geq 1/\beta$; in a kilomer (i.e., n = 1000), just one binding site of 1000 rings would need to be occupied, meaning the equilibrium concentration [M]* would be $10^{-3}/\beta$ to sustain this state). For chitin, n is about 1500 [61]. Previous experiments showed [15] that native chitin reached saturation concerning $Ni^{2+}{}_{aq}$ already at 1 nM/L, while it was only reached at much higher levels for insoluble salts including Ni minerals and for some complexes placed next to a chitin surface. Here, $\Sigma\sigma = -0.74$ as these materials are essentially free of N, or -0.54 for pectin (higher σ of methoxy groups [60]).

Data concerning the comparison of binding of Eu(III) to polysaccharides cellulose, chitin, and chitosan can be obtained from the literature [62]. Certain amino acids and peptides made thereof are involved in cross-linking chitin strands in arthropod chitin [63]. They readily undergo consumption by bacteria when stored moist and produce typical residues of H transfer upon pyrolysis, including benzene, toluene, and indole (from aromatic amino acids), way before the typical degradation products of polysaccharides, namely 3-acetamidofurane and its rearrangement products acetylpyrrole, and pyridine is formed (i.e., at 300 °C or less) [63]. TLRFS data show that whereas with the adsorption of this metal ion to cellulose or chitosan only $\approx$ two water molecules in the coordination polyhedron of Eu(III) are replaced vs. the aquaion; binding to chitin replaces six or more water molecules from the Eu coordination polyhedron. Accordingly, the binding of Eu will link two parallel chitin strands in both native biomass and solution. Cp. Figure 14 (from [64]).

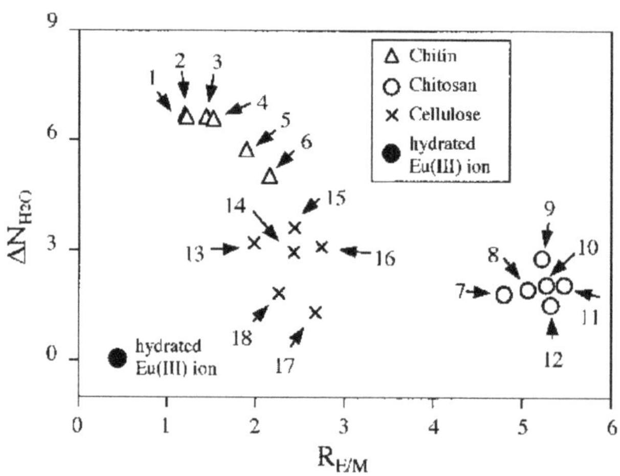

Figure 14. From [64]. pH between 6.5 and 7.9; highest pH means least replacement of H_2O by polysaccharide. Accordingly, Eu will stick much better to arthropods or lichens than to plant leaves, influencing rates of photooxidation of embedded substrates (sugars and amino acids) as well as increasing leaching rates of Eu^{2+}.

There is a linear correlation between the σ of the sidechain and the number of Eu binding positions occupied by something other than water molecules, combining data from [60] and [64] (Table 3):

Table 3. Sidechain electron acceptor activity (Hammett´s constant vs. replacement of water ligands from Eu(III) by polysaccharides.

Substance (Biopolymer)	L	$\sigma_{sidechain}$	Δn_{H2O}
cellulose, alkalinic conditions	O^-	−0.8	≤2
chitosan	NH_2	−0.66	about 2
cellulose	OH	−0.37	3–3.5
chitin	NH-$COCH_3$	0	6–6.5

Accordingly, pectin ($\sigma_{OCH3} = -0.27$) should display Eu(III) retention properties in between those of cellulose (in acidic conditions) and of chitin; the haloacetate treatment of chitosan would provide yet another intermediate-strength sorbent. Eu(III) photochemistry modifies chitin in the long run, also meaning that some structural motifs might be turned into radicals. The observed modification of chitin rendering Eu(III) retention almost irreversible indicates that $\sigma_{sidechain}$ is even higher than zero. The value for chitin indicates, in accordance with another suggestion by Ozaki et al. (2010) [62], that part of Eu(III) becomes locked away in some "cavity" between parallel chitin strands, replacing an accordingly larger number of water molecules.

Unlike other metals, the preheating of chitin by immersion into boiling *aqua dest.* does reduce the retrieval of REEs such as Gd and La [32]. An alternative would be the formation of some oxidation products of the acetamido group increasing Eu complexation, too, such as conversion into hydroxamic acid. σ for hydroxamates or ethylideneamines is not given in the literature, but it is for -NH(OH) [−0.34], benzylideneamines, or hexafluoroisopropylideneamine [60]. However, using Equation (1) and its formal rearrangement to yield:

$$\sigma_{Lig} = (\log \beta_{MLig} - b_{M}x^{+})/a_{M}x^{+} \text{ [bidentate binding]} \quad (3)$$

and data for acetohydroxamatocomplexes of Nd(III) and Eu(III) [65], namely $\log \beta_{Nd} \approx 5.17$ and $\log \beta_{Eu(III)} = 4.332$ irrespective of nitrate- or perchlorate-based ionic media, yields $\sigma_{CH3CO\text{-}N(H)O^-} \approx -0.2$ for acetohydroxamate, that is, a value which is much lower than for acetamido or formamido groups. Similar considerations hold for other plausible photooxidation products. Accordingly, Eu(III) at high concentrations and illumination intensities tends to migrate to some position "below" the surface where neither CH bond cleavage from co-adsorbed or dissolved organics nor Eu^{2+} release can readily occur (if chitin thus "spent up" is dissolved in DMF/Li^+, however, some Eu^{2+} gets into the liquid phase, which can be extracted by filtration and displays green–turquoise fluorescence when excited by blue/violet LED light). However, it is unlikely that this state of matters is ever reached with arthropods or lichens in ambient conditions, except perhaps when there is strong pollution such as in mine-tailing waters. There are few larger water bodies distinguished by $[Eu]_{aq} \geq 20$ nMol/L, such as Baikal (Siberia), Masurian lakes (Poland), or Lake Michigan (Figure 15):

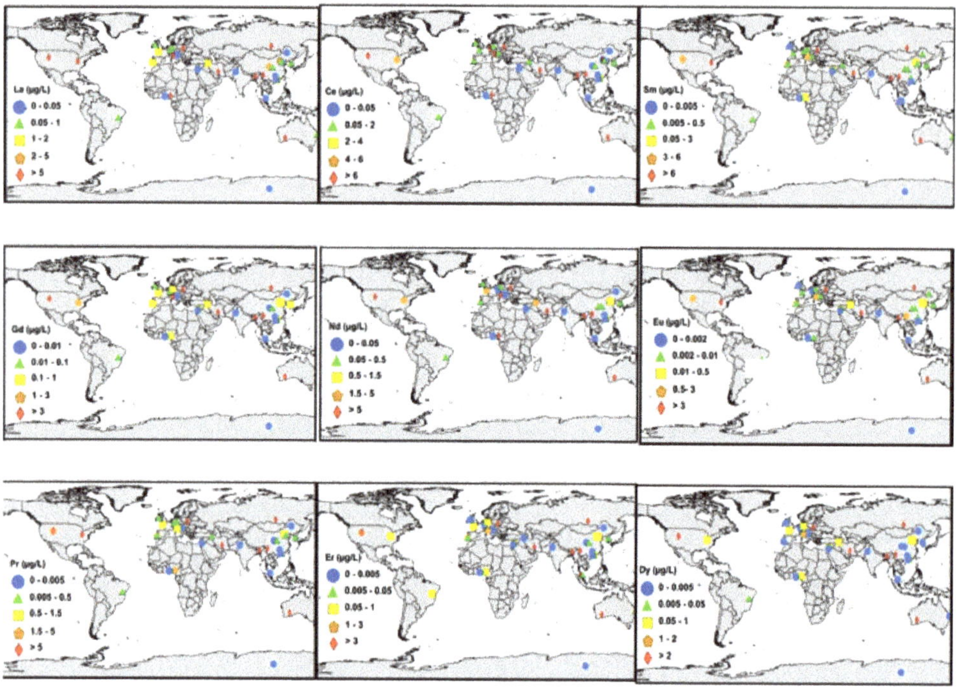

Figure 15. REE levels omitting Pm for many larger waterbodies around the globe [46]. Levels in ocean are much lower (<1 [odd-Z-REEs Tb–Lu] ... ≈ 50 [La, Ce, Nd] ng/L), similar to blue dots in this diagram [66].

With respect to non-electrode processes related to photoproduct Eu^{2+}, kinetics were studied for just few among the sizable number of inorganic species which might be reduced by Eu^{2+}, noting that reductions of I_3^- and V^{3+}, V(III) complexes and of $[Co^{III}L(NH_3)_5]^{2+}$ (L = simple anions, acrylate, NCS^-, N_3^-) are rather fast, while that of ClO_4^- is rather slow and that of Cr^{3+} is even reversible [67,68]. Apparently, electron transfer or H-H coupling next to electrodes involving surface hydroxide and $[Eu^{II}(OH)]^+$ prevails over electron transfer even on Cu with its sizable H_2 formation overpotential [58], as H_2 is formed both in gelatin (Figure 6) and in the bottom part of test cells (Figure 4). Accordingly, arthropods displaying the above effect should be almost "silent" towards predators in command of electroreceptors, such as sharks or platypuses [69], except of course for signals betraying nerve or muscle activity.

The DOC/DIC ratio is closely correlated in quite diverse lakes from China except for Karst waters [70]. Whereas the DIC levels were comparable among oligotrophic (100–300 μM/L) and eutrophic waters (60–200 μM/L, respectively), DOC—and by inference the level of substrates for photooxidation then causing Eu mobilization from chitinous interfaces—behave in a different way:

$$\text{Log DIC}_{\text{oligotr}} = \log \text{DOC} + 1.13\text{, that is, DOC} \approx \text{DIC}/13 \tag{4}$$

and

$$\text{Log DIC}_{\text{eutroph}} = 0.63 \log \text{DOC} + 0.81 \tag{5}$$

There is no such correlation for sulfate, which likewise forms rather stable REE(III) complexes [26,34], meaning that at rather high pH and water hardness, the availability of both non-chitin Eu and its substrates grow with carbonate content due to the solubility of

$[Eu(CO_3)]^+$. For both this ion and its excited state, log $\beta \approx 6.6$ [71], while log $\beta = 3.64$ for sulfatocomplex [34].

Irmerteich and Hartauer Lache are eutrophic waters, and additionally, at some spots in either pond, lignite is locally exposed, thus being subject to direct photooxidations. Accordingly, Eu/M´ ratios observed on chitin are lower than in "pure" (i.e., oligo- or mesotrophic) waters. Even though (estimated) Eu/M´ ratios in water are not that different in the four waterbodies, the re-uptake of Eu by the chitin interface during the night cannot compensate for enhanced photochemically induced desorption. Sampling was conducted in the morning at dawn, meaning that the Eu levels observed on crayfish chitin (all data are given where it was possible to detect and quantify Eu) should have been close to the maximum, as shown in the following graph (Figure 16).

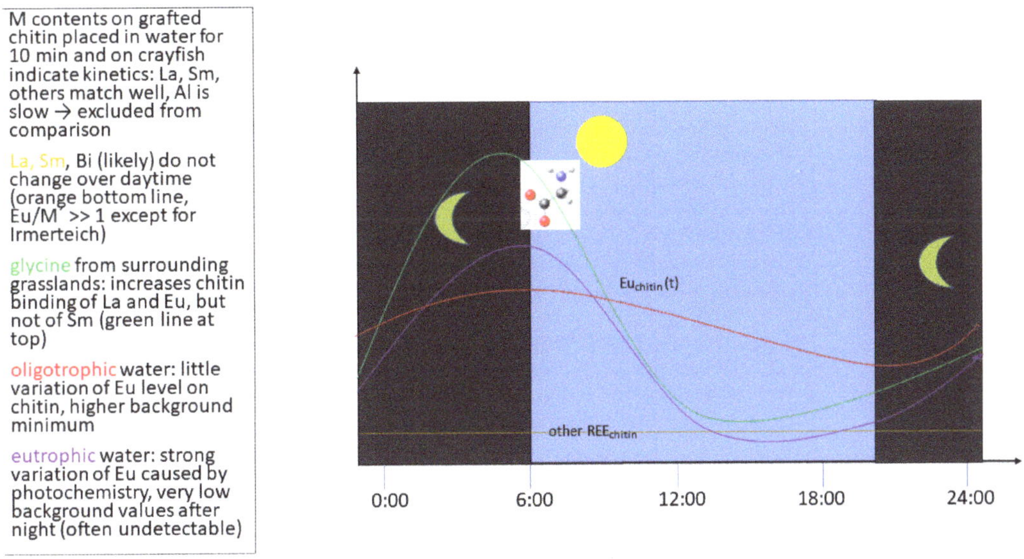

Figure 16. Predicted variation of Eu bound to chitin with and without glycine, potential substrates of Eu-mediated photooxidation over daytime. Sampling was conducted int the morning at dawn [32].

The high partition ratios of La for grafted chitin placed on either site of the water–sediment transition (Table 2) might indicate La use by methanol oxidases in sediment [72,73].

The general statements hold for all water volumes inhabited by arthropods or their larvae (insects, crabs, spiders, and water scorpions) provided there is a significant P level, available organic matter, and sufficient light influx. Sizes might range from phytotelmata like those of pitcher plants or bog pools up to sizable lakes and shallow rivers. The diurnal vertical migration of arthropods is typical in all of them, while living plant tissue or peat (providing citrate, malate, and humic acids) are substrates readily photooxidized by Eu(III) [11,74] and are involved in the supply or exchange of organic matter acting as ion exchangers. Main contribution would be from phytogenic glucose in addition to the organics mentioned before. Phytotelmata or bog pools are distinguished by very high levels of bacteria involved in P dynamics, whereas there are no larger predators feeding on the arthropods, except sometimes for frogs (tadpoles are herbivorous in common circumstances) [75]; particularly, fishes are absent. LREEs are—in addition to Bi, Ni, and V—among the elements producing the largest transport effects associated with chitin. Like Ni, LREEs are involved in the metabolism of (here: aerobic) microorganisms [72,73], while Eu undergoes selective bioaccumulation, too [76], without becoming a phototoxic agent [9,36]. Insofar, the hypothesis summarized in Figures 1, 2 and 12 can be considered tested and reaffirmed in order to understand the actual behavior of Eu in eutrophic conditions which tend to prevail both with time and increasing human and agricultural activities around them [28]. Phosphate levels and P speciation thus become significant in the biomonitoring of ponds and lakes by this method at least insofar as biomonitoring employing chitin is concerned.

4. Conclusions

Understanding processes which occur in chitin in ambient conditions is important for planning its use in biomining as well as estimating its role in the transport of elements on mobile animals in aquatic environments. The results obtained from both grafted chitin and different interfaces of arthropods (crayfish and ants) and the electrochemical data on the behavior of free and adsorbed ions/complexes on chitin agree that metal transport on mobile chitin (i.e., animals partly or completely covered by it) can produce significant effects with regard to metal distribution in ecosystems, whether the animals move in a periodic manner or chitin parts are discarded during or after life. Thus, eutrophication in the water bodies populated by chitin-covered organisms such as water fleas, crayfish, shrimps, or gammarids will change Eu dynamics by producing a higher share and level of phosphorylated compounds and in turn speed up their photooxidation. The extended illumination of such substrates along with Eu(III) will not cause $EuPO_4$ to precipitate. Specific reactions of certain elements, such as the photochemical reduction of Eu(III) or the biomethylation of elements [77] accumulating on chitin (especially Bi) then cause fractionation across the water column. Animals predating [28,78] aquatic or semi-aquatic arthropods (larvae and submerging water insects) thus are exposed to a different "cocktail" of metals when feeding at different sites from surface to benthos.

Author Contributions: Conceptualization S.F., methodology, validation S.F. and F.B., formal analysis S.F., investigation F.B., writing- original draft preparation S.F., review and editing: both S.F. and F.B., visualization S.F. and F.B. (figs.8, 13), project administration and funding acquisition do not apply. All authors have read and agreed to the published version of the manuscript.

Funding: No-third-party funding was acquired; work was carried out using the department budget alone.

Institutional Review Board Statement: Not applicable.

Informed Consent Statement: Not applicable.

Data Availability Statement: Not applicable.

Conflicts of Interest: The authors declare no conflict of interest.

Abbreviations

β	complex formation constant. Depends on T and solvent for a given complex such as $[LaCl]^{2+}$ or $[Fe(glyc]^{2+}$
DMF	solvent N,N-dimethyl formamide. Fully miscible with water and aromatics but not with aliphatic hydrocarbons, perfluorinated organics, or long-chain ethers. Best solvent for performing photoreduction of Eu(III) when combined with organic matter.
DOC	Dissolved organic carbon. Usually < 10% of total limnetic C content. Main component is polysaccharide
DON	dissolved organic nitrogen. Can be both highly reactive in biological terms (amino acids) or persist over centuries.
ε	Redox potential vs. some standard electrode, often → SCE.

H_0	Chemical potential of solvated proton in non-aq. media. Concerning fully dissociated acids such as $HClO_4$ or HSO_3CF_3 in CH_3CN or → DMF, $H_0 = 0$ for a 1 M solution. The most acidic solutions which can be prepared ($H_0 < -20$) correspond to almost bare protons, with species such as H_2F^+, HCO^+, $H_2SO_3F^+$, or even H_4O^{2+}, H_3^+ just weakly interacting with anions such as SbF_6^- and its SO_3 solvates in such (viscous, highly associated) media, such as "magic acid" or HF/SbF_5 (the main problems in making such solutions are their corrosive properties and the fact that few strong fluoride acceptors do readily dissolve in either HF_{liq} or HSO_3F [79]). In water, H_0 = pH. Buffer systems in non-aq. solutions will have a H_0 value different from pH in water, e.g., glycine in DMF (11.0 [39] rather than 6.1)
HM	heavy metals. Here, denominating metals have an elemental density $\rho \geq$ **6** g/cm^3, thus deliberately excluding elements which differ massively from "typical" HM in their chemical properties while their densities are 5–6 g/cm^3, like Eu (unlike all other REEs), Ra, Ga, and V. Some non-metals sometimes are mis-labeled as HM for their toxic properties, especially As, Sb, and Te.
LM	Light metals. All metal-forming elements with low(-er) densities (see HM).
LREE	Light rare-earth elements. Unprecise definition but often used to summarize elements Z = 57–63 (La–Eu).
POC	Particulate organic carbon. In lakes, POC sometimes forms "lake snow". Consists of suspended biopolymers (chitin, cellulose, and insoluble salts with organic anions) and dead plankton.
REE	Rare-earth elements. All the metals with Z = 57–71 (La–Lu), sometimes including yttrium. Highly reactive, electropositive metals (Eu and Yb even dissolve in liquid ammonia [80], like alkali metals and heavy alkaline earths) among which, Eu has unique photochemical properties. Mostly trivalent, while Eu, and to a lesser extent Yb (also in water), Sm, Tm, Nd, and Dy (in solvents such as THF, series of decreasing stability) form M^{2+} ions and M(IV) can be prepared for Ce (various media), Pr, Nd, Tb, and Dy (in solvents HF, BrF_5, or solid state only). Peculiar spectroscopic properties due to partly vacant f orbitals are used in solid-state lasers and fluorescent agents. REE alloys (Nd, Sm, and Dy with Fe or Co) form the most powerful ferromagnetic materials known.
σ	L.P. Hammett´s (1893–1987) substituent constant which originally described changes in the reactivity of benzene or benzoic acid towards electrophiles; oxidantsor side chain reactants are observed when the aromatic ring is endowed with some substituent $\neq$ H, COO^-. The Hammett parameter can also be used to define binding in and electrochemical properties of metal complexes, with σ describing binding centers in the ligand [12].
SCE	saturated (6 M KCl solution) calomel electrode; redox couple

References

1. Pinto, P.X.; Al-Abed, S.R.; Reisman, D.J. Biosorption of heavy metals from mining influenced water onto chitin products. *Chem. Eng. J.* **2011**, *166*, 1002–1009. [CrossRef]
2. Anastopoulos, I.; Bhatnagar, A.; Bikiaris, D.N.; Kyzas, G.Z. Chitin adsorbents for toxic metals: A review. *Int. J. Mol. Sci.* **2017**, *18*, 114. [CrossRef]
3. Muzzarelli, R.A.A.; Rocchetti, R.; Marangio, G. Separation of zirconium, niobium, cerium and ruthenium on chitin and chitosan columns for the determination of cesium in nuclear fuel solutions. *J. Radioanal. Chem.* **1972**, *10*, 17–25. [CrossRef]
4. Ishay, J.S.; Benshalom-Shimony, T.; Ben-Shalom, A.; Kristianpoller, N. Photovoltaic effects in the Oriental hornet, *Vespa orientalis*. *J. Insect Physiol.* **1992**, *38*, 37–48. [CrossRef]
5. Dahmane, E.M.; Taourirte, M.; Eladlani, N.; Rhazi, M. Extraction and characterization of chitin and chitosan from *Parapenaeus longirostris* from Moroccan local sources. *Int. J. Polym. Anal. Charact.* **2014**, *19*, 342–351. [CrossRef]
6. Agarwal, S.; Mast, C.; Dehnicke, K.; Greiner, A. Rare earth metal initiated ring-opening polymerization of lactones. *Macromol. Rapid Commun.* **2000**, *21*, 195–198. [CrossRef]
7. Bauer, A. Orientierende Untersuchungen zur Bindung von Metallionen an Chitin und zur Davon Abhängigen Eignung von Arthropoden zur Bestimmung von Metallionenkonzentrationen in der Umwelt. Master's Thesis, IHI Zittau/TU Dresden, Zittau, Germany, 2014.
8. Vijayaraghavan, K.; Balasubramanian, R. Single and binary biosorption of cerium and europium onto crab shell particles. *Chem. Engin. J.* **2010**, *163*, 337–343. [CrossRef]
9. Fränzle, S.; Erler, M.; Blind, F.; Ariuntsetseg, L.; Narangarvuu, D. Chitin adsorption in environmental monitoring: Not an alternative to moss monitoring but a method providing (lots of) bonus information. *J. Sci. Arts Univ.* **2019**, *19*, 659–674.

10. Erler, M. Untersuchung des Bindungsverhaltens Ausgewählter Elemente und Ihrer Bodenrelevanten Komplexe an Chitin. Master's Thesis, IHI Zittau/TU Dresden, Zittau, Germany, 2020.
11. Blind, F. Orientierende Untersuchungen zur Platinmetall Freien Aktivierung von CH-Bindungen für Europium Basierte Brennstoffzellenanwendungen. Master's Thesis, IHI Zittau/TU Dresden, Zittau, Germany, 2018.
12. Blind, F.; Fränzle, S. Chitin as a sorbent superior to other biopolymers: Features and applications in environmental research, energy conversion, and understanding evolution of animals. *Polysaccharides* **2021**, *2*, 773–794. [CrossRef]
13. Luoga, Y.F. Potential Influence of Exposed Lignite on Local Biota and Biodiversity. Master's Thesis, IHI Zittau/TU Dresden, Zittau, Germany, 2020.
14. Nickens, K.P.; Patierno, S.R.; Ceryak, S. Chromium genotoxicity: A double-edged sword. *Chem. Biol. Interact.* **2010**, *188*, 276–288. [CrossRef]
15. Retschke, D. Orientierende Untersuchungen zur Adsorption von Schwermetallen (Nickel) unter dem Einfluss Ausgewählter Komplexliganden Sowie in Arealen Potenzieller und Manifester Methanogenese. Master's Thesis, IHI Zittau/TU Dresden, Zittau, Germany, 2017.
16. Schieritz, R. (IHI (International Institute) Zittau, TU Dresden, Zittau, Germany); Fränzle, S. (IHI (International Institute) Zittau, TU Dresden, Zittau, Germany). Investigation into adsorption of neutral biogenic organometal entities (pnictogen trimethyls, Fe, Mo, and W carbonyls) on chitin and chemical (I2, dioxygen complexes) photochemical modification of sorbate films thus produced. unpublished results. 2015.
17. Feldmann, J.; Cullen, W.R. Occurrence of volatile transition metal compounds in landfill gas: Synthesis of molybdenum and tungsten carbonyls in the environment. *Environ. Sci. Technol.* **1997**, *31*, 2125–2129. [CrossRef]
18. Yang, X.; Yin, D.; Sun, H.; Wang, X.; Dai, L.; Chen, Y.; Cao, M. Distribution and bioavailability of rare earth elements in aquatic microcosm. *Chemosphere* **1999**, *39*, 2443–2450. [CrossRef]
19. Takeno, N. Atlas of Eh-pH diagrams: Intercomparison of thermodynamic databases. *Geol. Surv. Jpn. Open File Rep.* **2005**, *419*, 102.
20. Bolsunovsky, A.; Melgunov, M.; Chuguevskii, A.; Lind, O.C.; Salbu, B. Unique diversity of radioactive particles found in the Yenisei River floodplain. *Sci. Rep.* **2017**, *7*, 11132. [CrossRef] [PubMed]
21. Curragh, E.F.; Henbest, H.B.; Thomas, A. Amine oxidation. Part VII. Cleavage of tetra-N-substituted 1,2-diamines by manganese dioxide. *J. Chem. Soc.* **1960**, 3559–3565. [CrossRef]
22. Tanner, S.P.; Choppin, G.R. Lanthanide and actinide complexes of glycine. *Inorg. Chem.* **1968**, *7*, 1268–1271.
23. Matsumoto, A.; Azuma, N. Photodecomposition of europium(III) acetate and formate in aqueous solutions. *J. Phys. Chem.* **1988**, *92*, 1830–1835. [CrossRef]
24. Jiang, J.; Gao, Y.; Pang, S.-Y.; Lu, X.-T.; Zhou, Y.; Ma, J.; Wang, Q. Understanding the role of manganese dioxide in the oxidation of phenolic compounds by aqueous permanganate. *Env. Sci. Technol.* **2015**, *49*, 520–528. [CrossRef] [PubMed]
25. Karraker, R.H. Stability Constants of Some Rare-Earth Metal Chelates. Ph.D. Thesis, Iowa State University of Science and Technology, Ames, IA, USA, 1961.
26. Moeller, T.; Martin, D.F.; Thompson, L.C.; Ferrus, R.; Feistel, G.R.; Randall, W.J. The coordination chemistry of yttrium and the rare earth metal ions. *Chem. Rev.* **1965**, *65*, 1–50. [CrossRef]
27. Sinsabaugh, R.L.; van Horn, D.J.; Shah, J.J.F.; Findlay, S. Ecoenzymatic stoichiometry in relation to productivity for freshwater biofilm and plankton communities. *Microb. Ecol.* **2010**, *60*, 885–893. [CrossRef] [PubMed]
28. Schwoerbel, J.; Brendelberger, H. *Einführung in die Limnologie*, 10th ed.; Springer: Berlin, Germany, 2013.
29. Thompson, A.J.; Sinsabaugh, R.L. Matrix- and particulate phosphatase and aminopeptidase activity in limnetic biofilms. *Aquat. Microb. Ecol.* **2000**, *21*, 151–159. [CrossRef]
30. Loppi, S.; Vannini, A.; Monaci, F.; Dagodzo, D.; Blind, F.; Erler, M.; Fränzle, S. Can chitin and chitosan replace the lichen *Evernia prunastri* for environmental biomonitoring of Cu and Zn air contamination? *Biology* **2020**, *9*, 301. [CrossRef]
31. Jurkowski, W.; Paper, M.; Brück, T.B. Isolation and Investigation of natural rare earth metal chelating agents from *Calothrix brevissima*-A Step Towards Unraveling the Mechanisms of Metal Biosorption. *Front. Bioeng. Biotechnol.* **2022**, *10*, 833112. [CrossRef] [PubMed]
32. Budelmann, P. Verbreitung der Flusskrebse (Decapoda) in der Südlichen Oberlausitz und die Eignung des Invasiven Kamberkrebses (*Orconectes Limosus*) für Chitin-Basiertes Monitoring von Schwermetallen in Limnischen Ökosystemen. Master's Thesis, IHI Zittau/TU Dresden, Zittau, Germany, 2021.
33. Brandys, M.; Stein, G. Photochemical evolution of hydrogen from aqueous solutions of europium ions at lambda > 310 nm. The role of europium(II). *J. Phys. Chem.* **1978**, *82*, 852–854. [CrossRef]
34. Wood, S.A. The aqueous geochemistry of the rare-earth elements and yttrium. 2. Theoretical predictions of speciation in hydrothermal solutions to 350 °C at saturation water vapor pressure. *Chem. Geol.* **1990**, *88*, 99–125. [CrossRef]
35. Gebauer, T. Methodische Optimierung des Übertrags von Metallionen aus Umweltprobenmodellen auf Chitinoberflächen und von Diesen zu Zwecken Analytischen Biomonitorings sowie Untersuchungen zur Diffusion/Ausbreitung von Analyten in Chitinproben. Master's Thesis, IHI Zittau/TU Dresden, Zittau, Germany, 2016.
36. Pulsfort, J. Biopolymere Als Potenzielle Substrate der Photooxidation Durch f-Block-Metalle. Master's Thesis, IHI Zittau/TU Dresden, Zittau, Germany, 2021.
37. Malhotra, N.; Hsu, H.-S.; Liang, S.-T.; Jmelou, M.; Roldan, M.; Lee, J.-S.; Ger, T.-R.; Hsiao, C.-D. An Updated Review of Toxicity Effect of the Rare Earth Elements (REEs) on Aquatic Organisms. *Animals* **2020**, *10*, 1663. [CrossRef] [PubMed]

38. Sjöberg, S.; Callac, N.; Allard, B.; Smittenberg, R.H.; Dupraz, C. Microbial communities inhabiting a rare earth element enriched birnessite-type manganese deposit in the Ytterby Mine, Sweden. *Geomicrobiol. J.* **2018**, *35*, 657–674. [CrossRef]
39. Junold, M. Untersuchung der Eignung von Bipolartransistor und MOSFET (Metall-Oxid-Feldeffekttransistor) zur Identifikation von Kurzkettigen Carbonsäuren. Master's Thesis, IHI Zittau/TU Dresden, Zittau, Germany, 2021.
40. Giroux, S.; Rubini, P.; Aury, S. Complexes of praseodymium(III) with d-gluconic acid. *Polyhedron* **2000**, *17*, 422–428. [CrossRef]
41. Gutmann, V.; Peychal-Heiling, G. Das polarographische Verhalten von Europium in Abhängigkeit vom Lösungsmittel und Leitsalz. *Mon. Chem.* **1969**, *100*, 1423–1433. [CrossRef]
42. Couffin, F. *Potentiels d 'Oxydo-Reduction des Elements Lanthanides et Actinides dans les Solvants Organiques*; CEA-BIB 233; 1980.
43. Farkas, E.; Enyedy, E.A.; Zekany, L.; Deak, G. Fe hydroxamatocomplexes, natural: Interaction between iron(II) and hydroxamic acids: Oxidation of iron(II) to iron(III) by desferrioxamine B under anaerobic conditions. *J. Inorg. Biochem.* **2001**, *83*, 107–114. [CrossRef]
44. Holleck, L. Zur Komplexchemie der seltenen Erden. Die polarographischen Strompotentialkurven des Europiums als Nachweismittel komplexer Bindung. *Z. für Naturf. B* **1947**, *2*, 81–89. [CrossRef]
45. Blind, F. (IHI (International Institute) Zittau, TU Dresden, Zittau, Germany); Fränzle, S. (IHI (International Institute) Zittau, TU Dresden, Zittau, Germany). Influence of CH bond energies and substrate ionization potentials, orbital sequences on the chance to accomplish photooxidation by Eu(III). Unpublished results. 2019.
46. Adeel, M.; Lee, J.Y.; Zain, M.; Rizwan, M.; Nawab, A.; Ahmad, M.A.; Shafiq, M.; Yi, H.; Jilani, G.; Javed, R.; et al. Cryptic footprints of rare earth elements on natural resources and living organisms. *Environ. Int.* **2019**, *127*, 785–800. [CrossRef]
47. Markert, B.; de Li, Z. Natural background concentrations of rare-earth elements in a forest ecosystem. *Sci. Total Environ.* **1991**, *103*, 27–35. [CrossRef]
48. Köllner, K.E.; Carstens, D.; Keller, E.; Vazquez, F.; Schubert, C.J.; Zeyer, J.; Bürgmann, H. Bacterial chitin hydrolysis in two lakes with contrasting trophic statuses. *Appl. Environ. Microbiol.* **2012**, *78*, 695–704. [CrossRef] [PubMed]
49. Dhar, N.R.; Mukherjee, S.K. Photosynthesis of amino acids in vitro. *Nature* **1934**, *134*, 499. [CrossRef]
50. Anonymous. NIST Table of Chemicals Properties. Available online: https://webbook.nist.gov/cgi/cbook.cgi?ID=116-09-6 (accessed on 15 January 2022).
51. Klapiszewski, Ł.; Wysokowski, M.; Majchrzak, I.; Szatkowski, T.; Nowacka, M.; Siwińska-Stefańska, K.; Szwarc-Rzepka, K.; Bartczak, P.; Ehrlich, H.; Jesionowski, T. Preparation and Characterization of Multifunctional Chitin/Lignin Materials. *J. Nanomat.* **2013**, *2013*, 425726. [CrossRef]
52. Hennig, H.; Rehorek, D. *Photochemische und Photophysikalische Reaktionen von Koordinationsverbindungen*; Deutscher Verlag der Wissenschaften: Berlin, Germany, 1987.
53. Horvath, O.; Stevenson, K.L. *Charge Transfer Photochemistry of Coordination Compounds*; VCH: New York, NY, USA; Weinheim, Germany, 1993.
54. Araki, K.; Sakuma, M.; Shiraishi, S. Photooxidation of d-fructose with iron(III) chloride under aerobic conditions. *Chem. Lett.* **1983**, 665–666. [CrossRef]
55. Sterner, R.W.; Elser, J.J. *Ecological Stoichiometry: The Biology of Elements from Molecules to the Biosphere*; Princeton Univ. Press: Princeton, NJ, USA, 2002.
56. Wagner, G.P.; Lo, J.; Laine, R.; Almeder, M. Chitin in the epidermal cuticle of a vertebrate *(Paralipophrys trigloides, Blenniidae, Teleostei)*. *Experientia* **1993**, *49*, 317–319. [CrossRef]
57. Weiss, M.; Simon, M. Consumption of labile dissolved organic matter by limnetic bacterioplankton: The relative significance of amino acids and carbohydrates. *Aquat. Microb. Ecol.* **1999**, *17*, 1–12.
58. Ruetschi, P.; Delahay, P. Hydrogen overvoltage and electrode material. A theoretical analysis. *J. Chem. Phys.* **1955**, *23*, 195–199. [CrossRef]
59. Digby, P.S.B. Semi-conduction and electrode processes in biological material. *Proc. Royal Soc. B* **1964**, *161*, 504–525.
60. Hansch, C.; Leo, A.; Taft, R.W. A Survey of Hammett Substituent Constants and Resonance and Field Parameters. *Chem. Rev.* **1991**, *97*, 165–195. [CrossRef]
61. Funahashi, R.; Ono, Y.; Qi, Z.-D.; Saito, T.; Isogai, A. Molar Masses and Molar Mass Distributions of Chitin and Acid-Hydrolyzed Chitin. *Biomacromolecules* **2017**, *18*, 4357–4363. [CrossRef]
62. Ozaki, T.; Kimura, T.; Yoshida, Z.; Francis, A.J. Speciation study of actinides and lanthanides by time-resolved laser-induced fluorescence spectroscopy. *Chem. Lett.* **2003**, *32*, 560–561. [CrossRef]
63. Stankiewicz, B.A.; Mastalerz, M.; Hof, C.H.J.; Bierstedt, A.; Flannery, M.B.; Briggs, D.E.G.; Evershed, R.P. Biodegradation of the chitin-protein complex in crustacean cuticle. *J. Org. Geochem.* **1998**, *28*, 67–76. [CrossRef]
64. Ozaki, T.; Kimura, T.; Ohnuki, T.; Kirishima, A.; Yoshida, T.; Isobe, H.; Francis, A.J. Association of europium(III), americium(III), and curium(III) with cellulose, chitin, and chitosan. *Environm. Toxicol. Chem.* **2006**, *25*, 2051–2058. [CrossRef] [PubMed]
65. Pati, A.; Bhattacharyya, A.; Pujari, P.K.; Pal, S.; Kundu, T.K. Investigation of lanthanide complexation with acetohydroxamic acid in nitrate medium: Experimental and DFT studies. *J. Chem. Sci.* **2021**, *133*, 60. [CrossRef]
66. Nozaki, Y. A Fresh Look at Element Distribution in the North Pacific. *EOS Trans.* **1997**, *78*, 221–226.
67. Adin, A.; Sykes, A.G. The Kinetics of the oxidation of europium(II) with vanadium(III) and chromium(III) in aqueous perchloric acid solutions. *J. Chem. Soc.* **1966**, 1230–1236. [CrossRef]

68. AAdegite; Egboh, H. Kinetics and Mechanism of the Oxidation of Europium(II) by Aqueous Solution of Iodine and Triiodide. *Inorg. Chim. Acta* **1977**, *21*, 1–4. [CrossRef]
69. Scheich, H.; Langner, G.; Tidemann, C.; Coles, R.B.; Guppy, A. Electroreception and electrolocation in platypus. *Nature* **1986**, *319*, 401–402. [CrossRef] [PubMed]
70. Song, K.; Wen, Z.; Xu, Y.; Yang, H.; Lyu, L.; Zhao, Y.; Fang, C.; Shang, Y.; Du, J. Dissolved carbon in a large variety of lakes across five limnetic regions in China. *J. Hydrol.* **2018**, *563*, 143–154. [CrossRef]
71. Kim, J.I.; Klenze, R.; Wimmer, H.; Runde, W.; Hauser, W. A study of the carbonate complexation of Cm(III) and Eu(III) by time-resolved laser fluorescence spectroscopy. *J. Alloy. Compd.* **1994**, *213*, 333–340. [CrossRef]
72. Pol, A.; Barends, T.R.M.; Dietl, A.; Khadem, A.F.; Eygensteyn, J.; Jetten, M.S.M.; den Camp, H.J.M.O. Rare earth metals are essential for methanotrophic life in volcanic mudpots. *Environ. Microbiol.* **2014**, *16*, 255–261. [CrossRef]
73. Cook, E.C.; Featherston, E.R.; Showalter, S.A.; Cotruvo, J.A., Jr. Structural basis for rare earth element recognition by *Methylobacterium extorquens* lanmodulin. *Biochemistry* **2019**, *58*, 120–125. [CrossRef]
74. Monsallier, J.-M.; Scherbaum, F.J.; Buckau, G.; Frimmel, F.H. Influence of photochemical reactions on the complexation of humic acid with europium(III). *J. Photochem. Photobiol. A Chem.* **2001**, *138*, 55–56. [CrossRef]
75. Kitching, R.L. Food webs in phytotelmata: "bottom-up" and "top-down" explanations for community structure. *Annu. Rev. Entomol.* **2001**, *46*, 729–760. [CrossRef]
76. Maleke, M.; Valverde, A.; Vermeulen, J.-G.; Cason, E.; Gomez-Arias, A.; Moloantoa, K.; Coetsee-Hugo, L.; Swart, H.; van Heerden, E.; Castillo, J. Biomineralization and bioaccumulation of europium by a thermophilic metal resistant bacterium. *Front. Microbiol.* **2019**, *10*, 81–88. [CrossRef]
77. Ridley, W.P.; Dizikes, L.; Cheh, A.; Wood, J.M. Recent studies on biomethylation and demethylation of toxic elements. *Environ. Health Perspect.* **1977**, *19*, 43–46. [CrossRef]
78. Michalke, K.; Schmidt, A.; Huber, B.; Meyer, J.; Sulkowski, M.; Hirner, A.V.; Boertz, J.; Mosel, F.; Dammann, P.; Hilken, G. Role of intestinal microbiota in transformation of bismuth and other metals and metalloids into volatile methyl and hydride derivatives in humans and mice. *Appl. Environ. Microbiol.* **2008**, *74*, 3069–3075. [CrossRef] [PubMed]
79. Waddington, N.C. *Nichtwässrige Lösungsmittel*; utb: Heidelberg, Germany, 1972.
80. Golub, A.A.; Köhler, H. *Pseudohalogenide*; Deutscher Verlag der Wissenschaften: Berlin, Germany, 1978.

polysaccharides

MDPI

Article

Low-Denaturazing Glucose Oxidase Immobilization onto Graphite Electrodes by Incubation in Chitosan Solutions

Mireia Buaki-Sogó [1,*], Laura García-Carmona [1,*], Mayte Gil-Agustí [1], Marta García-Pellicer [1] and Alfredo Quijano-López [1,2]

[1] Instituto Tecnológico de la Energía (ITE), Avenida Juan de la Cierva 24, 46980 Paterna, Spain; mayte.gil@ite.es (M.G.-A.); marta.garcia@ite.es (M.G.-P.); alfredo.quijano@ite.es (A.Q.-L.)

[2] Instituto de Tecnología Eléctrica, Universitat Politècnica de València, Camino de Vera s/n Edificio 6C, 46022 Valencia, Spain

* Correspondence: mireia.buaki@ite.es (M.B.-S.); laura.garcia@ite.es (L.G.-C.); Tel.: +36-961-366-670 (L.G.-C.)

Abstract: In this work, glucose oxidase (GOx) has been immobilized onto graphite rod electrodes through an assisted-chitosan adsorption reaching an enzyme coverage of 4 nmol/cm^2. The direct and irreversible single adsorption of the Flavine Adenine Dinucleotide (FAD) cofactor has been minimized by electrode incubation in a chitosan (CH) solution containing the enzyme GOx. Chitosan keeps the enzyme structure and conformation due to electrostatic interactions preventing FAD dissociation from the protein envelope. Using chitosan, both the redox cofactor FAD and the protein envelope remain in the active form as demonstrated by the electrochemistry studies and the enzymatic activity in the electrochemical oxidation of glucose up to a concentration of 20 mM. The application of the modified electrodes for energy harvesting delivered a power density of 119 µW/cm^2 with a cell voltage of 0.3 V. Thus, chitosan presents a stabilizing effect for the enzyme conformation promoted by the confinement effect in the chitosan solution by electrostatic interactions. Additionally, it facilitated the electron transfer from the enzyme to the electrode due to the presence of embedded chitosan in the enzyme structure acting as an electrical wiring between the electrode and the enzyme (electron transfer rate constant 2.2 s^{-1}). This method involves advantages compared with previously reported chitosan immobilization methods, not only due to good stability of the enzyme, but also to the simplicity of the procedure that can be carried out even for not qualified technicians which enable their easy implementation in industry.

Keywords: glucose oxidase; adsorption; FAD; denaturation; chitosan; enzyme immobilization; glucose biofuel cell

Citation: Buaki-Sogó, M.; García-Carmona, L.; Gil-Agustí, M.; García-Pellicer, M.; Quijano-López, A. Low-Denaturazing Glucose Oxidase Immobilization onto Graphite Electrodes by Incubation in Chitosan Solutions. *Polysaccharides* **2022**, *3*, 388–400. https://doi.org/10.3390/polysaccharides3020023

Academic Editor: Azizur Rahman

Received: 14 February 2022
Accepted: 28 April 2022
Published: 3 May 2022

1. Introduction

The entire field of enzyme-based electrodes can trace its origin back to the enzymatic glucose-based electrode fabricated by Clark and Lyons in 1962 with the publication of the original patent of an amperometric enzyme electrode for glucose sensing [1]. Currently, the importance of the glucose biosensors market is clear, covering about 85% of the entire biosensor market in the world, mainly due to the notable biomedical significance of the rapid and convenient assay of blood glucose in diabetes [2–4]. In addition, other relevant applications, such as biofuel cells for energy harvesting using molecules such as glucose as an attractive new power source, take advantage of technological advances related with GOx enzyme for self-powered, wearable, and implantable biomedical devices [5].

The enzyme GOx catalyzes the oxidation of β-D-glucose to gluconic acid with molecular oxygen as an electron acceptor with the production of hydrogen peroxide (H_2O_2) following reactions 1 and 2. In these reactions the fully oxidized form of Flavine Adenine Dinucleotide (FAD) catalyzes the oxidation of glucose while the reduced form ($FADH_2$) is regenerated in the presence of O_2 producing H_2O_2. Reaction 3 represents another mechanism in which the electronic transference from the reduced form of the enzyme ($FADH_2$)

occurs in the absence of mediators or electron acceptors such as oxygen [6–8]. However, this mechanism, usually called Direct Electron Transfer (DET), has not been proved to drive the reaction in the absence of mediators or electron acceptors. Currently, reaction 1 and 2 are the most accepted mechanisms to explain GOx redox mechanism since the two flavin active sites are deeply buried within the structure of the homodimer [9]. For this reason, efforts regarding GOx immobilization, traditionally focused on achieving DET, should be focused on stabilization issues related to the loose of the structure instead on the orientation of the enzyme on the electrodic surface.

$$\text{Glucose} + \text{GOx(FAD)} \rightarrow \text{Gluconic acid} + \text{GOx(FADH}_2\text{)} \ (1)$$

$$\text{GOx(FADH}_2\text{)} + \text{O}_2 \rightarrow \text{GOx(FAD)} + \text{H}_2\text{O}_2 \ (2)$$

$$\text{GOx(FADH}_2\text{)} \rightarrow \text{GOx(FAD)} + 2\text{e}^- + 2\text{H}^+ \ (3)$$

Particularly regarding bioelectrodes development, GOx should be immobilized onto suitable conductive support to carry out the enzymatic reaction. For real industrial use of this technology in long-term devices, the enzyme stability should be preserved to keep activity in the working medium and enhance enzyme stability towards denaturation of the enzyme and FAD irreversible adsorption on the electrode in storage and operational conditions [5,6]. Among the different enzyme immobilization strategies [5,10,11] adsorption is one of the preferred techniques due to its simplicity and limited decrease in enzymatic activity despite the disadvantages related to desorption processes, denaturation, and the method's low specificity [12].

The use of graphite electrodes as supports in the development of bioelectrodes for biodevices has been widely spread due to the exceptional electrical properties of such materials. Additionally, graphite electrodes present a surface able to adsorb biomolecules due to the affinity of graphite surface for organic entities, so it has been used in several works as a support for enzyme immobilization using different strategies [13–16]. When performing GOx adsorption onto graphite electrodes, one of the main drawbacks comes from enzyme denaturation due to the most preferable and irreversible adsorption of FAD included into GOx structure, losing the quaternary structure, and, therefore, the catalytic activity [17–19]. The denaturation process could be mainly related to enzyme concentration or ionic strength of incubation solution, temperature, and/or incubation time [19]. This is a key aspect for the assessment of the enzyme activity since if FAD is irreversibly adsorbed in a denaturized form of GOx, can lead to misunderstanding when trying to assess the direct electrochemistry of the enzyme and to evaluate the bioelectrocatalytic activity in the presence of glucose [9,20], since no response would be detected.

Inactivation due to denaturation is mainly related to changes in quaternary structure due to subunit dissociation [21,22]. Hence, preventing subunit dissociation is an important strategy for multimeric enzyme stabilization [23]. Some methods based on this idea have been proposed, such as protein engineering or crosslinking [24–27]. However, these methods are quite complex and present limitations dealing with the stability of engineered enzymes or the selection of reaction conditions and reagents for crosslinking, which is usually a sensitive procedure that can lead to irreversible damage in the enzyme structure [28]. Therefore, the stabilization of multimeric enzymes against inactivation is considered a particular challenge.

Biopolymers can be used to immobilize enzymes belonging to various catalytic classes with the retention of good catalytic properties. Moreover, the produced biocatalytic systems offer improved thermal stability and, in general, are noted for their good reusability [29]. In this sense, several biomaterials are frequently used in research and industry due to the biocompatibility characteristics very convenient for their extensive potential in industrial applications. In this sense, biopolymers, such as Polyhydroxyalkanoates, linear polyesters containing 3-hydroxy fatty acid monomers, or polysaccharides, such as chitosan, are commonly employed [30,31]. However, sometimes it is necessary to modify these polymers in certain applications which can lead to modify their mechanical properties.

Specifically, Chitosan is a widely used polysaccharide obtained from the deacetylation of chitin, which is the major component in crustacean shells such as shrimps, lobsters, and crabs. Chitosan properties such as hydrophilicity, low toxicity, biocompatibility, biodegradability, film-forming ability, and bioactivity attract the interest of biologists and chemists to explore novel uses of this biopolymer. Chitosan molecules possess a positive charge in solution due to the amino groups present in their structure, so the low isoelectric point of some multimeric enzymes, as is the case for GOx, leads to a fast interaction between proteins and chitosan molecules [31]. For all these reasons, chitosan is the most frequently used biopolymer for enzyme immobilization [31,32]. Thus, although several methods using chitosan as the main scaffold for the adsorption immobilization strategy have been developed so far, several disadvantages have been detected, hindering their applicability in real-world applications. Specifically, these methods involved multilayer assemblies, crosslinking steps, acid treatments, and deposition of polymeric films containing GOx enzyme on the electrode surface [33–36].

Hence, the procedure here presented describes a method for GOx immobilization onto graphite rod electrodes avoiding enzyme denaturation through adsorption by incubation in chitosan solutions. The suitability of the method to keep enzyme activity, which is the key aspect regarding immobilization procedures for bioelectrodes development, is evaluated in this paper through cyclic voltammetry in oxygen-free solutions to assess electron transfer between the graphite surface and the active centre of GOx. Moreover, the catalytic response of the enzyme for energy harvesting in a glucose BFC has been evaluated to confirm the proper stability of the enzyme using this immobilization procedure.

2. Materials and Methods

2.1. Reagents and Instrumentation

Chitosan of medium molecular weight with a deacetylation degree > 85% and graphite rod electrodes were purchased from Sigma-Aldrich (St. Louis, MO, USA). Glucose oxidase (GOx) from Aspergillus niger Type VII, (110 U/mg and a ratio GOx/catalase higher or equal to 100, used without extra purification), D-Glucose biotechnology grade standard, and Hydroquinone (HQ) for mediated electron transfer were from VWR LIFE SCIENCE (Radnor, PA, USA). For the preparation of Phosphate Buffer Saline (PBS) solutions, Na_2HPO_4 and NaH_2PO_4 from VWR LIFE SCIENCE (Radnor, PA, USA), and KCl from Scharlab S.L. (Barcelona, Spain) were employed. Acetic acid for chitosan solutions was also acquired from Scharlab S.L. (Barcelona, Spain).

2.2. Apparatus and Electrodes

A Zhaner IM6 electrochemical workstation provided with a PP241 module for high currents was the equipment used to perform Cyclic Voltammetry (CV) and BFC polarization curve determination. Electrochemical characterization in a three-electrode cell configuration was performed using the graphite rod as the working electrode, an Ag/AgCl (3M KCl) electrode as the reference electrode, and a Pt wire as the counter electrode. Prior to all the cyclic voltammetry experiments, the electrolyte solution was purged with nitrogen for 30 min and then the inert atmosphere was kept during the recording of the voltammogram.

2.3. GOx Immobilization Procedure

For the immobilization of GOx onto graphite rod electrodes, the adsorption method was performed through the incubation of the graphite rods in a chitosan solution of GOx at 4 °C for 72 h. The low-density graphite (LDG) electrodes (Sigma-Aldrich. St. Louis, MO, USA) were first treated with coarse and fine emery paper. Then, the graphite rod was sonicated for 15 min in distilled water. Once the electrode was dried, it was cut into two rods of 7.5 cm and covered with Teflon tape leaving 1 cm in one end for the electrical connection and 2 cm in the other end as the immobilization surface.

For the immobilization procedure of GOx, 250 mg of chitosan were dissolved in 25 mL of a 0.3 M acetic acid solution and filtered through a 0.45 μm Nylon filter (VWR LIFE

SCIENCE) before enzyme addition. Then, 5 mL of a solution of 10 mg/mL of GOx in 1% (*w*/*w*) chitosan solution was prepared and stirred for 30 min at room temperature to allow the self-assembly of the enzyme in the chitosan solution. Graphite rods were incubated in such solutions for 72 h at 4 °C. After 72 h, graphite rods were removed from the solution and washed in PBS pH 5.5 for 30 min. The washing procedure was repeated three times with fresh PBS solutions. The modified bioelectrode (LDG-CH10) was kept in PBS pH 7.55 at 4 °C until use.

FAD adsorption onto the graphite rod electrodes was performed in PBS pH 7 solutions in the same conditions as reported above. The graphite rod electrode was incubated in a FAD solution of 10 μg/mL during 72 h at 4 °C. After the 72 h, the modified graphite electrode (FAD@LDG) was removed from the solution and washed in PBS pH 5.5 solution for 30 min. This washing procedure was also repeated three times and the FAD-modified graphite electrodes were kept at 4 °C until use.

For the assessment of enzyme denaturation during the adsorption, immobilization was performed in PBS as the incubation medium. A volume of 5 mL of a solution of 10 mg/mL GOx in PBS pH 7.55 was prepared. Graphite rods were incubated in such solutions for 72 h at 4 °C. After 72 h of incubation, the same washing procedure as in the case of immobilization in chitosan was carried out. The bioelectrode thus modified (LDG-PBS10), was kept in PBS pH 7.55 at 4 °C until use.

The evaluation of FAD irreversible adsorption on the graphite electrode surface was evaluated by placing the modified graphite electrode in 5 mL of a KCl 3M solution overnight with magnetic stirring. Later, cyclic voltammetry experiments were performed at 0.1 V/s between −0.7 V and 0.4V under inert atmosphere before and after the KCl treatment to be compared.

2.4. Electrochemical Characterization

The direct electrochemistry of immobilized GOx was evaluated by cyclic voltammetry. The electrolyte for the cyclic voltammetry experiments was PBS KCl 0.1 M pH 7.55 that was purged with nitrogen, at least for 30 min, before cyclic voltammogram recording. Cyclic voltammetry was performed at different scan rates: 25, 50, 100, 150, 200, 250, 300, 400, and 500 mV/s, with a potential window between −0.7 V and 0.4 V.

For activity assays, 1 M glucose solution was prepared before the experiment and left to mutarrotate overnight. Glucose concentration in the electrochemical cell was increased by a factor of 5 mM in each addition. Cyclic voltammograms were recorded between −0.7 V and 0.4 V at 0.1 V/s.

For the polarization curve, a two-electrode configuration cell with a Pt wire as cathode and the GOx-modified graphite electrode as anode were employed. The power density was calculated using the product of voltage and intensity divided by the electroactive surface area obtained from the Randles–Sevcik equation, as previously reported [37].

For the inhibition assays, a solution of $AgNO_3$ 0.1M was used. A suitable amount of the $AgNO_3$ solution was poured into the electrochemical cell to obtain a concentration of Ag^+ ions of 1 μM. Then, bioelectrocatalytic activity for glucose oxidation was evaluated by cyclic voltammetry following the procedure described above.

3. Results and Discussion

3.1. Bioelectrode Modification with Glucose Oxidase

The modification of the LDG electrodes was carried out using chitosan as an immobilization medium where, apart from hydrophobic interactions, electrostatic interactions between chitosan and the enzyme also act. As a control, the immobilization in PBS, where only hydrophobic interactions between the enzyme and the support were involved, was also performed to assess FAD dissociation and irreversible adsorption. The general procedure for the immobilization procedure is schematized in Figure 1. The electrode surface was first incubated in glucose oxidase (GOx) chitosan-based solution for 72 h at 4 °C for the selected working concentration of 10 mg/mL of GOx. This incubation time allowed the enzyme to be bound to LDG electrodes due to the affinity of carbon-based surfaces for biomolecules with chitosan entities embedded in the enzyme structure being part of the whole system [38]. Later, to remove the loosely bound enzyme, graphite rods were washed using a fresh PBS solution.

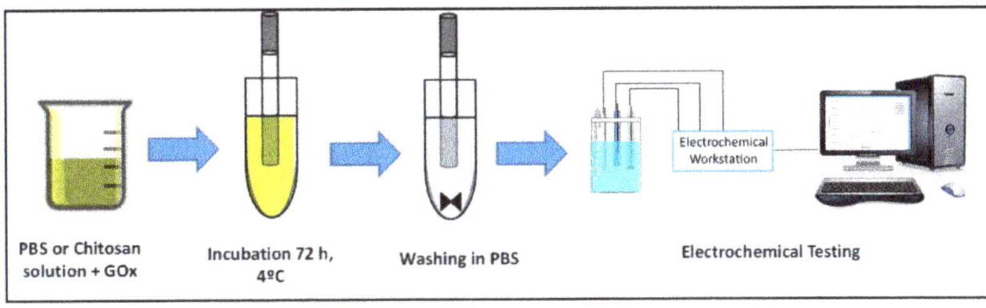

Figure 1. GOx immobilization procedure and working operation.

The immobilization of the enzyme by incubation in chitosan solutions was confirmed due to the presence of two redox peaks in the cyclic voltammogram corresponding to the oxidation-reduction of the active center of the GOx enzyme FAD at −0.22 V and −0.35 V respectively at 0.2 V/s under N_2 atmosphere (solid blue line in Figure 2A). The results indicate that FAD cofactor was undergoing a reversible electron transfer with the graphite electrode as shown in reaction 3.

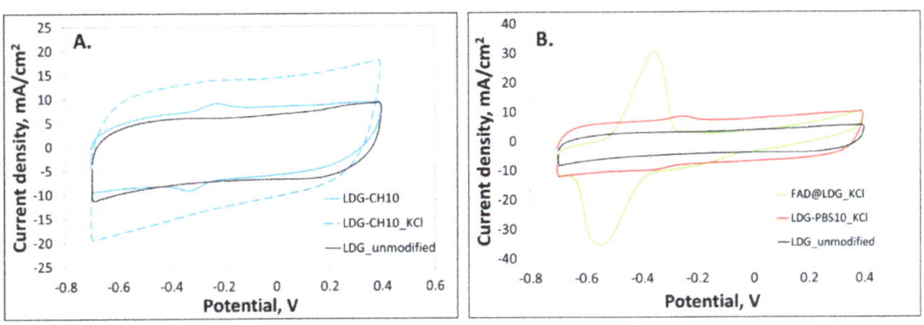

Figure 2. Bioelectrode evaluation for FAD irreversible adsorption. (**A**) chitosan-based GOx immobilization (solid blue line), chitosan-based GOx modified electrode treated with KCl (blue dashed line). (**B**) KCl treatment of the electrodes modified by GOx adsorption using PBS (solid red line) and FAD direct adsorption (solid green line). Unmodified LDG electrodes (solid black line). All the voltammograms were acquired at 0.2 V/s in O_2 free atmosphere.

As mentioned before, bioelectrodes prepared through GOx adsorption on carbon-based materials can suffer enzyme denaturation during the adsorption [18–20]. The mechanism for this process starts with FAD dissociation from the protein structure (during the immobilization process or directly from free FAD that commercial GOx can contain as an impurity) where it is embedded and the subsequent migration towards graphite surface, leading to FAD cofactor irreversible adsorption due to the affinity of graphite for organic moieties [39]. The immobilization by incubation of the LDG rod in a biopolymeric solution of chitosan was carried out to avoid the denaturation of the enzyme taking advantage of the stabilizing effect of the biopolymeric medium by electrostatic interactions [29,30,40]. However, before the electrochemical characterization of the GOx-bioelectrode prepared in chitosan solutions, it is necessary to ascertain whether the redox peaks observed in Figure 2 are due or not to denaturized FAD irreversible adsorption.

For this purpose, the bioelectrode prepared by incubation in chitosan solution was submitted to a treatment with concentrated KCl solutions to assess the enzyme denaturation process. It has been reported that treatment with concentrated salt solutions can strip FAD from flavoenzymes, but the same treatment is ineffective in removing adsorbed FAD from graphite surfaces [41–43]. Thus, if the specie adsorbed on the graphite surface is the whole GOx enzyme with its protein embedding structure, basic treatment will strip it from the electrode surface and the corresponding redox peaks will disappear from the cyclic voltammogram. In contrast, if the peaks observed come from FAD direct irreversible adsorption due to denaturation, the redox peaks will remain in the cyclic voltammogram after the treatment.

Thus, as can be seen in Figure 2A (blue dashed line), redox peaks of chitosan-based GOx modified electrodes disappear after the treatment with KCl, indicating that signals were originated from non-denaturized GOx and that no dissociation and irreversible adsorption of FAD cofactor has occurred.

On the other hand, in the case of immobilization of GOx in PBS (LDG-PBS10), the redox peaks do not disappear after the treatment with KCl (Figure 2B red solid line), which points to the fact that FAD has been irreversibly adsorbed on the graphite surface. This behavior has been confirmed by an experiment of direct adsorption of FAD on the graphite electrode (FAD@LDG). This modified electrode was also submitted to the KCl treatment and as expected, the redox peaks remained in the cyclic voltammogram after the basic treatment, which confirms that FAD has been irreversibly adsorbed on the graphite surface (see Figure 2B and Figure S1A in Supplementary Materials). Therefore, we can conclude that the incubation in chitosan solutions favors the immobilization of GOx without FAD dissociation.

It is worth highlighting that FAD redox peaks in Figure 2B at −0.22 V and −0.35 V for the GOx enzyme adsorbed on the LDG electrode in PBS, showed a positive shift of about 140 mV with respect to the redox peak potential of FAD directly adsorbed on the LDG electrode also in PBS medium. This result suggests that immobilization conditions and assembly may have distorted the enzyme structure making the flavine center more accessible [9]. Additionally, it is worth mentioning that KCl treatment can also modify the electrode surface, which contributes to an increase in the capacitive current after the KCl treatment, as can be seen in Figure S1B.

3.2. *Electrochemistry of Immobilized GOx in Chitosan Solutions*

The electrochemical characterization of the GOx bioelectrodes obtained by incubation in chitosan and PBS solutions was carried out by cyclic voltammetry at different scan rates in an O_2 free atmosphere to detect the oxidation-reduction reactions of the FAD cofactor on the graphite surface (Figure S2). The peak current increases with the scan rate demonstrating that the redox reaction of the FAD cofactor is a typical surface-confined process. Enzyme coverage in mol/cm^2 was obtained from the Equation (2) derived from Laviron expression (1) [44] for the bioelectrodes prepared by incubating the modified graphite rod electrode in chitosan and PBS solutions at two different concentrations (see Table S1).

$$I_p = \frac{n^2F^2vA\Gamma}{4RT} = \frac{nFQv}{4RT} \tag{1}$$

$$Q = nFA\Gamma \tag{2}$$

In these equations, Q is the charge obtained from the integration of the cathodic peak in the cyclic voltammograms recorded at different scan rates; n corresponds to the number of transferred electrons (n = 2); F is the Faraday (96,400 C/mol) constant; A is the electrode electroactive area in cm^2 obtained from cyclic voltammetry and Randles–Sevcik equation; and Γ is the enzyme coverage in mol/cm^2. To explore the efficiency of the bioelectrodes developed, the electron transfer rate constant was also determined from the scan rate dependence of ΔEp using the method of Laviron [44]. Values for the electron transfer rate constant in the LDG electrodes prepared (LDG-CH and LDG-PBS) are presented in Table S1 together with enzyme coverage.

Enzyme coverage was higher for the electrode LDG-CH10 prepared with the higher GOx concentration in the chitosan solution with a value near 4 nmol/cm^2. The electron transfer rate constant for the electrodes prepared by incubation in PBS and chitosan solutions of GOx were 1.1 s^{-1} and 2.2 s^{-1}, respectively. These values are in the same order of magnitude than similar systems that can be found in literature [18]. Moreover, the values for enzyme coverage obtained for the electrodes incubated in PBS are lower than those obtained for the bioelectrodes prepared in chitosan solutions (Table S1). Furthermore, it is worth mentioning that the values of the electron transfer rate constants obtained for electrodes modified in PBS solutions are in the same range than those previously reported for FAD adsorption on graphite electrodes [41].

Several works have been published over the years dealing with Direct Electron Transfer (DET) between GOx and carbon-based electrodes aimed at the development of glucose biosensors and biofuel cells [7,21,45,46] but there is no clear evidence that this phenomenon actually occurs [9,21].

Electron transfer of the bioelectrodes mediated by the flavoprotein FAD embedded into the GOx structure can be evaluated from the analysis of cyclic voltammetry experiments acquired in oxygen-free solutions. According to reaction 3, oxidation and reduction of FAD is a two-electron and two-proton coupled reaction. The cathodic peak obtained in the cyclic voltammetry is attributed to the reduction of GOx(FAD), while the anodic peak current is attributed to oxidation of GOx($FADH_2$). The pH value of the solution will influence electron transfer of GOx, and both the anodic and cathodic peak potentials will be shifted to more negative values as the solution pH increases as is observed in Figure S3. Moreover, formal potential exhibits a linear dependence of pH ranging from 5 to 9 (Inset of Figure S3) with a slope around to −50 mV/pH (R^2 = 0.9964), a value close to the theoretical −59.2 mV/pH for the two-electron and two-proton coupled reaction [11]. In addition, the linear relationship between peak current and scan rate implies that the bioelectrode presents the typical behavior of a surface-confined redox process with anodic and cathodic peak separation characteristic of quasi-reversible redox systems with a mean value lower than 100 mV [47]. In addition, anodic and cathodic peak currents are on the same order of magnitude, this being associated with the stability of the adsorbed redox species (Figure S2 Supplementary Materials).

The redox potential of GOx in solution determined by Vogt and co-workers by UV/vis spectroelectrochemistry was −0.385 V vs Ag/AgCl at pH = 7.4 [48]. According to cyclic voltammograms, a formal potential ($E_{1/2}$) of −0.307 V is obtained for chitosan-based electrodes. This potential is near the values reported for GOx and out of the range for O_2 electrochemical reduction [49]. Moreover, this value is in accordance with the values found in the literature for some GOx-based bioelectrodes where chitosan has been employed as an immobilization matrix [38,50,51]. In particular, Zhao and co-workers [38] stated that chitosan, in a protonated state in acidic solutions, can embed a part in the protein acting as a molecular "wire" that connects the active center of the enzyme with the electrode surface resulting in an enhancement of electron transfer process. Thus, chitosan with hydroxyl groups and positively charged amino groups behaves as a polyelectrolyte and can embed in the protein acting as a molecular wire connecting the graphite electrode and the GOx enzyme. Then, we assume that, in addition to the "wire" effect, chitosan as incubation medium modulates the adsorption allowing the migration of the enzyme towards the electrode surface avoiding unfolding and denaturation due to the electrostatic interactions that would prevent structure distortion [40]. Thus, the behavior detected for the bioelectrodes prepared incubating in chitosan solutions points to the successful immobilization of GOx enzyme onto LDG electrodes surface, and not from FAD irreversible adsorption due to denaturation, demonstrating a favoring effect of the incubation in chitosan solutions in the enzyme structure.

3.3. Evaluation of Immobilized GOx Activity

3.3.1. GOx Activity and Inhibition Assays

To prove the bioelectrocatalytic performance of the GOx adsorbed onto LDG electrodes, the activity of the enzyme for glucose catalysis has been proved. The bioelectrodes prepared by incubation in chitosan solutions have been proved not to be denatured and to retain the quaternary structure of the enzyme and consequently, the bioelectrocatalytic activity as demonstrated from the results of the KCl assays. Cyclic voltammetry was recorded in oxygen-free PBS pH 7.55 solutions with increasing concentrations of glucose (Figure 3A) to check the proper activity of GOx in the electrochemical oxidation of glucose.

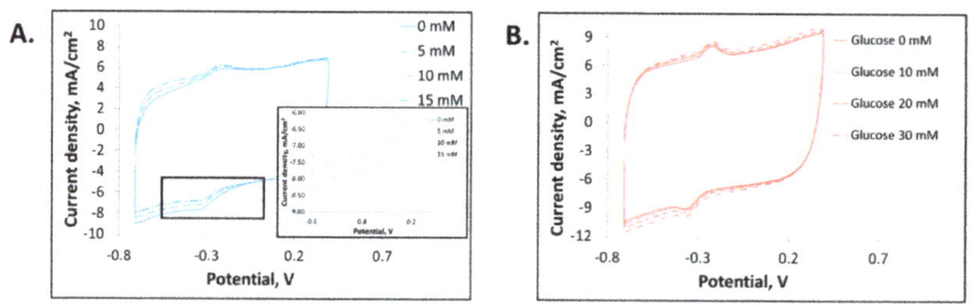

Figure 3. GOx activity evaluation in the presence of increasing concentrations of glucose using (**A**) Chitosan-modified bioelectrode (LDG-CH10) and (**B**) PBS modified bioelectrode (LDG-PBS10) treated with KCl. Voltammograms acquired at 0.1 V/s.

For the bioelectrodes prepared starting from a GOx concentration of 10 mg/mL (Figure 3), a reduction in the cathodic current density and a small increase in the anodic current density were observed as a response of the reaction of GOx and glucose. Thus, as explained in reaction 1, GOx(FAD) is reduced by glucose during oxidation. In addition, in KCl treated electrodes, there is little or negligible variation in FAD currents in presence of glucose, which means that there is no active GOx on the electrodic surface (Figure 3B). The effect observed in the voltammograms is more probably related to the modification of the double layer due to the addition of glucose to the working solution.

As observed in Figure 3A, the current corresponding to the reduction of FAD to $FADH_2$ decreases with the increase of glucose concentration up to saturation at a concentration of 15 mM glucose. This phenomenon is due to the simultaneous action of two processes: enzymatic catalysis and electrode reaction [52]. In the absence of glucose, the total amount of FAD that is reduced remains constant, being always reduced the same amount of FAD. However, in the presence of glucose, enzymatic catalysis also occurs; therefore, the amount of FAD that is reduced (monitored by cyclic voltammetry) is decreased, since it is participating in the enzymatic catalysis, thus modifying the current obtained.

The bioactivity of the modified graphite electrode was further investigated by employing mediators to enhance electrochemical currents [7,11]. The cyclic voltammetry in oxygen-free PBS pH 7 solutions containing 2 mM hydroquinone (HQ) as a redox mediator showed, after the addition of glucose, remarkable enzyme activity in presence of the mediator (Figure 4A) while no current increase was observed for the bare graphite rod electrode when increasing amounts of glucose are added in the presence of HQ (Inset in Figure 4A). This increase indicates that the presence of a mediator promotes GOx activity. Specifically, GOx transfers electrons and, thus, regenerates the hydroquinone, which is oxidized to benzoquinone in the electrode. The result can be explained following reactions 1, 4, and 5:

$$\text{GOx(FADH}_2) + 2\text{HQ}^+ \rightarrow \text{GOx(FAD)} + 2\text{HQ} + 2\text{H}^+ \quad (4)$$

$$2\text{HQ} \rightarrow 2\text{e}^- + 2\text{HQ}^+ \quad (5)$$

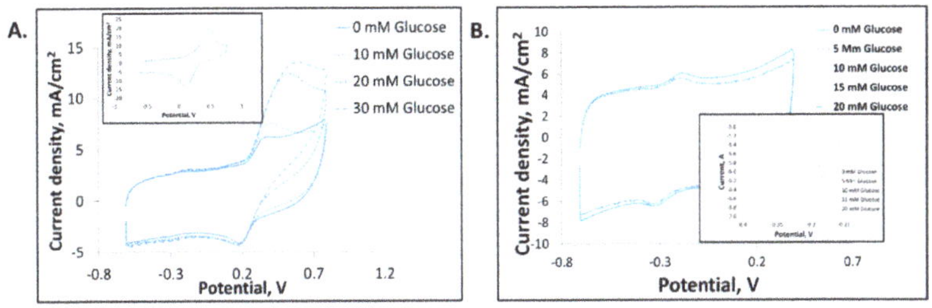

Figure 4. Cyclic voltammetry in oxygen-free PBS solution of LDG-CH10 bioelectrode using increasing glucose concentrations (**A**) in the presence of HQ 2 mM (inset: bare graphite rod) and (**B**) in the presence of the inhibitor Ag^+ 1 µM. Voltammograms acquired at 0.1 V/s.

On the other hand, in order to ascertain whether the bioelectrocatalytic activity observed comes from the immobilized enzyme, an inhibition assay was performed (Figure 4B). In this way, it can be confirmed that current response obtained is coming from the bioelectrocatalytic activity of the enzyme. One of the inhibitors of GOx enzyme activity is the Ag^+ ion [53,54] and its presence in low concentrations prevents, totally or partially, the oxidation of glucose by the enzyme GOx. The chitosan-modified electrodes showing activity towards glucose oxidation were assayed in an inhibition experiment with silver. Thus, when Ag^+ in concentration 1 µM is added to the working solution prior to glucose addition, no changes were observed when the glucose concentration in the electrolyte solution is increased. This means that the biocatalytic activity of the enzyme has been inhibited by the Ag^+ ions.

In Figure 4B, a partial response is observed up to a glucose concentration of 5 Mm in the presence of silver as inhibitor; further additions return the current density values to the initial value found when the glucose concentration in the working electrolyte was zero. It was determined that this behavior is related to the concentration of GOx in the bioelectrode and its response time, since it is known that as the glucose concentration increases, the response time required in an inhibition assay with silver also does [55]. Moreover, inhibition with silver occurs due to the interaction of Ag^+ ions with the functional groups of the enzyme; therefore, the higher the concentration of enzyme in the electrode, the greater the amount of Ag^+ ions necessary to totally block the catalytic activity [54]. However, in the case of a bioelectrode prepared with lower concentration of GOx in chitosan (4 mg/mL, LDG-CH4) the effect of the inhibition with silver is clearly observed (Figure S4).

3.3.2. GOx Activity for Energy Harvesting

Finally, the modified graphite electrode by incubation in chitosan solutions was tested as proof of the concept of the suitability of the bioelectrode in a glucose biofuel cell for energy harvesting. For the quantification of the power density achieved in the presence of glucose, the LDG-CH4 electrode was set as anode due to its faster response as observed in the inhibition assay. A Pt wire electrode was used as a cathode and 0.1 M PBS pH 7.4 was used as the working electrolyte. For an electrochemical surface area of 0.07 cm^2, obtained from the Randles–Sevcik equation, the power density delivered using GOx chitosan-based modification was 119 $\mu W/cm^2$ for a concentration of 20 mM glucose with a cell voltage of 0.3 V (Figure 5). This result shows the proper enzyme immobilization by chitosan presence and the potential of the bioelectrode prepared for energy harvesting applications.

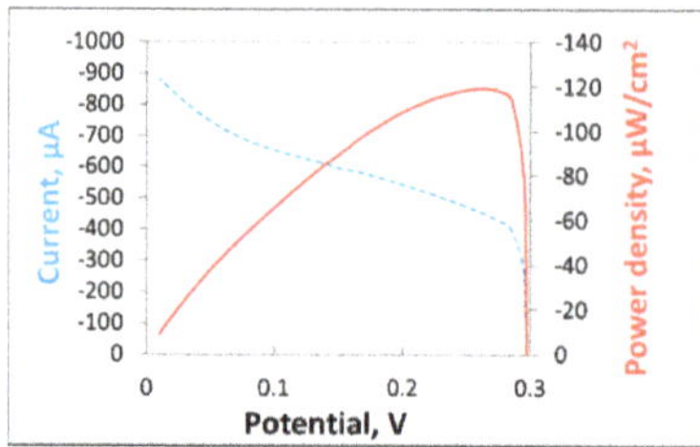

Figure 5. Polarization and power density curves of the graphite bioelectrode modified with chitosan solutions of GOx (4 mg/mL) in the presence of 20 mM glucose.

The power density of 119 μW/cm^2 delivered by the biofuel cell using the modified graphite rod as anode is in the range of the values that can be found in the literature for glucose biofuel cells for similar electrode configurations [5,56]. Thus, these power densities and currents are more appropriate for energy harvesting, where the high OCP and currents are desired for the proper operation of biofuel cells. Nonetheless, this approach for enzyme immobilization would also apply to the development of biosensors, but, in the latter case, it is convenient to work at lower potentials to avoid possible interferences.

4. Conclusions

The proposed method for GOx immobilization using a chitosan solution allows extremely simple and highly versatile enzyme attachment on graphite supports. This strategy allows keeping the stability of the enzyme in terms of quaternary structure, avoiding loss of FAD and enhancing electron transfer kinetics, leading to a good catalytic activity of the enzyme for glucose catalysis. By incubating graphite electrodes in chitosan solutions, FAD adsorption is minimized due to the beneficial effect of the electrostatic interactions in the chitosan medium. Moreover, chitosan is embedded in the enzyme structure promoting electron transfer between the enzyme and the electrode. Bioelectrocatalytic activity towards glucose is retained as demonstrated by the electrochemical response obtained for the electrode when contacted with increasing glucose concentrations in energy harvesting applications.

Thus, the proposed method to immobilize GOx onto graphite electrodes using chitosan as an immobilization medium constitutes a simple and easy-to-use approach for GOx adsorption compared with similar systems reported in the literature so far. To our best knowledge, the most important achievement of the method proposed is based on its extreme simplicity since chitosan does not need any modification for the application of the method but also the ease of the process. These facts allow the use of this strategy in any kind of application related with bioelectrodes development, such as sensing or energy harvesting, and could be further extended to other enzymes of similar features. However, further work will be necessary to evaluate the long-term stability of the immobilization, and deep evaluation of its features for sensing and energy harvesting for power supply.

Supplementary Materials: The following supporting information can be downloaded at: https://www.mdpi.com/article/10.3390/polysaccharides3020023/s1. Table S1: Enzyme coverage and electron transfer rate constant obtained for PBS- and Chitosan-incubated bioelectrodes. Figure S1: KCl treatment effects in electrodes. Figure S2: GOx immobilization in PBS for (A) 4mg/mL GOx (LDG-PBS4) and (B) 10 mg/mL GOx, (LDG-PBS10) and in chitosan for (C) 4 mg/mL GOx (LDG-CH4) and (B) 10 mg/mL GOx, (LDG-CH10). Figure S3: pH dependence of redox peaks in LDG_CH10 electrode. Figure S4: Cyclic voltammetry in oxygen-free PBS solution of LDG-CH4 bioelectrode with increasing glucose concentration in the absence (A) and in the presence (B) of the inhibitor Ag^+. Voltammograms acquired at 0.1 V/s.

Author Contributions: Conceptualization, M.B.-S.; methodology, M.B.-S.; validation, M.B.-S. and L.G.-C.; formal analysis, M.B.-S. and L.G.-C.; investigation, M.B.-S.; resources, M.G.-A., M.G.-P. and A.Q.-L.; writing—original draft preparation, M.B.-S. and L.G.-C.; writing—review and editing, M.B.-S., L.G.-C., M.G.-A. and A.Q.-L.; visualization, M.B.-S.; L.G.-C.; M.G.-A.; M.G.-P.; and A.Q.-L.; supervision, M.G.-A. and A.Q.-L.; project administration, M.B.-S.; funding acquisition, M.G.-P. All authors have read and agreed to the published version of the manuscript.

Funding: This research was funded by the Spanish Ministry of Science, Innovation, and University, grant number PTQ-14-07145 (Bio2 project) and by the Instituto Valenciano de Competitividad Empresarial-IVACE.

Institutional Review Board Statement: Not applicable.

Informed Consent Statement: Not applicable.

Acknowledgments: The authors acknowledge the Instituto Valenciano de Competitividad Empresarial (IVACE), and M.B.-S. is gratefully acknowledged for a Torres Quevedo research contract through the State Program for Talent and Employability Promotion 2013–2016.

Conflicts of Interest: The authors declare no conflict of interest.

References

1. Clark, L.C., Jr. Electrochemical Device for Chemical Analysis. U.S. Patent 2913386A, 17 November 1959.
2. Bruen, D.; Delaney, C.; Florea, L.; Diamond, D. Glucose Sensing for Diabetes Monitoring: Recent Developments. *Sensors* **2017**, *17*, 1866. [CrossRef] [PubMed]
3. Khor, S.M.; Choi, J.; Won, P.; Ko, S.H. Challenges and Strategies in Developing an Enzymatic Wearable Sweat Glucose Biosensor as a Practical Point-Of-Care Monitoring Tool for Type II Diabetes. *Nanomaterials* **2022**, *12*, 221. [CrossRef] [PubMed]
4. Kim, J.; Campbell, A.S.; Wang, J. Wearable non-invasive epidermal glucose sensors: A review. *Talanta* **2018**, *177*, 163–170. [CrossRef] [PubMed]
5. Buaki-Sogó, M.; García-Carmona, L.; Gil-Agustí, M.; Zubizarreta, L.; García-Pellicer, M.; Quijano-López, A. Enzymatic Glucose-Based Bio-batteries: Bioenergy to Fuel Next-Generation Devices. *Top. Curr. Chem.* **2020**, *378*, 49. [CrossRef] [PubMed]
6. Bankar, S.B.; Bule, M.V.; Singhal, R.S.; Ananthanarayan, L. Glucose oxidase—An overview. *Biotechnol. Adv.* **2009**, *27*, 489–501. [CrossRef]
7. Yin, Z.; Ji, Z.; Zhang, W.; Taylor, E.W.; Zeng, X.; Wei, J. The glucose effect on direct electrochemistry and electron transfer reaction of glucose oxidase entrapped in a carbon nanotube-polymer matrix. *ChemistrySelect* **2020**, *5*, 12224–12231. [CrossRef]
8. Huang, J.; Zhang, Y.; Deng, X.; Li, J.; Huang, S.; Jin, X.; Zhu, X. Self-encapsulated enzyme through in situ growth of polypyrrole for high-performance enzymatic biofuel cell. *Chem. Eng. J.* **2021**, *429*, 132148. [CrossRef]
9. Bartlett, P.N.; Al-Lolage, F.A. There is no evidence to support literature claims of direct electron transfer (DET) for native glucose oxidase (GOx) at carbon nanotubes or graphene. *J. Electroanal. Chem.* **2018**, *819*, 26–37. [CrossRef]
10. Homaei, A.A.; Sariri, R.; Vianello, F.; Stevanato, R. Enzyme immobilization: An update. *J. Chem. Biol.* **2013**, *6*, 185–205. [CrossRef]
11. You, C.; Li, X.; Zhang, S.; Kong, J.; Zhao, D.; Liu, B. Electrochemistry and biosensing of glucose oxidase immobilized on Pt-dispersed mesoporous carbon. *Microchim. Acta* **2009**, *167*, 109–116. [CrossRef]
12. Jesionowski, T.; Zdarta, J.; Krajewska, B. Enzyme immobilization by adsorption: A review. *Adsorption* **2014**, *20*, 801–821. [CrossRef]
13. Haghighi, B.; Gorton, L.; Ruzgas, T.; Jönsson, L.J. Characterization of graphite electrodes modified with laccase from Trametes Versicolor and their use for bioelectrochemical monitoring of phenolic compounds in flow injection analysis. *Anal. Chim. Acta.* **2003**, *487*, 3–14. [CrossRef]
14. Jarosz-Wilkołazka, A.; Ruzgas, T.; Gorton, L. Use of laccase-modified electrode for amperometric detection of plant flavonoids. *Enzyme Microb. Technol.* **2004**, *35*, 238–241. [CrossRef]
15. Onay, A.; Dogan, Ü.; Ciftci, H.; Cetin, D.; Suludere, Z.; Tamer, U. Amperometric glucose sensor based on the glucose oxidase enzyme immobilized on graphite rod electrode modified with Fe3O4-CS-Au magnetic nanoparticles. *Ionics* **2018**, *24*, 4015–4022. [CrossRef]
16. Bandapati, M.; Krishnamurthy, B.; Goel, S. Fully Assembled Membraneless Glucose Biofuel Cell With MWCNT Modified Pencil Graphite Leads as Novel Bioelectrodes. *IEEE Trans. NanoBioscience* **2019**, *18*, 170–175. [CrossRef] [PubMed]
17. Chi, Q.; Zhang, J.; Dong, S.; Wang, E. Direct electrochemistry and surface characterization of glucose oxidase adsorbed on anodized carbon electrodes. *Electrochimica Acta* **1994**, *39*, 2431–2438. [CrossRef]
18. Ianniello, R.M.; Lindsay, T.J.; Yacynych, A.M. Differential pulse voltammetric study of direct electron transfer in glucose oxidase chemically modified graphite electrodes. *Anal. Chem.* **1982**, *54*, 1098–1101. [CrossRef]
19. Goran, J.M.; Mantilla, S.M.; Stevenson, K.J. Influence of Surface Adsorption on the Interfacial Electron Transfer of Flavin Adenine Dinucleotide and Glucose Oxidase at Carbon Nanotube and Nitrogen-Doped Carbon Nanotube Electrodes. *Anal. Chem.* **2013**, *85*, 1571–1581. [CrossRef]

20. Pereira, S.; Santos, N.; Carvalho, A.; Fernandes, A.; Costa, F. Electrochemical Response of Glucose Oxidase Adsorbed on Laser-Induced Graphene. *Nanomaterials* **2021**, *11*, 1893. [CrossRef]
21. Pilipenko, O.S.; Atyaksheva, L.F.; Poltorak, O.M.; Chukhrai, E.S. Dissociation and catalytic activity of oligomer forms of β-galactosidases. *Russ. J. Phys. Chem. A* **2007**, *81*, 990–994. [CrossRef]
22. Poltorak, O.M.; Chukhrai, E.S.; Kozlenkov, A.A.; Chaplin, M.F.; Trevan, M.D. The putative common mechanism for inactivation of alkaline phosphatase isoenzymes. *J. Mol. Catal. B Enzym.* **1999**, *7*, 157–163. [CrossRef]
23. Fernandez-Lafuente, R. Stabilization of multimeric enzymes: Strategies to prevent subunit dissociation. *Enzym. Microb. Technol.* **2009**, *45*, 405–418. [CrossRef]
24. Bolivar, J.M.; Cava, F.; Mateo, C.; Rocha-Martín, J.; Guisán, J.M.; Berenguer, J.; Fernandez-Lafuente, R. Immobilization–stabilization of a new recombinant glutamate dehydrogenase from Thermus thermophilus. *Appl. Microbiol. Biotechnol.* **2008**, *80*, 49–58. [CrossRef] [PubMed]
25. Kabashima, T.; Li, Y.; Kanada, N.; Ito, K.; Yoshimoto, T. Enhancement of the thermal stability of pyroglutamyl peptidase I by introduction of an intersubunit disulfide bond. *Biochim. Biophys. Acta Protein Struct. Mol. Enzymol.* **2001**, *1547*, 214–220. [CrossRef]
26. Fuentes, M.; Segura, R.L.; Abian, O.; Betancor, L.; Hidalgo, A.; Mateo, C.; Fernandez-Lafuente, R.; Guisan, J.M. Determination of protein-protein interactions through aldehyde-dextran intermolecular cross-linking. *Proteomics* **2004**, *4*, 2602–2607. [CrossRef] [PubMed]
27. Yamazaki, T.; Tsugawa, W.; Sode, K. Increased thermal stability of glucose dehydrogenase by cross-linking chemical modification. *Biotechnol. Lett.* **1999**, *21*, 199–202. [CrossRef]
28. Tao, Q.; Li, A.; Zhang, Z.; Ma, R.; Shi, L. Stabilization of Multimeric Enzymes against Heat Inactivation by Chitosan-graft-poly(N-isopropylacrylamide) in Confined Spaces. *ACS Biomater. Sci. Eng.* **2017**, *3*, 3141–3145. [CrossRef]
29. Dosadina, E.E.; Belov, A.A. Interaction Between Chitosan Solutions, Cellulose Carriers and Some of the Multi-enzyme Complexes. *Int. J. Bioorg. Chem.* **2017**, *2*, 51–60.
30. Raza, Z.A.; Riaz, S.; Banat, I.M. Polyhydroxyalkanoates: Properties and chemical modification approaches for their functionalization. *Biotechnol. Prog.* **2018**, *34*, 29–41. [CrossRef]
31. Riaz, S.; Rhee, K.Y.; Park, S.J. Polyhydroxyalkanoates (PHAs): Biopolymers for Biofuel and Biorefineries. *Polymers* **2021**, *13*, 253. [CrossRef]
32. Verma, M.L.; Kumar, S.; Das, A.; Randhawa, J.S.; Chamundeeswari, M. Chitin and chitosan-based support materials for enzyme immobilization and biotechnological applications. *Environ. Chem. Lett.* **2020**, *18*, 315–323. [CrossRef]
33. Dabhade, A.; Jayaraman, S.; Paramasivan, B. Development of glucose oxidase-chitosan immobilized paper biosensor using screen-printed electrode for amperometric detection of Cr(VI) in water. *3 Biotech* **2021**, *11*, 183. [CrossRef] [PubMed]
34. El Ichi, S.; Zebda, A.; Alcaraz, J.-P.; Laaroussi, A.; Boucher, F.; Boutonnat, J.; Reverdy-Bruas, N.; Chaussy, D.; Belgacem, M.N.; Cinquin, P.; et al. Bioelectrodes modified with chitosan for long-term energy supply from the body. *Energy Environ. Sci.* **2015**, *8*, 1017–1026. [CrossRef]
35. Nasar, A.; Rahman, M.M. Applications of chitosan (CHI)-reduced graphene oxide (rGO)-polyaniline (PAni) conducting composite electrode for energy generation in glucose biofuel cell. *Sci. Rep.* **2020**, *10*, 10428.
36. Rassas, I.; Braiek, M.; Bonhomme, A.; Bessueille, F.; Rafin, G.; Majdoub, H.; Jaffrezic-Renault, N. Voltammetric glucose biosensor based on glucose oxidase encapsulation in a chitosan-kappa-carrageenan polyelectrolyte complex. *Mater. Sci. Eng. C* **2019**, *95*, 152–159. [CrossRef]
37. Bai, L.; Yuan, R.; Chai, Y.; Yuan, Y.; Wang, Y.; Xie, S. Direct electrochemistry and electrocatalysis of a glucose oxidase-functionalized bioconjugate as a trace label for ultrasensitive detection of thrombin. *Chem. Commun.* **2012**, *48*, 10972–10974. [CrossRef]
38. Zhao, C.; Meng, Y.; Shao, C.; Wan, L.; Jiao, K. Unadulterated Glucose Biosensor Based on Direct Electron Transfer of Glucose Oxidase Encapsulated Chitosan Modified Glassy Carbon Electrode. *Electroanalysis* **2008**, *20*, 520–526. [CrossRef]
39. Kanda, M.; Brady, F.O.; Rajagopalan, K.V.; Handler, P. Studies on the Dissociation of Flavin Adenine Dinucleotide from Metalloflavoproteins. *J. Biol. Chem.* **1972**, *247*, 765–770. [CrossRef]
40. Xie, Y.; Li, Z.; Zhou, J. Hamiltonian replica exchange simulations of glucose oxidase adsorption on charged surfaces. *Phys. Chem. Chem. Phys.* **2018**, *20*, 14587–14596. [CrossRef]
41. Gorton, L.; Johansson, G. Cyclic voltammetry of FAD adsorbed on graphite, glassy carbon, platinum, and gold electrodes. *J. Electroanal. Chem. Inter. Electrochem.* **1980**, *113*, 151–158. [CrossRef]
42. Ianniello, R.M.; Lindsay, T.J.; Yacynych, A.M. Direct electron transfer in immobilized flavoenzyme chemically modified graphite electrodes. *Anal. Chim. Acta* **1982**, *141*, 23–32. [CrossRef]
43. Coughlan, M.P.; Johnson, D.B. Preparation and properties of immobilised xanthine oxidase. *Biochim. Biophys. Acta Enzymol.* **1973**, *302*, 200–204. [CrossRef]
44. Laviron, E. General expression of the linear potential sweep voltammogram in the case of diffusionless electrochemical systems. *J. Electroanal. Chem. Inter. Electrochem.* **1979**, *101*, 19–28. [CrossRef]
45. Liang, B.; Guo, X.; Fang, L.; Hu, Y.; Yang, G.; Zhu, Q.; Wei, J.; Ye, X. Study of direct electron transfer and enzyme activity of glucose oxidase on graphene surface. *Electrochem. Commun.* **2015**, *50*, 1–5. [CrossRef]
46. Kang, Z.; Jiao, K.; Yu, C.; Dong, J.; Peng, R.; Hu, Z.; Jiao, S. Direct electrochemistry and bioelectrocatalysis of glucose oxidase in CS/CNC film and its application in glucose biosensing and biofuel cells. *RSC Adv.* **2017**, *7*, 4572–4579. [CrossRef]

47. Nicholson, R.S. Theory and Application of Cyclic Voltammetry for Measurement of Electrode Reaction Kinetics. *Anal. Chem.* **1965**, *37*, 1351–1355. [CrossRef]
48. Vogt, S.; Schneider, M.; Schafer-Eberwin, H.; Noll, G. Determination of the pH dependent redox potential of glucose oxidase by spectroelectrochemistry. *Anal. Chem.* **2014**, *86*, 7530–7535. [CrossRef]
49. Wang, Y.; Yao, Y. Direct electron transfer of glucose oxidase promoted by carbon nanotubes is without value in certain mediator-free applications. *Microchim. Acta* **2012**, *176*, 271–277. [CrossRef]
50. Ivnitski, D.; Branch, B.; Atanassov, P.; Apblett, C. Glucose oxidase anode for biofuel cell based on direct electron transfer. *Electrochem. Commun.* **2006**, *8*, 1204–1210. [CrossRef]
51. Gao, Y.-F.; Yang, T.; Yang, X.-L.; Zhang, Y.-S.; Xiao, B.-L.; Hong, J.; Sheibani, N.; Ghourchian, H.; Hong, T.; Moosavi-Movahedi, A.A. Direct electrochemistry of glucose oxidase and glucose biosensing on a hydroxyl fullerenes modified glassy carbon electrode. *Biosens. Bioelectron.* **2014**, *60*, 30–34. [CrossRef]
52. Zhang, D.; Chen, X.; Ma, W.; Yang, T.; Li, D.; Dai, B.; Zhang, Y. Direct electrochemistry of glucose oxidase based on one step electrodeposition of reduced graphene oxide incorporating polymerized l-lysine and its application in glucose sensing. *Mater. Sci. Eng. C* **2019**, *104*, 109880. [CrossRef] [PubMed]
53. Dhekra, A.; Khaled, H.; Mina, S.; Mustapha, M.; Ali, O. Immobilization of Glucose Oxidase in Anthracene-Based Semi-Conducting Polymer: Application on Glucose Biosensing. *J. Bioeng. Biomed. Sci.* **2017**, *7*, 5.
54. Wang, S.; Su, P.; Yang, Y. Online immobilized enzyme microreactor for the glucose oxidase enzymolysis and enzyme inhibition assay. *Anal. Biochem.* **2012**, *427*, 139–143. [CrossRef] [PubMed]
55. Rust, I.M.; Goran, J.M.; Stevenson, K.J. Amperometric Detection of Aqueous Silver Ions by Inhibition of Glucose Oxidase Immobilized on Nitrogen-Doped Carbon Nanotube Electrodes. *Anal. Chem.* **2015**, *87*, 7250–7257. [CrossRef]
56. Yu, S.; Myung, N.V. Recent Advances in the Direct Electron Transfer-Enabled Enzymatic Fuel Cells. *Front. Chem.* **2021**, *8*, 620153. [CrossRef] [PubMed]

polysaccharides

Article

Development and In Vitro Cytotoxicity of *Citrus sinensis* Oil-Loaded Chitosan Electrostatic Complexes

Antonio M. N. de Toledo *, Adriana R. Machado and Leonor A. de Souza-Soares

School of Chemistry and Food, Federal University of Rio Grande, Rio Grande 96203-900, Brazil; adririzo85@gmail.com (A.R.M.); leonor.souzasoares@gmail.com (L.A.d.S.-S.)
* Correspondence: amatinato@gmail.com

Abstract: Electrostatic complexes based on chitosan, lecithin, and sodium tripolyphosphate were produced and evaluated with respect to their encapsulation capacity and cytotoxicity. Physical chemical properties were determined by zeta potential values and size distributions. For encapsulation assays, the emulsification method was followed, and *Citrus senensis* peel oil was utilized as volatile compound model. Morphology of complexes with oil incorporated was observed by scanning electron microscopy. The cytotoxicity of complexes was related to cell viability of zebrafish hepatocytes. The complexes produced presented positive Zeta potential values and size distributions dependent on the mass ratio between compounds. Higher concentrations of sodium tripolyphosphate promote significant changes ($p < 0.05$) in zeta values, which did not occur at smaller concentrations of the crosslinking agent. These complexes were able to encapsulate Citrus sinensis peel oil, with encapsulation efficiency higher than 50%. Cytotoxicity profiles showed that in a range of concentrations (0.1–100 μg/mL) studied, they did not promote cellular damage in zebrafish liver cells, being potential materials for food and pharmaceutical applications.

Keywords: electrostatic association; encapsulation; cytotoxicity

check for updates

Citation: de Toledo, A.M.N.; Machado, A.R.; de Souza-Soares, L.A. Development and In Vitro Cytotoxicity of *Citrus sinensis* Oil-Loaded Chitosan Electrostatic Complexes. *Polysaccharides* **2022**, *3*, 347–355. https://doi.org/10.3390/polysaccharides3020020

Academic Editor: Azizur Rahman

Received: 28 December 2021
Accepted: 25 March 2022
Published: 6 April 2022

1. Introduction

Colloids based on the bonding between opposing charges of different compounds are one of the most versatile colloid research materials. By using low-cost processes, without the use of high energy or toxic solvents, these become promising in the pharmaceutical and food industries. In the development of these complexes, their physicochemical characteristics can be modulated, defining their application as texture modifiers in food products, active agent carriers, and antimicrobial agents [1–3].

Among the compounds of interest for production of these complexes, chitosan has shown their relevance. Due to its cationic character, this polysaccharide has the availability of charges to bind to anionic compounds [4–6], forming electrostatic complexes that are promising for the encapsulation of volatile active compounds, such as orange essential oil [7–9]. This oil originates from the secondary metabolism of *Citrus sinensis* and easily oxidizes in the presence of air, light, and humidity contributing to the oxidation of the products that have it in its formulation [10,11].

Ionotropic gelation using sodium tripolyphosphate as a polyanion is the most widespread formation protocol of chitosan nanoparticles and colloids [12]. Despite the good potential of such particles, studies have demonstrated their in vitro cytotoxic potential in several cell lines [13–15]. However, there are few studies of their action on liver cell lines. Cultures of hepatocytes, such as zebrafish liver (ZFL cell line), are interesting because they reproduce in vitro the action of xenobiotics that are metabolized by this organ [16].

Aiming to reduce the chitosan-tripolyphosphate particles cytotoxicity, the addition of soybean lecithin phospholipids may be an alternative since they are present in living organisms and integrate the cell membrane. In addition, in acidic media, they have a negative net charge, being suitable for electrostatic interaction with chitosan. Finally, the surfactant character of lecithin may enhance the encapsulation of lipophilic compounds such as orange essential oil [17,18].

In this way, the objective of this work was the development of a new electrostatic complex based on lecithin and chitosan crosslinked with sodium tripolyphosphate, its use for the encapsulation of orange essential oil, and subsequent evaluation of its cytotoxicity in zebrafish liver cells (ZFL cell line).

2. Materials and Methods

2.1. Raw Materials and Chemical Reagents

Chitosan (CHI) (molecular weight of 74.03 kDa, deacetylation degree of 95% and polydispersity of 0.03) (Polymar Ltd., Fortaleza, Brazil), soy lecithin (LIP) (food grade) (Delaware S/A, Porto Alegre, Brazil), orange essential oil (OEL) (Phytotherapica, São Paulo, Brazil), and sodium tripolyphosphate (STPP) (Synth, Broomfield, CO, USA) were used as raw materials. Acetic acid (PA, purity > 99.7%, Sigma-Aldrich, Saint Louis, Missouri, MO, USA) and Milli-Q water (Millipore Corporation, Burlington, MA, USA) were used as solvents. Culture medium RPMI 1640 (Gibco), trypsin 0.1% (w/v) (Gibco), 3-4,5-dimethylthiazol-2-yl, 2,5-diphenyltetrazolium (MTT, Sigma-Aldrich, Saint Louis, Missouri, MO, USA), and dimethylsulfoxide (DMSO, purity > 99.9%, Sigma-Aldrich, Saint Louis, Missouri, MO, USA) were used for the cytotoxicity assays.

2.2. Stock Solutions Preparation

Chitosan 0.5% (w/v) was solubilized in acetic acid solution (0.1% v/v). Sodium tripolyphosphate 1% (w/v) and lecithin 0.7% (w/v) were solubilized in Milli-Q water. The pH was adjusted to 3.5 for all solutions previously the surfactant-polyelectrolyte complexes (SPECs) production.

2.3. Production of Surfactant-Polyelectrolyte Complexes (SPECs)

SPECs were prepared according to Calvo et al. [12] based on the ionic gelation of chitosan, with some modifications. In this study, two anionic compounds were utilized for SPECs production: Initially, CHI and STPP solutions were homogenized in a rotor stator device (Ultraturrax T-18, IKA, Königswinter, North Rhine-Westphalia, Germany) by direct mixing at 10,000 rpm for 2 min. Subsequently, LIP was slowly added at the dispersion in permanent stirring (2000 rpm). The formulations (Table 1) were produced varying the ratio between chitosan and sodium tripolyphosphate and keeping the mass ratio CHI:LIP constant at 10:1. Figure 1 represents the schematic production of SPECs.

Table 1. Formulations of surfactant-polyelectrolyte complexes.

Formulation	CHI (mg/mL)	LIP (mg/mL)	STPP (mg/mL)	Ratio STPP:CHI
F1	1.2	0.12	0.15	1:8
F2	1.0	0.10	0.15	1:6.7
F3	1.2	0.12	0.20	1:6
F4	1.0	0.10	0.20	1:5

2.4. Size Distribution and Zeta Potential Analysis

Size distributions of SPECs were measured by dynamic light scattering (DLS). These measurements were performed using Zetasizer Nano ZS (Malvern Instrument Ltd., Malvern, Worcestershire, UK) equipped with 633 nm laser and with a 2 mL rectangular cuvette (path length 10 mm). Zeta potentials were carried out by microelectrophoresis in the same equipment, operating with a He–Ne (633 nm) laser, 4.0 mV as a light source and using disposable zeta cells (DTS 1060).

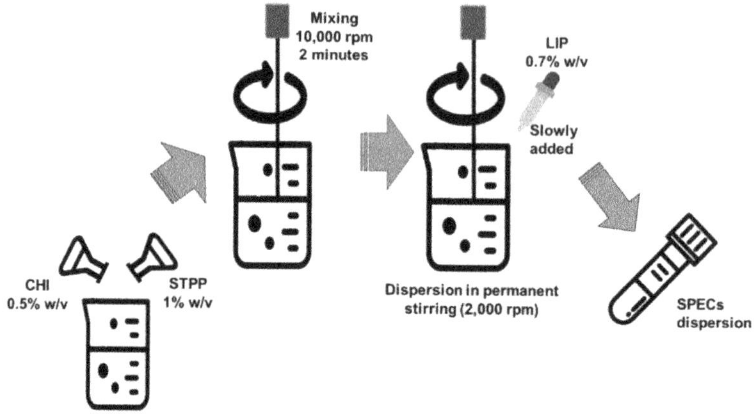

Figure 1. Schematic production of SPECs.

2.5. Incorporation of Orange Essential Oil into SPECs

Orange essential oil was added by dripping in SPECs dispersion under stirring. Afterward, the suspensions were submitted to lyophilization. The SPECs:oil mass ratios ranged from 1:1 to 5:1 (E1 = 1:1; E2 = 3:1; E3 = 5:1). Size distributions and zeta potentials of these formulations were measured as described in Section 2.4.

2.5.1. Encapsulation Efficiency

The encapsulation efficiency was carried out as described by Férnandez-Urrusuno et al. [19] with modifications. The dispersions were centrifuged at 15,500× g for 30 min. To the supernatant was added chloroform (1:10 volume ratio), and liquid–liquid extraction was carried out in separation funnels. Afterward, chloroform solvent was evaporated in a rotary evaporator (60 °C), and then, the flasks were weighed. The encapsulation efficiency (EE) was calculated from Equation (1):

$$EE = (mi - mf)/mi \quad (1)$$

where mi is the initial mass of essential oil added, and mf is the oil-free in the supernatant solution.

2.5.2. Microstructure of Lyophilized Dispersions

In order to gain information about the encapsulation process, scanning electron microscopy was carried out in two magnitudes (40× and 1000×) for lyophilized suspensions of SPECs incorporated or not with essential oil. The microscopies were performed in a scanning electron microscope (Jeol, model JSM—6610LV, Pleasanton, CA, USA) with an acceleration voltage of 20 KV.

2.6. In Vitro Cytotoxicity Assessments

For these assays, the ZFL strain of D. rerio hepatocytes (CRL2643—American Type Culture Collection—ATCC) was used. The cells culture was maintained with RPMI 1640 culture medium, supplemented with 10% fetal bovine serum and 1% antibiotic and antimycotic, in glass culture bottles at 28 °C.

Cell Viability Analysis: MTT Method

Cell viability was assessed by the 3-4,5-dimethylthiazol-2-yl, 2,5-diphenyltetrazolium (MTT) method at 0, 24, 48 and 72 h post-exposure. The MTT test is colorimetric and relies on the reduction of the MTT by the mitochondrial enzymes of the viable cells. When this compound is reduced, its coloration and physical state change from yellow (liquid) to violet (solid). The difference between the stains was then determined by ELISA plate reader absorbance, and subsequently, viability was calculated by assigning the control group absorbance as 100% viable cells [20].

Two distinct experiments were carried out with ZFL cells: empty SPECs dispersions and oil loaded SPECs dispersions. For each group, independent triplicate, represented by three 96-well culture plates, was conducted.

Cells of the ZFL strain were incubated for 24 h in 96-well culture plates at 28 °C for adherence (3×10^5 cells/mL). Subsequently, the cells were treated with different concentrations of SPECs (empty SPECs and oil loaded SPECs). The evaluated concentrations ranged from 0.1 µg/mL to 100 µg/mL. Control cells received the same volume as the vehicle used for the solubilization, in this case, immersion water (Novafarma, Brazil).

After exposition, cells were washed with PBS buffer and 180 µL of culture medium, and 20 µL of MTT solution (5 mg/mL) was added to each well. The plates were incubated for 3 h at 28 °C. Thus, the MTT-containing medium was removed, and the formazan crystals were dissolved in 200 µL of dimethylsulfoxide (DMSO). Absorbance values at 490 nm wavelength were then determined in the ELISA (ELX 800 Universal Reader, Bio-TEK, Winooski, VT, USA).

2.7. Statistical Analysis

The experimental data were presented as means ± standard deviation. The results obtained were submitted to variance analysis (ANOVA), and Tukey's test was applied to evaluate significant differences using the statistical program PAST (v.5). Values of $p \leq 0.05$ were considered statistically significant.

3. Results and Discussion

3.1. Physicochemical Properties of SPECs

According to Figure 2, it was possible to notice that the ratio between CHI and STPP influenced the SPECs particle size. The consecutive decrease of the crosslinking agent in relation to chitosan promoted the extinction of the multimodal character (F1 in contrast to F2) and, subsequently, promoted the existence of a population of nanoparticles with mean size around 100 nm (Figure 2). This fact occurred because STPP promotes crosslinks in the fragments of free chitosan in solution. Therefore, a higher amount of STPP tends to promote more aggregates and then increase the system polydispersity [21].

Calvo et al. [12] studied the influence of chitosan and sodium tripolyphosphate concentrations on the mean size of the nanoparticles, similar to F3 and F4 formulations. These authors found lower mean values (263.8 ± 23.6 and 307.6 ± 14.6 nm) when compared to our work. This difference might be attributed to the presence of the lecithin fatty acids in the formulations. At higher concentrations of chitosan (1.2 mg/mL), present in formulations F1 and F3, the mean particle size (Table 2) increased when compared to the formulations with low chitosan content (F2 and F4). This fact was attributed to a small increase in the solution viscosity. The higher viscosity makes the interactions between the forming compounds of SPECs difficult, favoring the formation of larger particles [22].

Table 2. Physicochemical parameters of surfactant-polyelectrolyte complexes.

Formulation	Ratio STPP/CHI	Mean Size (nm)	PDI (–)	ζ-Potential (mV)
F1	1/8	814.05 ± 34.3 ^{b}	0.300 ± 0.02	21.1 ± 3.36 ^{a}
F2	1/6.7	718.53 ± 34.6 ^{a,b}	0.323 ± 0.03	20.7 ± 3.67 ^{a}
F3	1/6	668.93 ± 03.2 ^{a}	0.394 ± 0.03	15.6 ± 3.50 ^{a}
F4	1/5	587.50 ± 38.9 ^{a,c}	0.362 ± 0.01	17.4 ± 3.20 ^{a}

Different lowercase letters on the same column indicate a significant difference between values (p value < 0.05).

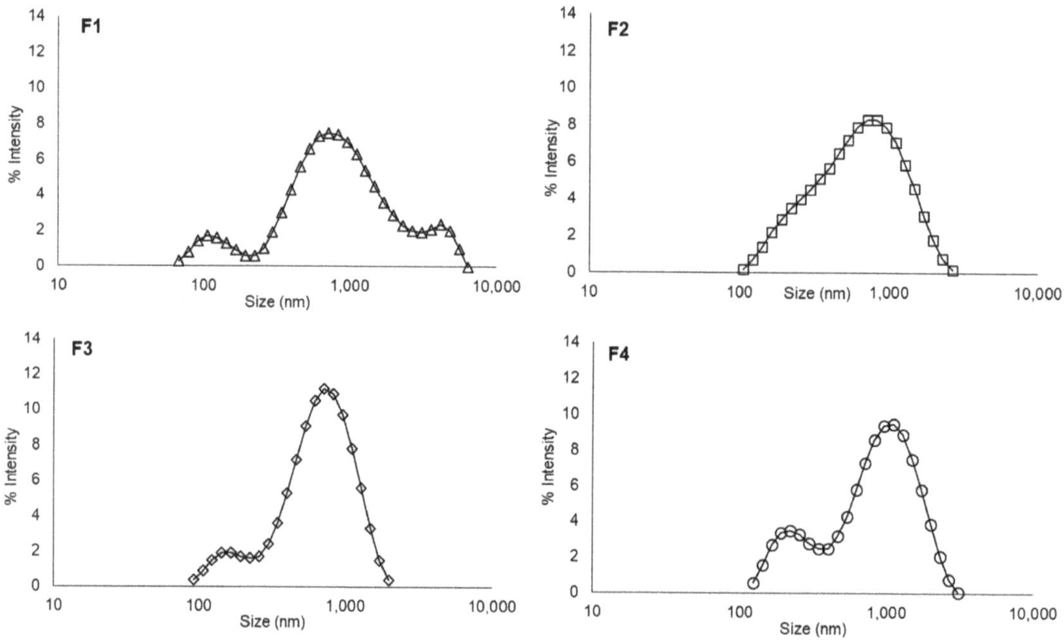

Figure 2. Size distributions of the four formulations of surfactant-polyelectrolyte complexes studied (F1, F2, F3 and F4).

Another parameter used to evaluate the size of the particles formed is the polydispersity index. All formulations presented good uniformity (Table 2). This index represents the particle diameter distribution range, where high values indicate heterogeneity in the diameter of the suspended particles. According to Malvern [23], PDI lower than 0.3 are ideal because they indicate that the particle diameter distribution is in a narrow range. Zeta potential values (Table 2) were affected by the amount of STPP added as well as the total chitosan concentration in the systems. Higher concentrations of STPP (F1 and F2) promote significant changes ($p < 0.05$) in zeta values, which did not occur at smaller concentrations of the crosslinking agent (F3 and F4). It is believed that the STPP amount added in formulations F1 and F2 promoted significant rearrangements in the structures, bonding more chitosan chains (which was in excess in all formulations). On the other hand, in F3 and F4 formulations, the addition of STPP just assisted in chitosan gelation, reducing their zeta potential and promoting not a reticulated network but rather nanometric particles.

3.2. Physicochemical Properties of SPECs Incorporated with Orange Essential Oil

For encapsulation studies, formulation F3 was used based on the previous tests (not shown) of solubility. Table 3 shows that, for all wall–core mass ratios, encapsulation efficiency was higher than 50%. Formulation E1 showed a lower retention capacity when compared to E2 and E3. This fact can be attributed to the greater amount of oil added in comparison with the others, saturating the system. This amount of oil added in the E1 formulation also influenced particles mean size: ANOVA results showed that the formulation E1 had a significantly (p value < 0.05) higher value than other formulations, while E2 and E3 did not differ significantly from each other. In addition to these results, all formulations had a larger size when compared to formulation F3 (668.93 ± 3.2 nm), which had no orange oil incorporated. This might suggest that the orange oil was successfully incorporated into the SPECs. Zeta potential values of the three encapsulated formulations was higher when compared with zeta potential of the empty SPECs (Table 3). It is believed that the incorporation of oil promoted a closure of the pores of the structure formed by CHI, LIP, and STPP, exposing some free amino groups present in the structures. The pores closure can be observed by the scanning electron microscopies of empty SPECs (F3) and loaded with orange oil (E1) (Figure 3).

Table 3. Encapsulation efficiency, mean size, PDI, and zeta potential of orange essential oil loaded SPECs.

Formulation	Mass Ratio Wall: Core	Encapsulation Efficiency (%)	Mean Size (nm)	PDI	Zeta Potential (mV)
F3 (empty)	-	-	668.93 ± 03.2 a	0.394 ± 0.2 a	15.6 ± 3.50 b
E1	1:1	61.05 ± 9.55 b	1069.4 ± 30.2 a	0.108 ± 0.8 a	24.3 ± 3.32 b
E2	3:1	97.79 ± 1.29 a	885.1 ± 22.4 b	0.08 ± 0.07 b	37.1 ± 4.20 a
E3	5:1	99.32 ± 0.23 a	801.9 ± 43.2 b	0.325 ± 0.2 a	29.9 ±3.51 b

Different lowercase letters on the same column indicate a significant difference between values (p value < 0.05).

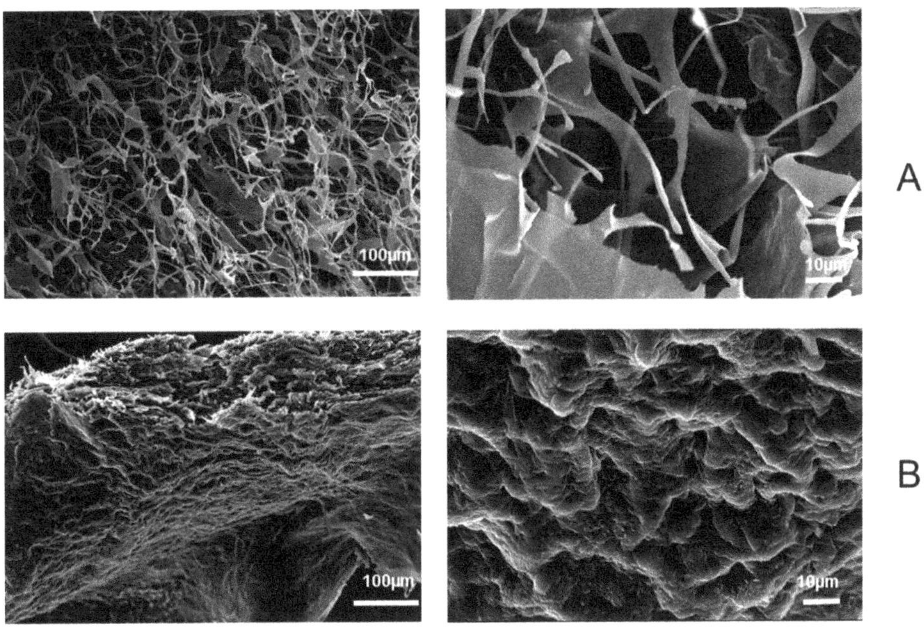

Figure 3. Scanning electron microscopies of (**A**): lyophilized empty SPECs; (**B**) orange essential oil loaded SPECs.

3.3. *In Vitro Cytotoxicity Assessments*

SPECs Dispersions and Orange Essential Oil Loaded SPECs

The cytotoxicity assays for dispersions of empty SPECs (F3) and essential oil loaded SPECs (E3 was chosen based on net charge) are shown in Figure 4. The glass bottles cultures and microscopic image of cell culture are shown in Figure 5. According to statistical results (p-value < 0.05) no significant differences between the treatments performed and the control group were found at the different time intervals. These results are interesting since several factors could promote injury in the cell culture such as the positive charge of the dispersions [24], the uptake by cells of submicron-sized particles present in the dispersions [25,26], or even the presence of orange essential oil [27,28]. Thus, some type of oxidative stress was expected at least after 72 h exposure.

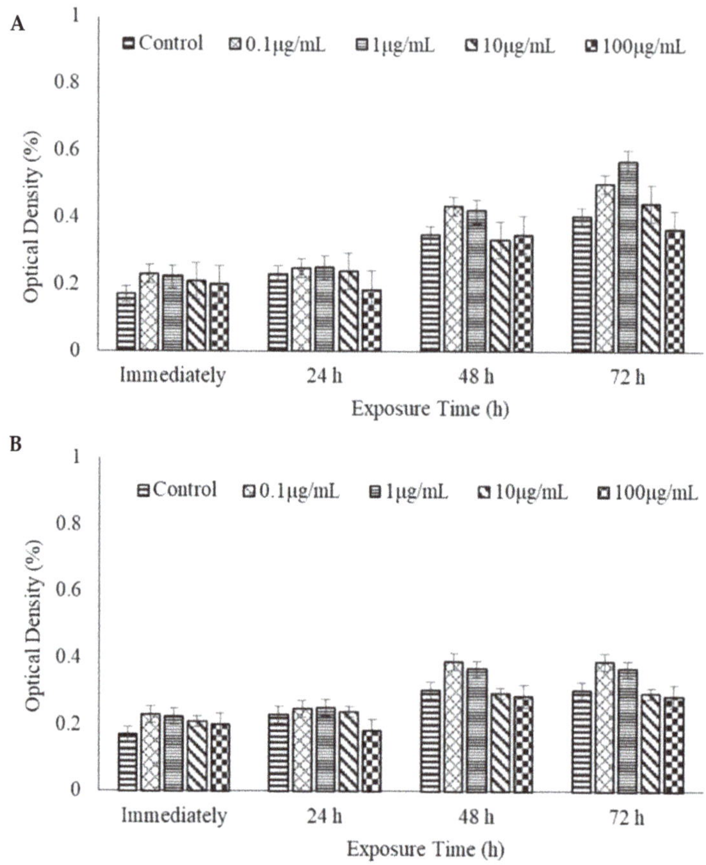

Figure 4. Results of cell viability as a function of the treatments performed with different concentrations of (**A**) empty SPECs and (**B**) oil loaded SPECs. Each value represents the mean of triplicate measurements and varied from the mean by not more than 10%.

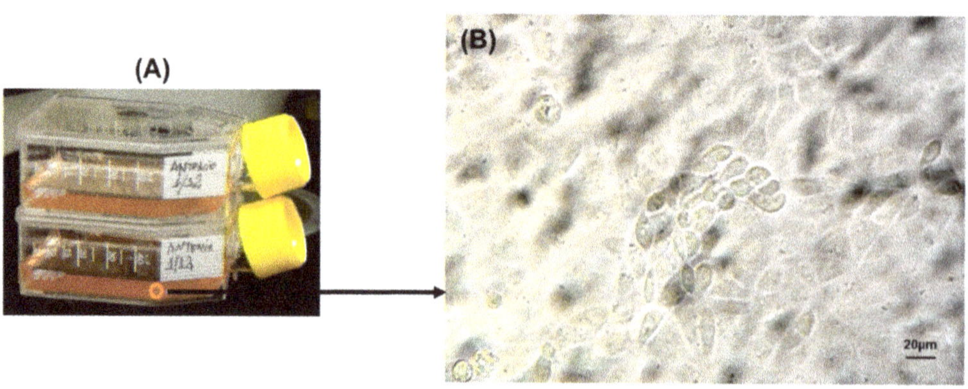

Figure 5. (**A**) Glass culture bottles with ZFL cells; (**B**) microscopic image of ZFL cells.

Many factors may have contributed to these results. Regarding the surface charge of dispersions (empty SPECs = 15.6 ± 3.50 mV; oil loaded SPECs = 29.9 ± 3.51 mV), it is believed that it was affected by lower acidification used for resuspension. Zeta potential, that is, surface charge, can greatly influence the particle stability in suspension through the electrostatic repulsion between the particles. The greater the zeta potential, the more stable the suspension is likely to be because the charged particles repel one another and thus overcome the natural tendency to aggregate [14]. The lower concentration of acetic acid did not promote the protonation of free amino groups present in chitosan, and it affected the mean sizes, forming aggregates. The large size of complexes did not allow the uptake by cells.

Regarding to different types of liver cell cultures, the results obtained are consistent with the literature and suggest a positive relation between chitosan nanoparticles and these types of cell culture. Qi et al. (2005) [14] studied the effects of chitosan nanoparticle size and surface charge on tumor cell cultures (BEL 7402, BGC 823, and Colo 320) and culture of human normal liver cells (L-02). These authors found that chitosan nanoparticles showed higher cytotoxicity in cancer cells but had effects on normal human liver cells. It was believed that the little cytotoxicity was due to the liver being the primary location of detoxification and showing the highest abundance of critical phase II enzymes, such as GSTs42 [29]. Loutfy et al. (2016) [29] also found little cytotoxicity by chitosan nanoparticles in culture of human tumor liver cells after 48 h of cell exposure. Despite these positive results, flow cytometry and cellular DNA fragmentation showed the accumulation of cells in the G2/M phase and a dramatic effect on DNA concentration after 48 h of cell exposure.

4. Conclusions

It was possible to produce a new complex based on chitosan, sodium tripolyphosphate, and lecithin. Its physicochemical properties were dependent on the mass ratio between the forming compounds, and a smaller amount of crosslinking agent promoted smaller particle sizes. The orange essential oil encapsulation by these complexes was successful and presented encapsulation efficiency values greater than 50%. The cytotoxicity results of complexes of essential oil added or not to ZFL cells did not show cellular damage in the studied concentrations.

Author Contributions: Conceptualization, Investigation, Validation, Writing–Review and editing, Visualization: A.M.N.d.T., A.R.M. and L.A.d.S.-S. Methodology: A.M.N.d.T. and L.A.d.S.-S. Formal analysis, Data curation, Writing–Original draft preparation: A.M.N.d.T. Resources, Supervision, Project Administration and Funding Acquisition: L.A.d.S.-S. All authors have read and agreed to the published version of the manuscript.

Funding: CAPES—Coordenação de Aperfeiçoamento de Pessoal de Nível Superior—for the research grant and FAPERGS: 2485.271.13638.270202013-PVSRS.

Data Availability Statement: The data presented in this study are available on request from the corresponding author.

Acknowledgments: The authors would like to thank Cell Culture Laboratory and Center of electron microscopy (CEME-SUL) at the Federal University of Rio Grande, Brazil.

Conflicts of Interest: The authors declare no conflict of interest.

References

1. Lankalapalli, S.; Kolapalli, V.R. Polyelectrolyte Complexes: A Review of their Applicability in Drug Delivery Technology. *Indian J. Pharm. Sci.* **2009**, *71*, 481–487. [CrossRef] [PubMed]
2. Siyawamwaya, M.; Choonara, Y.E.; Bijukumar, D.; Kumar, P.; Du Toit, L.C.; Pillay, V. A Review: Overview of Novel Polyelectrolyte Complexes as Prospective Drug Bioavailability Enhancers. *Int. J. Polym. Mater. Polym. Biomater.* **2015**, *64*, 955–968. [CrossRef]
3. Meka, V.S.; Sing, M.K.G.; Pichika, M.R.; Nali, S.R.; Kolapalli, V.R.M.; Kesharwani, P. A comprehensive review on polyelectrolyte complexes. *Drug Discov. Today* **2017**, *22*, 1697–1706; ISSN 1359. [CrossRef] [PubMed]
4. Picone, C.S.; Cunha, R.L. Chitosan-gellan electrostatic complexes: Influence of preparation conditions and surfactant presence. *Carbohydr. Polym.* **2013**, *94*, 695–703. [CrossRef] [PubMed]

5. Maciel, V.B.V.; Yoshida, C.M.P.; Franco, T.T. Chitosan/pectin polyelectrolyte complex as a pH indicator. *Carbohydr. Polym.* **2015**, *132*, 537–545. [CrossRef] [PubMed]
6. Aranaz, I.; Alcántara, A.R.; Civera, M.C.; Arias, C.; Elorza, B.; Heras Caballero, A.; Acosta, N. Chitosan: An Overview of Its Properties and Applications. *Polymers* **2021**, *13*, 3256. [CrossRef] [PubMed]
7. Yoksana, R.; Jirawutthiwongchai, J.; Arpo, K. Encapsulation of Ascorbyl Palmitate in Chitosan Nanoparticles By Oil-in-water Emulsion and Ionic Gelation Processes. *Colloids Surfaces B Biointerfaces* **2010**, *76*, 292–297. [CrossRef]
8. Jarudilokkul, S.; Tongthammachat, A.; Boonamnuayvittaya, V. Preparation of chitosan nanoparticles for encapsulation and release of protein. *Korean J. Chem. Eng.* **2011**, *28*, 1247–1251. [CrossRef]
9. Azevedo, M.A.; Bourbon, A.I.; Vicente, A.A.; Cerqueira, V.M.A. Alginate/chitosan nanoparticles for encapsulation and controlled release of vitamin B2. *Int. J. Biol. Macromol.* **2014**, *71*, 141–146. [CrossRef]
10. Gomes, M.S.; Cardoso, M.G.; Soares, M.J.; Batista, L.R.; Machado, S.M.F.; Andrade, M.A.; Azeredo, C.M.O.; Resende, J.M.V.; Rodrigues, L.M.A. Use of Essential Oils of the Genus Citrus as Biocidal Agents. *Am. J. Plant Sci.* **2014**, *5*, 299–305. [CrossRef]
11. Mahato, N.; Sharma, K.; Koteswararao, R.; Sinha, M.; Baral, E.; Cho, M.H. Citrus essential oils: Extraction, authentication and application in food preservation. *Crit. Rev. Food Sci. Nutr.* **2019**, *59*, 611–625. [CrossRef] [PubMed]
12. Calvo, P.; Remuñàn-Lopez, C.; Vila-Jato, J.L.; Alonso, M.J. Chitosan and Chitosan/Ethylene Oxide-Propylene Oxide Block Copolymer Nanoparticles as Novel Carriers for Proteins and Vaccines. *Pharm. Res.* **1997**, *14*, 1431–1436. [CrossRef] [PubMed]
13. Huang, M.; Khor, E.; Lim, L.Y. Uptake and Cytotoxicity of Chitosan Molecules and Nanoparticles: Effects of Molecular Weight and Degree of Deacetylation. *Pharm. Res.* **2004**, *21*, 344–353. [CrossRef]
14. Qi, L.; Xu, Z. In vivo antitumor activity of chitosan nanoparticles. *Bioorg. Med. Chem. Lett.* **2006**, *16*, 4243–4245. [CrossRef] [PubMed]
15. Loh, J.W.; Saunders, M.; Lim, L.Y. Cytotoxicity of monodispersed chitosan nanoparticles against the Caco-2 cells. *Toxicol. Appl. Pharmacol.* **2012**, *262*, 273–282. [CrossRef] [PubMed]
16. Gosh, C.; Zhou, Y.L.; Collodi, P. Derivation and characterization of a zebrafish liver cell line. *Cell Biol. Toxicol.* **1994**, *10*, 167–176. [CrossRef]
17. Dickinson, E. Towards more natural emulsifiers. *Trends Food Sci. Technol.* **1993**, *4*, 330–334. [CrossRef]
18. Whitehurst, R.J. *Emulsifiers in Food Technology*; Blackwell Publishing Ltd.: Hoboken, NJ, USA, 2004.
19. Fernández-Urrusuno, R.; Calvo, P.; Remuñán-López, C. Enhancement of Nasal Absorption of Insulin Using Chitosan Nanoparticles. *Pharm. Res.* **1999**, *16*, 1576–1581. [CrossRef]
20. Mosmann, T. Rapid colorimetric assay for cellular growth and survival: Application to proliferation and cytotoxicity assays. *J. Immunol. Methods* **1983**, *65*, 55–63. [CrossRef]
21. Harish-Prashanth, K.V.; Tharanathan, R.N. Crosslinked chitosan—Preparation and characterization. *Carbohydr. Res.* **2006**, *341*, 169–173. [CrossRef]
22. Malvern. Available online: http://www.malvern.com/br/support/resource-center/Whitepapers/WP111214DLS (accessed on 10 January 2018).
23. Fischer, D.; Li, Y.; Ahlemeyer, B.; Krieglstein, J.; Kissel, T. In vitro cytotoxicity testing of polycations: Influence of polymer structure on cell viability and hemolysis. *Biomaterials* **2003**, *24*, 1121–1131. [CrossRef]
24. Behzadi, S.; Serpooshan, V.; Tao, W.; Hamaly, M.A.; Alkawareek, M.Y.; Dreaden, E.C.; Brown, D.; Alkilany, A.M.; Farokhzad, O.C.; Mahmoudi, M. Cellular uptake of nanoparticles: Journey inside the cell. *Chem. Soc. Rev.* **2017**, *46*, 4218–4244. [CrossRef]
25. Huang, M.; Ma, Z.; Khor, E.; Lim, L.Y. Uptake of FITC-chitosan nanoparticles by A549 cells. *Pharm. Res.* **2002**, *19*, 1488–1494. [CrossRef]
26. Bayala, B.; Bassole, I.H.; Scifo, R.; Gnoula, C.; Morel, L.; Lobaccaro, J.M.; Simpore, J. Anticancer activity of essential oils and their chemical components—A review. *Am. J. Cancer Res.* **2014**, *4*, 591–607. [PubMed]
27. Monajemi, R.; Oryan, S.; Haeri-Roohani, A.; Ghannadi, A.; Jafarian, A. Cytotoxic Effects of Essential Oils of Some Iranian Citrus Peels. *Iran. J. Pharm. Res.* **2010**, *4*, 183–187. [CrossRef]
28. Lungu-Mitea, S.; Oskarsson, A.; Lundqvist, J. Development of an oxidative stress in vitro assay in zebrafish (Danio rerio) cell lines. *Sci. Rep.* **2018**, *8*, 12380. [CrossRef]
29. Loutfy, S.A.; Hanaa, M.; Alam, E.D.; Mostafa, H.E.; Allam, N.G.; Abdellah, A.M. Synthesis, characterization and cytotoxic evaluation of chitosan nanoparticles: In vitro liver cancer model. *Adv. Nat. Sci. Nanosci. Nanotechnol.* **2016**, *7*, 955–961. [CrossRef]

polysaccharides

MDPI

Article

Surface Property Modification of Collagen, Hyaluronic Acid, and Chitosan Films with the Neodymium Laser

Sylwia Grabska-Zielińska [1,*] and Alina Sionkowska [2]

1 Department of Physical Chemistry and Physicochemistry of Polymers, Faculty of Chemistry, Nicolaus Copernicus University in Toruń, 7 Gagarin Street, 87-100 Toruń, Poland
2 Department of Chemistry of Biomaterials and Cosmetics, Faculty of Chemistry, Nicolaus Copernicus University in Toruń, 7 Gagarin Street, 87-100 Toruń, Poland; as@chem.umk.pl
* Correspondence: sylwia.gz@umk.pl

Abstract: In this paper, surfaces of thin films prepared from blends of collagen, hyaluronic acid, and chitosan and modified by neodymium laser radiation were researched. To evaluate the laser beam effect on the surface structure, scanning electron microscopy (SEM) imaging and infrared spectroscopy (FTIR-ATR) were employed. The results demonstrated that during laser treatment the specimens lost water due to the evaporation process. SEM images revealed some changes in the biopolymer films structure. After laser treatment, the micro-foam formation was observed on the biopolymeric films. The micro-foaming in films based on ternary blends was more extensive than in those made of a single biopolymer. The results of this study indicate that collagen, hyaluronic acid, and chitosan materials can be modified with laser treatment. Such treatment can be used for material modification for potential biomedical purposes.

Keywords: hyaluronic acid; chitosan; collagen; surface properties; laser; surface modification

Citation: Grabska-Zielińska, S.; Sionkowska, A. Surface Property Modification of Collagen, Hyaluronic Acid, and Chitosan Films with the Neodymium Laser. *Polysaccharides* **2022**, 3, 178–187. https://doi.org/10.3390/polysaccharides3010008

Academic Editor: Azizur Rahman

Received: 15 November 2020
Accepted: 10 January 2022
Published: 19 January 2022

1. Introduction

Surface properties of new materials are of significant importance for several reasons [1]. The surface also plays a pivotal role in biomedical applications of new materials prepared with the use of biopolymers. The surface can be modified in a number of ways, namely, by physico-chemical, mechanical, and biological methods. Laser treatment of biopolymer materials is one of the physico-chemical methods of surface treatment [2]. Therefore, ultrafast lasers have been commonly used in biomedical sciences due to their ability to produce micro/nanostructures that improve the biomaterial biocompatibility. Surface topography is responsible for cell adhesion, migration, multiplication, and differentiation. In general, laser material treatment and processing may include welding, drilling, cutting, surface hardening, cladding, alloying, laser-assisted forming (bending), rapid prototyping, ablation, and shot peening. During material processing with the use of a laser, the laser light strikes the superficies of the film. A part of the beam energy is reflected from the surface material, while the residual of beam energy is transferred into it, causing a high reflectivity to the laser radiation that characterizes the majority of the materials. Conversion of the absorbed energy to heat leads to several process occurrences in the materials [3,4]. Lasers are used to design the surfaces of different types of materials and can manipulate roughness and wetting characteristics [2,3]. Applications for laser-structured surfaces are emerging in different types of areas, including biomedical field, water harvesting, spectroscopic analysis, antibacterial and antifungal, and oil/water separation, etc., which have been observed recently [2,5]. The effects of several lasers on the surface properties of biopolymer films have already been studied [6,7]. Structured surfaces of several biopolymeric materials allow the application of commonly used biopolymers in an extensive range of specialized fields.

It has been demonstrated that laser energy can lead to the surface foaming of silk fibroin, chitosan, collagen, and other biopolymer films [8,9]. However, to the best of our knowledge, the effect of laser radiation on the surface of hyaluronic acid films has not been studied yet. Moreover, nowadays, there is a growing interest in newly designed materials based on the mixtures of two or more polymers [10]. The effect of laser radiation on several biopolymer films was studied, but the laser modification of binary and ternary biopolymer blends has not been researched yet. In this work, films made of hyaluronic acid, collagen/hyaluronic acid, and collagen/hyaluronic acid with a 30% addition of chitosan were treated with the Nd:YAG laser beam.

Hyaluronic acid (HA) is a chemical compound, which belongs to the polysaccharides group. It is an anionic, nonsulfated glycosaminoglycan. It can also be called a hyaluronan [11,12]. HA is distributed widely throughout epithelial, connective, and neural tissues [13,14]. It is an endogenous component of human skin [15]. Additionally, it is naturally occurring in the extracellular matrix of human tissues [15]. Thanks to its strong water-binding capacity and the possibility for fibroblast stimulation, hyaluronic acid is highly recommend for the improvement of skin elasticity, moisture, and skin wrinkles [15–17].

Collagen (Coll) is the main structural protein in the extracellular matrix of connective tissues in the body. Thus, it is the most widely occurring protein in mammals. Collagen mostly occurs in fibrous tissues, such as ligaments, tendons, and skin [18].

Chitosan (CTS) is a linear polysaccharide [19]. Chitosan can be found by treating chitin shells of shrimps and other crustaceans with alkaline substances (e.g., sodium hydroxide). Chitosan is frequently used in biomedical applications [20,21].

The interactions between hyaluronic acid, collagen, and chitosan in mixtures and materials based on hyaluronic acid/collagen/chitosan blends have been studied within the last few years [22–24]. It was found that the interactions between collagen and chitosan in a solution led to the miscibility of these biopolymers. Partial miscibility was also observed for ternary blends made of collagen, chitosan, and hyaluronic acid in a solution [23,24].

The purpose of this work was to investigate the impact of laser radiation on the structure of biopolymeric films based on hyaluronic acid, collagen/hyaluronic acid, and collagen/hyaluronic acid, with the addition of chitosan. The influence of UV irradiation on the properties of such films has been studied previously [25]. However, to the best of our knowledge, the influence of laser light on the surface properties of hyaluronic acid-based films has not been studied yet.

2. Materials and Methods

2.1. Reagents

Hyaluronic acid (HA) (Mv = 1.8×10^6 g/mol) and chitosan (CTS) (degree of deacetylation DD = 78% Mv = 1.4×10^6 g/mol) were supplied by Sigma Aldrich (Poznań, Poland). Collagen (Coll) was prepared in our laboratory from rat-tail tendons, according to the procedure [26]. The rat-tail tendons were excised and washed in distilled water. Next, they were dissolved in acetic acid solution (0.1 M) for three days (fridge, 4 °C). The undissolved parts of the tendons were removed by centrifugation (10 min, 10,000 rpm). The obtained collagen solution was frozen (freezer, −18 °C) and lyophilized (−55 °C, 5 Pa, 48 h, ALPHA 1–2 LD plus, CHRIST, Osterode am Harz, Germany) [26].

2.2. Fabrication of Films

Collagen and chitosan were dissolved in acetic acid solution (0.1 M). Hyaluronic acid was dissolved in hydrochloric acid solution (0.1 M). The percentage concentration of each polymer was 1%. Biopolymers were blended together in suitable ratios and mixed with a magnetic stirrer (2-component mixture for 2 h, 3-component mixture for 3 h). The films were prepared by solvent evaporation (room temperature and humidity).

For laser treatment, the following films were prepared: (1) films made of hyaluronic acid; (2) films made of a mixture of collagen and hyaluronic acid (50/50 wt ratio); and (3) films made of a mixture of collagen and hyaluronic acid (50/50) with 30% chitosan addition.

2.3. Laser Treatment

Thin films were submitted to the fractional pulse neodymium laser radiation treatment (Nd:YAG (neodymium-doped yttrium aluminum garnet); Nd:$Y_3Al_5O_{12}$). In this study, the Nd:YAG laser generated 355 nm wavelength. The polymeric films were modified at a constant temperature and humidity. The films were radiated with a 10-fold and 20-fold pulse. The energy of the pulse was 100 mJ. The scheme of the sample preparation and laser treatment is shown in Figure 1.

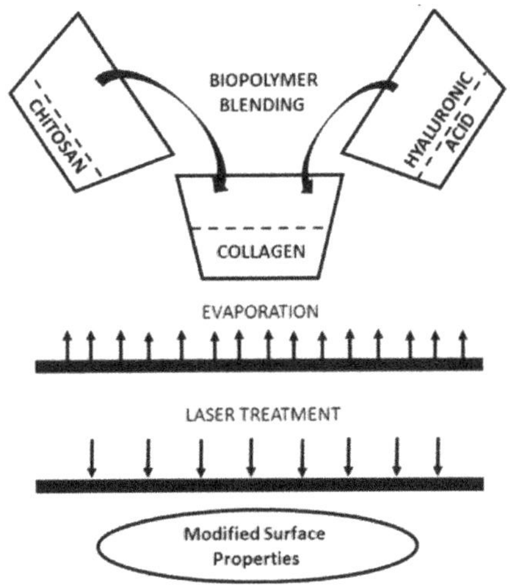

Figure 1. Preparation of biopolymer blend based on collagen, chitosan, and hyaluronic acid and laser treatment of biopolymer film.

2.4. The Characterization of Biopolymeric Films

The structures of the laser-treated samples were studied by FTIR-ATR spectroscopy and scanning electron microscopy (SEM) imaging.

Attenuated total reflection infrared spectroscopy was used to prepared the FTIR spectra. A Nicolet iS10 equipped with an ATR device with a diamond crystal was used as the instrument for recording each kind of sample. All of the spectra were recorded by the absorption mode at 4 cm^{-1} intervals and 64-times scanning. The absorption values were observed in the range of 400–4000 cm^{-1}. The IR spectra were recorded before and after laser treatment.

The morphology of the prepared films was characterized by scanning electron microscopy (SEM) using Quanta 3D FEG. The SEM images were recorded for the surface of films before and after laser treatment.

The roughness of the films was studied by atomic force microscopy and published earlier in our previous paper [25]. The roughness parameters of the laser-treated films was studied by atomic force microscopy using the Veeco SPM (digital instrument) microscope and calculated for the scanned area (5 μm × 5 μm) using Nanoscope software. The AFM calculations for laser-treated films were obtained for different sample places and the most typical results were presented in this paper.

The mechanical properties and contact angle measurements for native films were presented in our previous paper [25]. Performing these tests was not possible for the laser-treated samples because too small areas of the films were irradiated and the samples were not suitable for this type of research.

3. Results and Discussion

The properties of new materials, such as temperature, surface chemistry, wettability, hardness, surface roughness, surface reactivity, and surface charge, are some of the essential surface properties of materials. Each of the properties listed above plays a big role in cell adhesion and proliferation. To investigate the surface structure, attenuated total reflectance (ATR) spectroscopy can be used. ATR is a special accessory unit. It can be used with FTIR spectrometers. It enables taking measurements directly over the whole sample surface in the solid state by pressing that sample towards the ATR crystal. In this study, the structure of a hyaluronic acid, collagen/hyaluronic, and collagen/hyaluronic acid with a 30% addition of chitosan films was characterized by ATR-FTIR spectroscopy. ATR spectra of films based on HA, and mixtures of Coll/HA and Coll/HA/30CTS, have been shown in Figure 2. The obtained spectra are typical for hyaluronic acid, collagen, and chitosan.

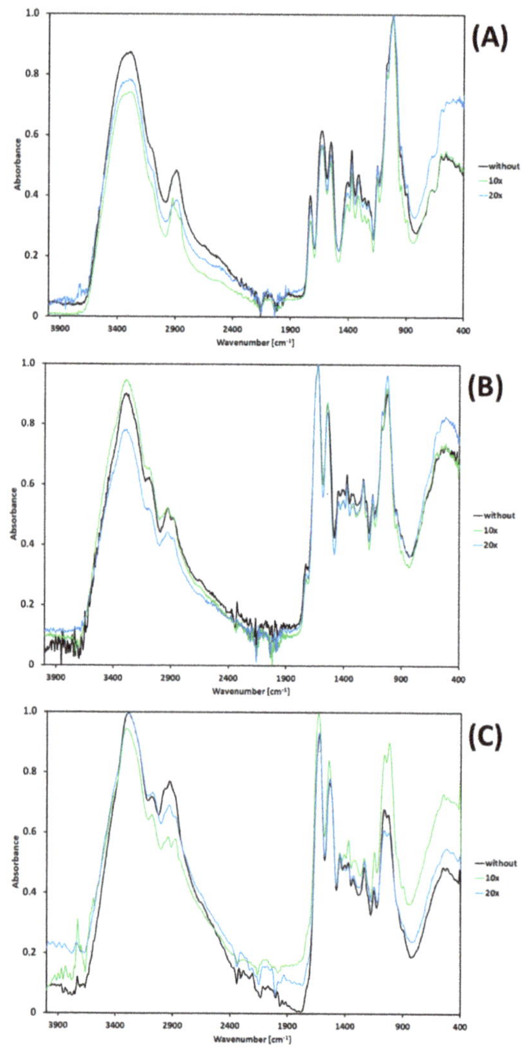

Figure 2. ATR-FTIR spectra of (**A**) HA, (**B**) Coll/HA, and (**C**) Coll//HA/30CTS films before and after dose of laser radiation.

As Figure 2 demonstrates, the strong band at about 3292 cm^{-1} is rather broad and can be assigned to hydrogen-bonded OH and NH stretching vibrations [27,28]. The band at around 2889 cm^{-1} can be observed due to CH stretching vibrations. The bands at 1632 and 1419 cm^{-1} can be attributed to the asymmetric CO and symmetric CO stretching modes of the planar carboxyl groups in the acid. The absorption bands around 1729, 1547, and 1320 cm^{-1} are characteristic of the amide I, II, and III bands, respectively [27,28]. Three signals centered near 1147, 1075, and 1020 cm^{-1} are assigned to the COC (O-bridge), CO (exocyclic), and COH groups, respectively. As observed from Figure 2, the overall spectral pattern was not significantly changed after treatment with laser radiation by various numbers of pulses in comparison to untreated specimens. However, the intensity of the strong and wide band at about 3292 cm^{-1}, assigned to the hydrogen-bonded OH, decreased after the laser treatment. This evidences the fact that during the laser treatment the specimens lost water due to the evaporation process. The collagen structure is strongly dependent on water content. Hence, the dynamics of water molecules in collagen play an important role in determining the structural and functional properties of collagen-based materials. The laser treatment led to the destruction of several hydrogen bonds, and the water bonded to collagen was released. Water is also bonded with chitosan and hyaluronic acid via hydrogen bonds; therefore, during the laser treatment the release of water from those biopolymers can also be observed. The release of water from the film based on a single polymer is higher than from the film made of biopolymer blends. There was also evidence of an increase in the concentration of oxygenated surface functional groups with the increase in the pulse frequency of the laser treatment, especially for ternary blends made of collagen, hyaluronic acid, and chitosan. In the blend, the intensity of the band at around 1630–1700 cm^{-1} is higher than for single biopolymers after laser treatment. This band is related to carbonyl groups in biopolymers. It suggests that oxidation processes occur with higher efficiency on the surface of a ternary blend than on the surface of a film made of a single biopolymer.

Scanning electron microscopy (SEM) was used to study the structure of films made of hyaluronic acid, Coll/HA, and Coll/HA/30CTS before and after laser radiation treatment. SEM images are shown in Figures 3 and 4.

Upon laser treatment, changes in the structure of biopolymeric films are observed. The material expansion into a new foam structure is visible, which is easily observed in the SEM images, especially for the Coll//HA/30CTS mixture. For films made of HA, only small alterations were observed, which suggests that those films are more resistant to laser treatment, and surface foaming under experimental conditions is not observed. A more efficient micro-foaming process in films based on ternary blends than those in films made of a single biopolymer is probably due to the higher absorption of laser energy by the blend components. The mechanism of micro-foam formation on some biopolymer films' surfaces was studied in our previous paper, which was regarding laser ablation [8,9,26]. Previously, we found that the femtosecond laser irradiation caused an expansion of materials above the surface of collagen films. The received results can be considered in terms of possible ablation mechanisms and compared with the last results obtained previously with nanosecond laser pulses [26].

It was found by Castillejo et al. [27] that ultraviolet laser irradiation with the use of pulses of the duration from the nanosecond to the femtosecond range generated a foam layer on films made of chitosan, starch, and their blends. The example of the possibility of a broader use of laser-induced biopolymer foaming structures in biology is also mentioned in this paper [26]. However, the low-level red laser induced small alterations in fluorescence emission and had a negligible effect on the topography of thin collagen films, as demonstrated by Stylianou et al. [28]. Although irradiation did not affect the nanotopography of collagen, it influenced the cell behavior. Daskalova et al. [29] found that the femtosecond laser treatment of thin films made of collagen, gelatin, and collagen-elastin resulted in the creation of micro/nanopores with different sizes of cavity formations; moreover, the pore sizes could be influenced by tuning the laser parameters. It has been demonstrated that any changes in the surface structure after laser treatment depend on laser type, pulse energy, and radiation wavelength. The interaction between a collagen film and UV laser radiation can be complex, and it is a result of the photomechanical regime, with low thermal degradation and combined with photochemical transformations [30].

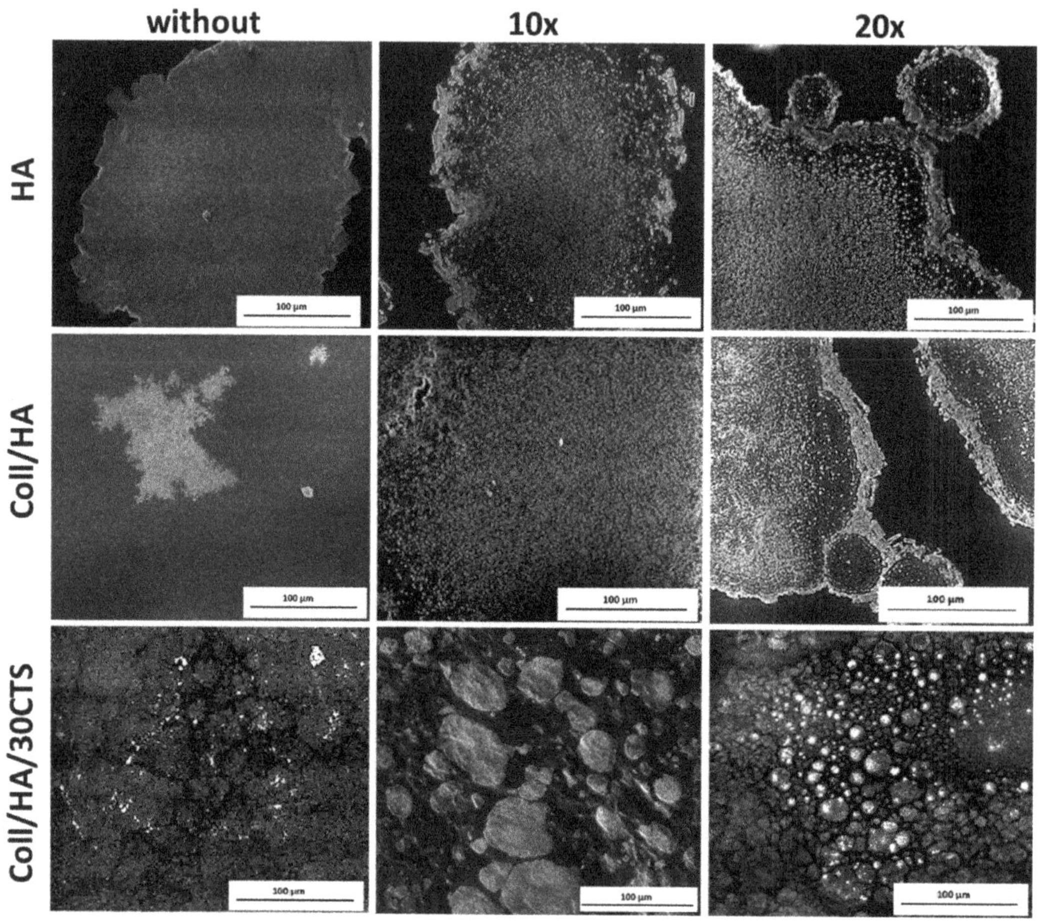

Figure 3. The SEM images of HA, Coll/HA, and Coll//HA/30CTS films before and after treatment of laser radiation (with resolution 100 µm).

In collagen materials, mainly thermal degradation and water evaporation occurs after the IR laser treatment. It was observed by Oh et al. [29] that the thermoelastic effect was generated and the mechanical stress was induced by laser irradiation on collagen with the Nd:YAG laser. The influence of laser radiation on films made of hyaluronic acid has not been widely studied. Preliminary studies on lasers and X-ray irradiation effects on hyaluronic acid dermal fillers have been carried out by Mladenova et al. [30]. It was demonstrated that the FTIR spectra of irradiated samples indicate no substantial changes of the spectral model in the characteristic absorption band regions corresponding to vibrations of the acetamido and carboxylate groups in the pyranose ring. In our study, we treated hyaluronic acid films with the Nd:YAG laser; therefore, results may differ, as hyaluronic acid dermal fillers are usually prepared as hydrogels.

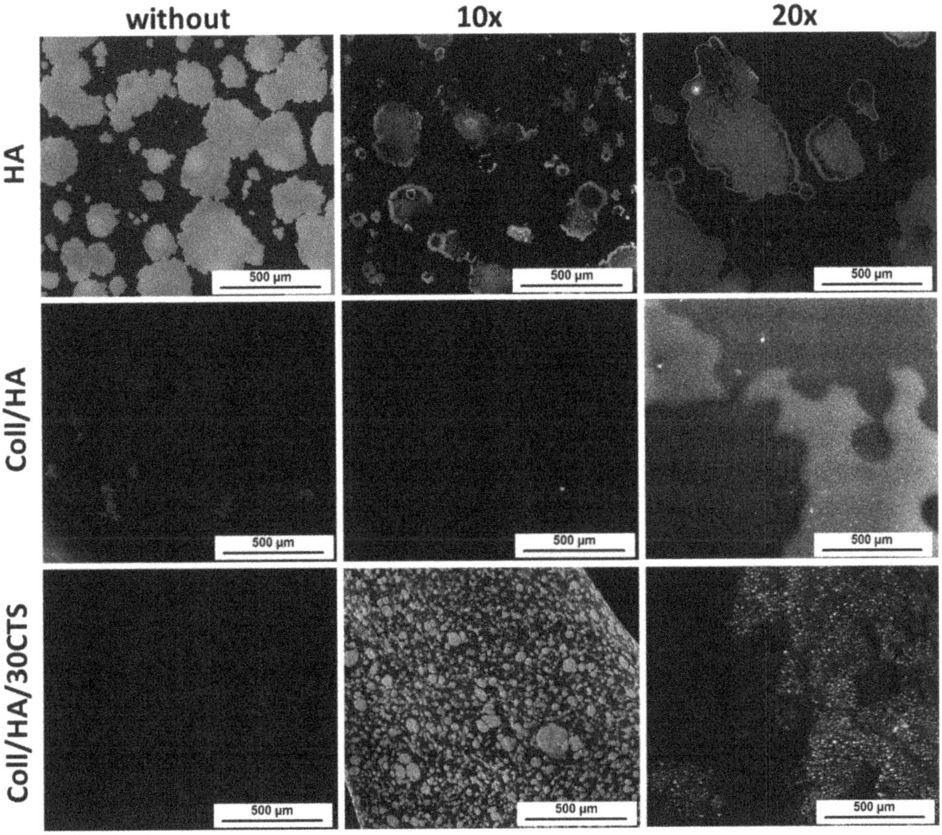

Figure 4. The SEM images of HA, Coll/HA, and Coll/HA/30CTS films before and after treatment of laser radiation (with resolution 500 µm).

The roughness parameters, AFM images, mechanical parameters, and contact angle parameters of native films were presented in our previous paper: Grabska, S.; Sionkowska, A. *The Influence of UV-Radiation on Hyaluronic Acid and Its Blends with Addition of Collagen and Chitosan*. Int. J. Polym. Anal. Char. 2019, 24, 285–294 [25]. The roughness parameters of laser-treated films are displayed in Table 1.

Table 1. The roughness parameters of laser-treated films based on HA, Coll/HA, and Coll/HA/30CTS.

Sample	Laser-Treated Films (20×)	
	Rq [nm]	Ra [nm]
HA	19.40	16.70
Coll/HA	112.50	99.30
Coll/HA/30CTS	9.06	7.99

Rq—mean square deviation of surface roughness; Ra—mean arithmetic deviation of the profile from the mean line.

Based on our previous study, it can be observed that AFM images display differences in the surface of films based on hyaluronic acid and its mixture with collagen and chitosan [25]. The AFM images demonstrate that the films based on hyaluronic acid and Coll/HA/30CTS are characterized by a smoother surface than the Coll/HA film [25]. The surface of the three-component film was similar to the hyaluronic acid film surface and these results are related to the interactions between hyaluronic acid, collagen, and chitosan [25].

Comparing the roughness parameters values for the films before and after the laser treatment, significant changes can be observed. The surface of the films after laser treatment was characterized by higher parameters of Rq (mean square deviation of surface roughness) and Ra (mean arithmetic deviation of the profile from the mean line). These observations are consistent with the SEM images (Figures 3 and 4).

Otherwise, in general, the roughness of the surface increases after the treatment of polymer films with lasers [6,31,32]. Daskalova et al. [32] report that from the AFM results pure chitosan film treated with laser, and the roughness parameter increases when the intensity of the laser increases. According to Volova et al. [31], pristine films based on poly-3-hydroxybutyrate demonstrated the lowest values of roughness parameters, such as: arithmetic mean surface roughness, root mean square roughness, or peak-to-valley height, relative to materials treated with the CO_2 laser LaserPro Explorer II. On the other hand, Michaljaničová et al. [6] compared the effect of the KrF and ArF lasers on the surface of poly(lactic acid) and polyhydroxybutyrate, and they reported an increase in surface roughness for polyhydroxybutyrate, regardless of the laser used. An increase in surface roughness for the poly(lactic acid) with the ArF laser treatment and a decrease for the KrF laser treatment was observed.

Overall, the roughness of film surface is a consequence of the laser treating and additional side effects, such as pores and foam formation, cracks, and damages [6,31,32].

Our previous paper, where the same materials were irradiated with UV light, demonstrates the parameters of the mechanical properties and the contact angle for the tested materials prior to laser treatment.

The films that are the subject of this work, unfortunately, after laser treatment, were not suitable for investigating the mechanical properties and the contact angle measurements. Therefore, we decided to quote only the previous results [25]. We dare to say that when the films are treated with laser, the mechanical properties of the materials becomes a little better, which was observed for the laser-deposited poly(methyl methacrylate) films [33], while the contact angle for the polar liquid (glycerin) of the films decreased, which was the case for the PLLA-plasma-treated materials [34].

4. Conclusions

The treatment of biopolymeric films made of hyaluronic acid, and those made of the blend of collagen and hyaluronic acid as well as the ternary Coll/HA/chitosan blend with pulse neodymium laser radiation, led to the alterations of the surface properties. After the laser treatment, oxidation processes occurred on the biopolymeric films, and the micro-foam formation was observed. Micro-foaming occurred with higher efficiency in the films based on ternary blends than in those made of a single biopolymer. It can be assumed that laser treatment can be used to tailor the material properties based on the ternary blends of biopolymers for potential biomedical applications.

Author Contributions: Conceptualization, S.G.-Z. and A.S.; methodology, S.G.-Z.; software, S.G.-Z. and A.S.; validation, S.G.-Z. and A.S.; formal analysis, S.G.-Z.; investigation, S.G.-Z.; resources, A.S.; data curation, S.G.-Z.; writing—original draft preparation, S.G.-Z.; writing—review and editing, S.G.-Z. and A.S.; visualization, S.G.-Z.; supervision, A.S.; project administration, A.S.; funding acquisition, A.S. All authors have read and agreed to the published version of the manuscript.

Funding: This work was supported by the Polish National Science Centre (NCN) under grant 2013/11/B/ST8/04444.

Institutional Review Board Statement: Not applicable.

Informed Consent Statement: Not applicable.

Data Availability Statement: The data presented in this study are available on request from the corresponding author.

Acknowledgments: Grateful acknowledgements to Agata Kołodziejczak-Czubachowska for the obtaining of biopolymeric films and help with the sample irradiation.

Conflicts of Interest: The authors declare no conflict of interest.

References

1. Jaganathan, S.K.; Balaji, A.; Vellayappan, M.V.; Subramanian, A.P.; John, A.A.; Asokan, M.K.; Supriyanto, E. Review: Radiation-Induced Surface Modification of Polymers for Biomaterial Application. *J. Mater. Sci.* **2015**, *50*, 2007–2018. [CrossRef]
2. Peethan, A.; Unnikrishnan, V.K.; Chidangil, S.; George, S.D. Laser-Assisted Tailoring of Surface Wettability-Fundamentals and Applications: A Critical Review. *Rev. Adhes. Adhes.* **2019**, *7*, 331–365. [CrossRef]
3. Krishnan, A.; Fang, F. Review on Mechanism and Process of Surface Polishing Using Lasers. *Front. Mech. Eng.* **2019**, *14*, 299–319. [CrossRef]
4. Kusinski, J.; Kac, S.; Kopia, A.; Radziszewska, A.; Rozmus-Górnikowska, M.; Major, B.; Major, L.; Marczak, J.; Lisiecki, A. Laser Modification of the Materials Surface Layer-a Review Paper. *Bull. Polish Acad. Sci. Tech. Sci.* **2012**, *60*, 711–728. [CrossRef]
5. Kurella, A.; Dahotre, N.B. Review Paper: Surface Modification for Bioimplants: The Role of Laser Surface Engineering. *J. Biomater. Appl.* **2005**, *20*, 5–50. [CrossRef] [PubMed]
6. Michaljaničová, I.; Slepička, P.; Heitz, J.; Barb, R.A.; Sajdl, P.; Švorčík, V. Comparison of KrF and ArF Excimer Laser Treatment of Biopolymer Surface. *Appl. Surf. Sci.* **2015**, *339*, 144–150. [CrossRef]
7. Michaljanicova, I.; Slepicka, P.; Rimpelova, S.; Sajdl, P.; Svorcík, V. Surface Properties of Laser-Treated Biopolymer. *Mater. Technol.* **2016**, *3*, 331–335. [CrossRef]
8. Lazare, S.; Sionkowska, A.; Zaborowicz, M.; Planecka, A.; Lopez, J.; Dijoux, M.; Louména, C.; Hernandez, M.C. Bombyx Mori Silk Protein Films Microprocessing with a Nanosecond Ultraviolet Laser and a Femtosecond Laser Workstation: Theory and Experiments. *Appl. Phys. A-Mater.* **2012**, *106*, 67–77. [CrossRef]
9. Lazare, S.; Tokarev, V.; Sionkowska, A.; Wiśniewski, M. Surface Foaming of Collagen, Chitosan and Other Biopolymer Films by KrF Excimer Laser Ablation in the Photomechanical Regime. *Appl. Phys. A-Mater.* **2005**, *81*, 465–470. [CrossRef]
10. Sionkowska, A. Current Research on the Blends of Natural and Synthetic Polymers as New Biomaterials: Review. *Prog. Polym. Sci.* **2011**, *36*, 1254–1276. [CrossRef]
11. Hemshekhar, M.; Thushara, R.M.; Chandranayaka, S.; Sherman, L.S.; Kemparaju, K.; Girish, K.S. Emerging Roles of Hyaluronic Acid Bioscaffolds in Tissue Engineering and Regenerative Medicine. *Int. J. Biol. Macromol.* **2016**, *86*, 917–928. [CrossRef]
12. Zhai, P.; Peng, X.; Li, B.; Liu, Y.; Sun, H.; Li, X. The Application of Hyaluronic Acid in Bone Regeneration. *Int. J. Biol. Macromol.* **2020**, *151*, 1224–1239. [CrossRef] [PubMed]
13. Necas, J.; Bartosikova, L.; Brauner, P.; Kolar, J. Hyaluronic Acid (Hyaluronan): A Review. *Vet. Med.* **2008**, *53*, 397–411. [CrossRef]
14. Maharjan, A.S.; Pilling, D.; Gomer, R.H. High and Low Molecular Weight Hyaluronic Acid Differentially Regulate Human Fibrocyte Differentiation. *PLoS ONE* **2011**, *6*, e26078. [CrossRef] [PubMed]
15. Choi, S.Y.; Kwon, H.J.; Ahn, G.R.; Ko, E.J.; Yoo, K.H.; Kim, B.J.; Lee, C.; Kim, D. Hyaluronic Acid Microneedle Patch for the Improvement of Crow's Feet Wrinkles. *Dermatol. Ther.* **2017**, *30*, e12546. [CrossRef] [PubMed]
16. John, H.E.; Price, R.D. Perspectives in the Selection of Hyaluronic Acid Fillers for Facial Wrinkles and Aging Skin. *Patient Prefer. Adher.* **2009**, *3*, 225.
17. Pavicic, T.; Gauglitz, G.G.; Lersch, P.; Schwach-Abdellaoui, K.; Malle, B.; Korting, H.C.; Farwick, M. Efficacy of Cream-Based Novel Formulations of Hyaluronic Acid of Different Molecular Weights in Anti-Wrinkle Treatment. *J. Drugs Dermatol.* **2011**, *10*, 990–1000.
18. Ferreira, A.M.; Gentile, P.; Chiono, V.; Ciardelli, G. Collagen for Bone Tissue Regeneration. *Acta Biomater.* **2012**, *8*, 3191–3200. [CrossRef]
19. Rinaudo, M. Main Properties and Current Applications of Some Polysaccharides as Biomaterials. *Polym. Inter.* **2008**, *57*, 397–430. [CrossRef]
20. Kumar, M.N.V.R.; Muzzarelli, R.A.A.; Muzzarelli, C.; Sashiwa, H.; Domb, A.J. Chitosan Chemistry and Pharmaceutical Perspectives. *Chem. Rev.* **2004**, *104*, 6017–6084. [CrossRef] [PubMed]
21. Jayakumar, R.; Prabaharan, M.; Nair, S.V.; Tokura, S.; Tamura, H.; Selvamurugan, N. Novel Carboxymethyl Derivatives of Chitin and Chitosan Materials and Their Biomedical Applications. *Prog. Mater. Sci.* **2010**, *7*, 675–709. [CrossRef]
22. Lewandowska, K.; Sionkowska, A.; Grabska, S.; Kaczmarek, B.; Michalska, M. The Miscibility of Collagen/Hyaluronic Acid/Chitosan Blends Investigated in Dilute Solutions and Solids. *J. Mol. Liq.* **2016**, *220*, 726–730. [CrossRef]
23. Lewandowska, K.; Sionkowska, A.; Grabska, S.; Kaczmarek, B. Surface and Thermal Properties of Collagen/Hyaluronic Acid Blends Containing Chitosan. *Int. J. Biol. Macromol.* **2016**, *92*, 371–376. [CrossRef] [PubMed]
24. Sionkowska, A.; Kaczmarek, B. Preparation and Characterization of Composites Based on the Blends of Collagen, Chitosan and Hyaluronic Acid with Nano-Hydroxyapatite. *Int. J. Biol. Macromol.* **2017**, *102*, 658–666. [CrossRef]
25. Grabska, S.; Sionkowska, A. The Influence of UV-Radiation on Hyaluronic Acid and Its Blends with Addition of Collagen and Chitosan. *Int. J. Polym. Anal. Char.* **2019**, *24*, 285–294. [CrossRef]
26. Sionkowska, A.; Lewandowska, K.; Grabska, S.; Kaczmarek, B.; Michalska, M. Physico-Chemical Properties of Three-Component Mixtures Based on Chitosan, Hyaluronic Acid and Collagen. *Mol. Cryst. Liq. Cryst.* **2016**, *640*, 21–29. [CrossRef]
27. Kim, J.K.; Srinivasan, P.; Kim, J.H.; Choi, J.-I.; Park, H.J.; Byun, M.W.; Lee, J.W. Structural and Antioxidant Properties of Gamma Irradiated Hyaluronic Acid. *Food Chem.* **2008**, *109*, 763–770. [CrossRef] [PubMed]

28. Wu, Y. Preparation of Low-Molecular-Weight Hyaluronic Acid by Ozone Treatment. *Carbohyd. Polym.* **2012**, *89*, 709–712. [CrossRef]
29. Oh, H.B.; Kim, J.S.; Jung, G.I.; Baek, J.Y.; Kim, J.; Jun, J.H. Change of Induced Stress Wave on Collagen Tissue for Biostimulation by Frequency-Doubled Nd:YAG Laser. *J. Mech. Med. Biol.* **2018**, *18*, 1840003. [CrossRef]
30. Mladenova, R.; Karakirova, Y.; Shopska, M.; Sabatinov, O.; Gyurova, M. Preliminary Study on Lasers and X-Ray Irradiation Effects on Hyaluronic Acid Dermal Fillers. *Comptes Rendus L'academie Bulg. Sci.* **2018**, *50*, 1451–1457.
31. Volova, T.G.; Golubev, A.I.; Nemtsev, I.V.; Lukyanenko, A.V.; Dudaev, A.E.; Shishatskaya, E.I. Laser Processing of Polymer Films Fabricated from PHAs Differing in Their Monomer Composition. *Polymers* **2021**, *13*, 1553. [CrossRef] [PubMed]
32. Daskalova, A.; Bliznakova, I.; Angelova, L.; Trifonov, A.; Declercq, H.; Buchvarov, I. Femtosecond Laser Fabrication of Engineered Functional Surfaces Based on Biodegradable Polymer and Biopolymer/Ceramic Composite Thin Films. *Polymers* **2019**, *11*, 378. [CrossRef] [PubMed]
33. Süske, E.; Scharf, T.; Krebs, H.U.; Panchenko, E.; Junkers, T.; Egorov, M.; Buback, M.; Kijewski, H. Tuning of Cross-Linking and Mechanical Properties of Laser-Deposited Poly (Methyl Methacrylate) Films. *J. Appl. Phys.* **2005**, *97*, 063501. [CrossRef]
34. Slepička, P.; Michaljaničová, I.; Švorčík, V. Controlled Biopolymer Roughness Induced by Plasma and Excimer Laser Treatment. *Express Polym. Lett.* **2013**, *7*, 950–959. [CrossRef]

polysaccharides

MDPI

Review

Polysaccharide Stalks in *Didymosphenia geminata* Diatom: Real World Applications and Strategies to Combat Its Spread

Esther Somanader [1,2], Roshini Sreenivas [1], Golnoosh Siavash [1], Nicole Rodriguez [1], Tingxiao Gao [1,2], Hermann Ehrlich [1,2,3,4] and M. Azizur Rahman [1,2,*]

[1] Centre for Climate Change Research, University of Toronto Entrepreneurship (ONRamp), Toronto, ON M5G 1L5, Canada; esther@climatechangeresearch.ca (E.S.); rosh.2992@gmail.com (R.S.); golnooshs@hotmail.com (G.S.); nicolerodri9@gmail.com (N.R.); tx.gao@mail.utoronto.ca (T.G.); Hermann.Ehrlich@esm.tu-freiberg.de (H.E.)
[2] A.R. Environmental Solutions, ICUBE-University of Toronto, Mississauga, ON L5L 1C6, Canada
[3] Institute of Electronic and Sensor Materials, TU Bergakademie Freiberg, 09599 Freiberg, Germany
[4] Center for Advanced Technology, Adam Mickiewicz University, 61614 Poznan, Poland
* Correspondence: mazizur.rahman@utoronto.ca or aziz@climatechangeresearch.ca; Tel.: +1-647-892-4221

Abstract: *Didymosphenia geminata* is a species of freshwater diatom that is known as invasive and is propagating quickly around the world. While invasive species are generally considered a nuisance, this paper attempts to find useful applications for *D. geminata* in the biomedical field and wastewater remediation. Here, we highlight the polysaccharide-based stalks of *D. geminata* that enable versatile potential applications and uses as a biopolymer, in drug delivery and wound healing, and as biocompatible scaffolding in cell adhesion and proliferation. Furthermore, this review focuses on how the polysaccharide nature of stalks and their metal-adsorption capacity allows them to have excellent wastewater remediation potential. This work also aims to assess the economic impact of *D. geminata*, as an invasive species, on its immediate environment. Potential government measures and legislation are recommended to prevent the spread of *D. geminata*, emphasizing the importance of education and collaboration between stakeholders.

Keywords: chitin; diatoms; didymo; invasive species; polysaccharide structure; drug delivery; wound dressing; wastewater treatment; government legislation; economic impact

Citation: Somanader, E.; Sreenivas, R.; Siavash, G.; Rodriguez, N.; Gao, T.; Ehrlich, H.; Rahman, M.A. Polysaccharide Stalks in *Didymosphenia geminata* Diatom: Real World Applications and Strategies to Combat Its Spread. *Polysaccharides* **2022**, *3*, 83–94. https://doi.org/10.3390/polysaccharides3010004

Academic Editor: Cédric Delattre

Received: 29 November 2021
Accepted: 3 January 2022
Published: 6 January 2022

1. Introduction

Diatoms are a type of unicellular microalgae that play a large role in the biosilica production and carbon fixation of the planet [1,2]. Additionally, they are also one of the largest producers of macronutrients such as nitrogen and phosphorus [1]. One species of freshwater diatom, *Didymosphenia geminata* (Lyngb.) has characteristic biomineralized polysaccharide stalks and blooms whose growth has recently proliferated throughout many aquatic ecosystems [3–6]. Although the invasive status of *D. geminata* may sometimes be uncertain, it is not the organisms themselves that have had the largest impact on the local environment but rather the adhesive stalks formed by *D. geminata* [7,8]. Unfortunately, the detailed chemistry of these fibrous structures has not been known until now. According to the modern view, these stalks are primarily composed of sulfated polysaccharides, proteins, and some uronic acid [3,4] and can spread and grow over 500 μm long and over multiple centimeters thick [8]. They are also able to divide during the cell reproduction process, leading to the production of large, branching, dense mats that can cover stream bottoms, covering areas of over 20 km [4,6,8]. Furthermore, the chemistry of the frustules of *D. geminata* has also not been well studied. Thus, the possible presence of a chitinous template within them, similar to that reported for the siliceous cell walls of marine diatoms [6,9–11], is still not confirmed. Research into this intriguing question is extremely important because chitin was not found in the stalks of this microalga and from this point of view, it would be illogical to find chitin inside the walls. Our recent investigation into the polysaccharide stalks of *D. geminata* (see Figure 1) shows how this diatom can build stalk-based layers on the rock surface and how the siliceous cells remain attached to biomineralized polysaccharide-containing adhesive stalks. The giant siliceous frustules of *D. geminata* may also have interesting biomedical and technological applications due to their highly specific micro- and nanoporous architecture (Figure 1B,C).

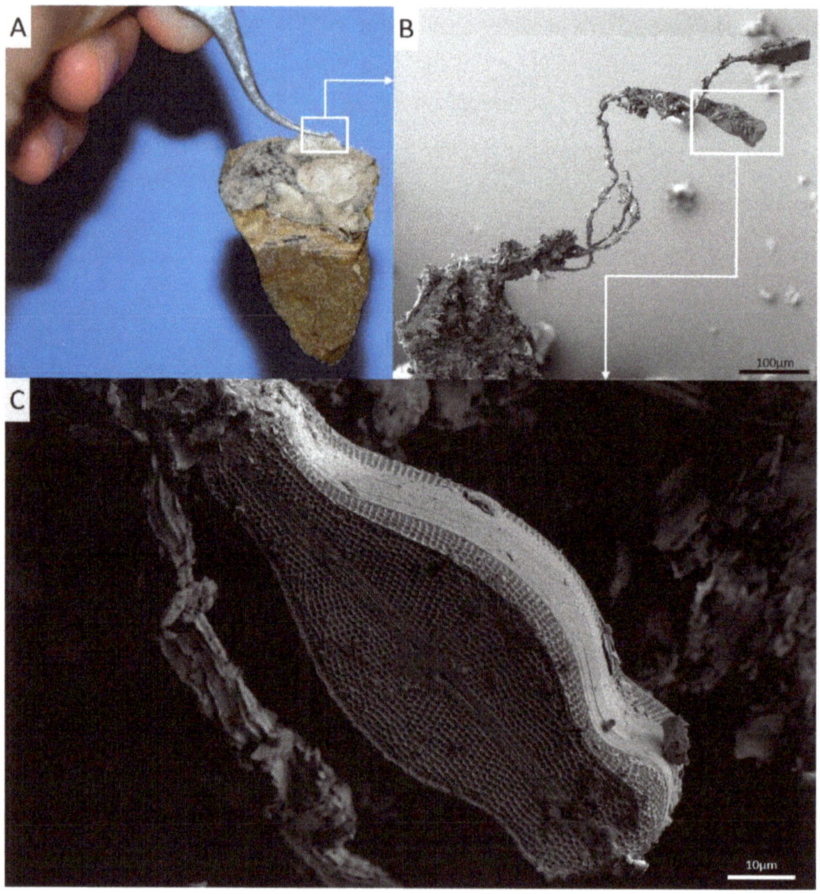

Figure 1. The diatom *D. geminata* can build stalk-based layers, which are responsible for very strong attachment to the rock surface (**A**). This felt-like layer has so firmly adhered to the surface of a hard substrate that a piece of stone can be held in the air, in a suspended state, without detaching from the ends of the tweezers. SEM images (**B**,**C**) show that siliceous cells remain attached to biomineralized polysaccharide-containing adhesive stalks.

D. geminata in benthic ecosystems has recently been a focus of research, in part due to the ability of blooms to manipulate and overwhelm the biodiversity of an ecosystem [12]. The presence of these large blooms alters local food-web structures by favoring smaller predators that can move between filaments and alter the composition of other diatoms [5]. The fish community is also affected by *D. geminata* blooms through changes to benthic invertebrate composition, with some New Zealand ecosystems experiencing a fish biomass decrease as a result [13]. The proliferation of *D. geminata* blooms also affects water quality, as diatom biodiversity and species composition are often used as measures of changes in environmental conditions [8]. Indeed, these blooms are caused by novel environmental conditions that are becoming increasingly more common due to climate change [14]. One of the largest predictors for the presence of polysaccharide blooms is the low concentration of dissolved phosphorus [7,15]. The formation of stalks is thought to be a way for *D. geminata* to elevate itself to reach the water column, where there is increased nutrient uptake [7,15,16]. While *D. geminata* bloom proliferation may have devastating effects on ecosystems, it can be suggested that its relatively large nanocalcite- and polysaccharide-based stalks may be used in many, more positive applications.

Here, we present a review of the recent literature focusing on the potential applications of *D. geminata*, including their use in wastewater treatment and as a biomaterial, based on the biological and chemical composition of its polysaccharide stalks. The motivation behind this emerging area of research is the unique structure of the microtubular stalks, which make them ideal reservoirs for small molecules, leading to their potential use in the aforementioned applications. Recently, *D. geminata* stalks have been shown to be effective in removing metallic ions such as Pb(II), Ni(II), and Cd(II) [4,17]. Because the stalks and frustules that are made of biogenic silica contain nontoxic, plant-based materials, they may also be ideal for biomedical applications or drug delivery [2,6,18,19]. Additionally, the environmental and economic impacts and environmental regulations of *D. geminata* are explored, and future policy recommendations are made. The research and recommendations we present in this review can be utilized for future studies regarding *D.geminata's* stalks. However, more research and progress is required with regard to the potential beneficial applications of *D. geminata* and the strategies to control its increasingly more common blooms.

2. Applications of *D. geminata* as a Biomaterial: Cell Adhesion, Proliferation, and Drug Delivery

D. geminata is a relatively new finding and has many potential uses in biomedicine and as a biomaterial for drug delivery, cell adhesion, and proliferation [6]. Principally, a biomaterial is defined as an inert material used in therapeutic or diagnostic procedures by interactions with components of living systems [20]. Polysaccharides are a great source of biomaterials, as they are naturally sourced, sustainable, and economically viable options due to their abundance in nature, including diverse aquatic niches. Chemically, polysaccharides are carbohydrates with repeating structures of monohydrides and are hence nontoxic by their very nature [18]. The polysaccharides in the stalk of *D. geminata* appear to be primarily sulfated xylogalactan, and the stalk was found to be intrinsically hydrophilic, a trait that is one of the first determiners of the extent of protein adsorption [21]. The presence of sulfate groups within polysaccharides shows an improvement in the immune system and would be beneficial to the nutraceutical industry [19]. The intrinsic biocompatible nature of *D. geminata* is attributed to its similarity in structure to glycosaminoglycans, which are a vital component in the tissue extracellular matrix [18].

Zglobicka in 2013 examined how previous studies had cast stalks of *D. geminata* as capable of adhering to multiple kinds of substrates because they were built in concentric circles of materials of variable compositions [22]. The chemical composition of the stalk seems to largely be unknown, but it is agreed that the main composition is amorphous silica [22,23]. The use of silica and biosilica in bone regeneration is well studied. It is understood to improve bone regeneration and increase bone density and is thought to play a role in the stabilization of collagen in the bone matrix [24]. The specific structure of *D. geminata* and its diatom frustules made of silica are especially beneficial with regard to drug delivery due to characteristics such as high permeability and low density [25].

D. geminata also exhibits anticancer properties [2]. As seen in Figure 2, the diatom has a nanosized porous silica capsule, which allows controlled drug release. The drug itself can be added to the external and internal surface of the diatom and then released based on need. The drug can be loaded onto the nanoparticles directly through noncovalent interactions like electrostatic bonding, hydrogen bonding, van der Waals forces, pi–pi stacking, and hydrophobic interactions [26].

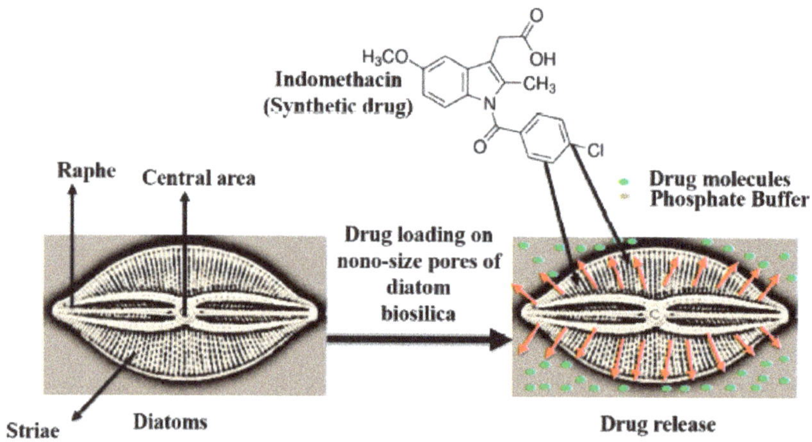

Figure 2. The mechanism of drug release from the porous diatom microshell [2].

The previous study also examined in some detail how the stalks formed by *D. geminata* can outlive the alga itself and are essentially a great substrate for cell adhesion [22]. Figure 3 shows a tubular structure of stalks. Further experimentation could involve examination of the interactions between a Didymo stalk as a substratum and a eukaryotic cell line in vitro. When testing a potential biomaterial for cell adhesion or cell proliferation in vitro, cellular models can be used to test the initial host response in addition to cell adhesion, proliferation, and cytotoxicity [27].

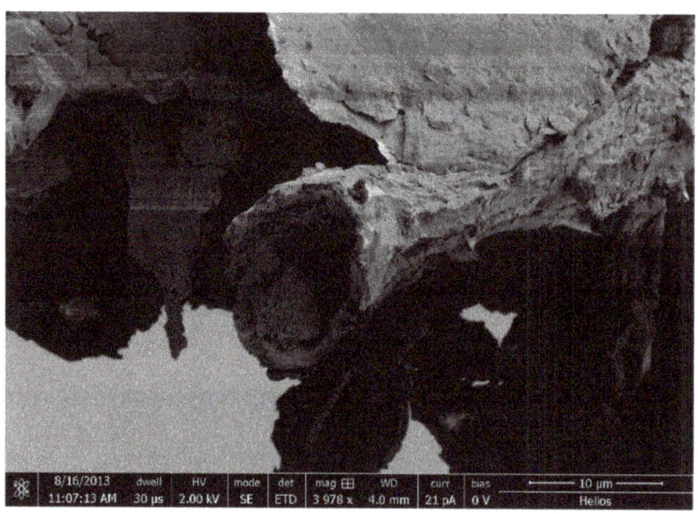

Figure 3. SEM image of the broken stalk of *D. geminata* clearly shows its tubular structure.

Wound healing is another potential use for a *D. geminata* stalk due to its unique sulfated polysaccharide composition. Recent studies have shown that sulfated polysaccharides from seaweeds are harvested and studied as promising sources of wound dressings due to their intrinsic biological activity [28]. Ideally, a wound dressing must be from an inexhaustible source, have the potential for low toxicity, be hypoallergenic, and have the potential for bacterial drug resistance [29]. *D. geminata* stalks are, in theory, a good contender for this application. They also exhibit antioxidant properties, which can be used to prevent the spread of disease [19]. Due to the physical properties of the *D. geminata* stalks, they can be easily pressed into a felt-like banding material. The capillary properties of such a material should determine its effectiveness in relation to blood, plasma, and other biological fluids [30]. Therefore, based on the well-known thermostability of polysaccharide-based materials, it is suggested that there should not be any issues with sterilizations of such kinds regarding *D. geminata*-based wound-dressing materials [31].

The recent pandemic has also been an eye-opener to the extent of our abilities to protect ourselves from deadly viruses. Climate change is also causing more health issues within communities that were previously not known to be vulnerable [32]. With its use in therapeutic purposes within the nutraceutical and pharmaceutical industries, *D. geminata* could have potential uses in the treatment of diseases such as COVID-19 as a filler for corresponding drugs.

Additionally, the authors agree that much work needs to be done to better understand *D. geminata*'s proficiency as a biomaterial. One could also utilize the services of the genome-splicing software 'CRISPR' to induce specific modifications at the genome level that would allow *D. geminata* to function in a more prolific manner as a naturally occurring large-scale biological material. A list of the references corresponding to the biomedical applications of *D.geminata* is provided here (Table 1). This information will help to easily find the reports of biomedical applications in the literature.

Table 1. References corresponding to the different applications of *D. geminata* as a biomaterial.

Applications of *D. geminata* as a Biomaterial	References
Cell Adhesion	[18,22,24,29]
Cell Proliferation	[18,24,29]
Drug Delivery	[2,6,18,19,23,25,29,30]
Wound Dressing	[18,28,29]

3. *D. geminata* Applications in Wastewater Treatment

The composition and structure of *D. geminata*, in particular its multiphase-biomineralized polysaccharide stalks [33], allow for potential valuable applications in wastewater treatment due to its metal-adsorption capacities [34]. The use of microalgae in wastewater treatment is not a novel concept, and research has revealed numerous advantages, such as low cost, high metal-ion uptake, and excellent metal selectivity [35]. Due to the availability of *D. geminata* in various regions, recent studies have emerged investigating the feasibility of using its biomass for heavy-metal remediation in water treatment [4,17]. Polysaccharide-containing biomass, which makes up the extracellular stalks of *D. geminata*, is considered to be an excellent adsorbent of heavy metals, predominantly via functional groups [34]. Wysokowski et al. [34] found that purified *D. geminata* stalks had considerable adsorption capabilities for both nickel(II) and cadmium(II) ions but were especially efficient in the adsorption of lead(II) ions, with a maximum adsorption capacity of 175.48 mg g^{-1} [4,34]. This Pb(II)-adsorption capacity is quite high in comparison to other microalgae biosorbents and, in fact, is almost comparable to the metal-sorption capabilities of some macroalgae (Table 2) [4,17,35,36]. The use of *D. geminata* for heavy-metal adsorption would be especially advantageous in treating industrial wastewater that has been contaminated with lead, which is a priority pollutant associated with high toxicity [37]. There is also potential for *D. geminata* stalks to act as an adsorbent for U(V) ions, with preliminary results showing a 96% decrease in U(V) concentration from adsorption via *D. geminata* nonwoven fabric [25].

Table 2. Pb(II)-sorption capacities for various micro- and macroalgae species.

	Algae Species	Pb(II) Sorption Capacity (mg g^{-1})	Original Reference
Microalgae	*Didymosphenia geminata*	129–175.48	[4,17]
	Chaetoceros sp.	8	[38]
	Chlorella sp.	10.4	[38]
	Phormidium sp.	2.3	[39]
	Rhizoclonium hookeri	81.7	[40]
Macroalgae	*Spirogyra* sp.	140	[41]
	Cladophora fascicularis	198.5	[42]
	Ascophyllum nodosum	178.6	[43]
	Fucus vesiculosus	215.5–259	[44]

Utilizing nonliving *D. geminata* biomass for heavy-metal remediation in water treatment has also proven to be advantageous in terms of sorbent regeneration and reusability [17,45]. Although dried *D. geminata* stalks appear to decrease in mechanical resistance upon adsorption of metals, Wysokowski et al. demonstrated the potential for the material still to go through multiple adsorption and desorption phases, with only a slight decrease in uptake capacity [4,46]. Reinoso-Guerra et al. [17] incorporated *D. geminata* biomass into a nanofibrillated cellulose membrane matrix with the addition of carbon nanotubes (CNTs) to reinforce the material and improve usability in wastewater applications. After evaluating the Pb(II)-adsorption ability of the matrix, it was discovered that the addition of CNTs improved the reusability of the membrane, with increased adsorption and decreased desorption in successive cycles. The boosted adsorption from CNT addition is associated with the enhanced porosity of the membrane and the additional available active sites [17]. Additionally, the presence of CNTs improved the mechanical resistance and antibacterial properties of the *D. geminata* membrane, making the material more viable and economical for water-treatment applications [17].

There is evidence to suggest that the presence of sulfur-based functional groups on *D. geminata* stalks specifically contributes to the excellent adsorption capabilities of the biomaterial [4,17]. The presence of either sulfate esters or sulfonic groups on the isolated polysaccharide stalks has been confirmed through infrared spectroscopy [19]. Metal adsorption on *D. geminata* stalks is thought to occur via a complex mechanism based on ion exchange between hydroxyl and sulfur-based functional groups and metal ions, forming complex ions and coordination bonds (Figure 4) [4,17]. In particular, Reinoso-Guerra et al. directly connected the adsorption of Pb(II) on a *D. geminata* stalk to sulfur-based functional groups through energy-dispersive X-ray spectroscopy (EDS) analysis [17]. Furthermore, experimentally determined maximum adsorption capacities at pH levels between 4 and 6 agree with the dissociation of these sulfur-based functional groups [4,17].

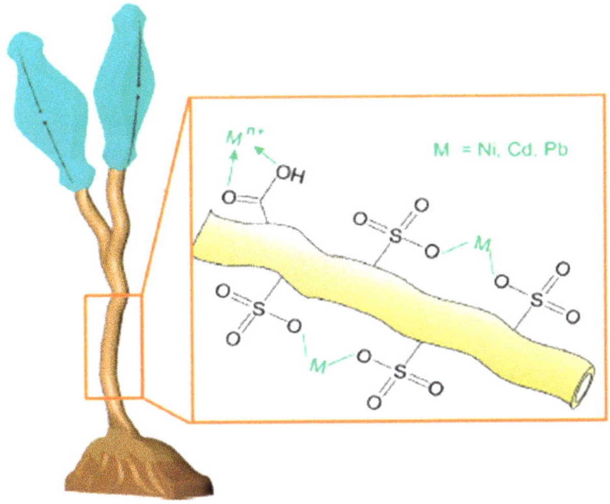

Figure 4. Schematic diagram for the adsorption of metal ions on *D. geminata* stalks [4]. The adsorption mechanism is based on ion exchange between groups on the adsorbent surface (hydroxyl and sulfonic groups) and metal ions, leading to the formation of coordination bonds, as one might observe in the diagram.

4. Environmental Regulations and Policies Pertaining to *D. geminata*

The containment and management of an invasive species requires effective environmental regulations, collaboration between stakeholders, and widespread educational initiatives. In the context of environmental management, it is recommended to focus predominantly on the impacts of a species on its surroundings instead of the geographical origin of the species [47]. Given the growing detrimental effects of *D. geminata*, local authorities have taken steps to contain this species.

Within Canada, to control the spread of *D. geminata*, the Government of Quebec recommends maintaining boats, clothing, and equipment in one waterway when possible [48]. If multiple waterways must be utilized, such items should be cleaned prior to relocation [48]. The Government of Quebec (2008) recommends guidelines developed in New Zealand [48]. Such cleaning protocols include treatment with dishwashing detergent, bleach, hot water, and freezing equipment [49]. Taylor and Bothwell (2014) argued that the current management practices and policies pertaining to decontamination efforts will not be effective in containing *D. geminata*. Instead, the researchers believe the focus should be redirected to identifying which environmental factors/conditions facilitate the spread and bloom of this species [8]. Stalk production for these algae proliferates under low nutrient conditions with a high amount of light [16]. Research by Kilroy and Bothwell (2012) has specifically demonstrated the impact of phosphorus levels on *D. geminata* [50]. Based on this information, local environmental policies should be developed to identify potential environmental areas that may be suitable for the spread of *D. geminata*. Such information can aid conservation authorities in selecting key areas to actively monitor. Credit Valley Conservation has categorized *D. geminata* as an aquatic species on their "watch list" of potentially invasive species [51]. With the aid of the public, conservation authorities should continually monitor areas within their jurisdiction for the emergence of these algae.

Further policy development in this area should focus on providing enhanced educational resources for the public and relevant stakeholders (e.g., outdoor organizations). The government of New Brunswick has recommended improved education efforts for the containment of *D. geminata* [52]. Educational posters and infographics pertaining to *D. geminata* (i.e., cleaning protocols) should be well distributed in priority areas such as locations with increasing blooms. Individuals are encouraged to report sightings of these algae to local authorities [53].

Although *D. geminata* has been labeled invasive, it is important to recognize its potential beneficial applications for biotechnology and biomaterials. We have discussed these fruitful applications throughout this paper. To improve research efforts, scientists should obtain permits for working with this invasive species. Specifically, strict protocols must be implemented to ensure that the species is contained. Researchers and conservation authorities should work collectively to isolate samples of *D. geminata* without any contamination. Leakproof containers must be utilized when transporting the collected samples. Such samples must only be handled in a certified and contained lab setting and discarded appropriately. Furthermore, legal repercussions can assist in ensuring effective compliance. In a global context, New Zealand has taken concrete legal action and has labeled its South Island as a controlled area for these algae [49]. Consequently, individuals are legally obligated to follow outlined cleaning guidelines before changing waterways to ensure that South Island remains free of *D. geminata* [49]. Following New Zealand's initiatives, provinces managing this invasive species should establish specific policies/laws that ensure effective enforcement. Such policies should accommodate research efforts by permitting scientists to work on this species in an isolated setting.

Collectively, the management of *D. geminata* must be implemented via effective environmental regulations, coordination between stakeholders, and enhanced education. Current efforts have focused on implementing cleaning protocols to contain these algae. Alongside such measures, future efforts should also accommodate researchers currently working on invasive species.

5. *D. geminata*'s Economic Impacts

Taking advantage of algae species such as *D. geminata* in industrial practices could be advantageous in reducing the negative impact of invasive species while reusing them as sustainable and natural resources. As mentioned in our previous review [6], *D. geminata*'s ability to treat wastewater and transform it into agricultural fertilizers could bring great economic benefits with its usage; however, it has yet to be determined whether the accompanying side effects of *D. geminata* application would potentially suppress the profitability of the water area. Thus, considerations of its negative impact and growth-controlling plans should be made before *D. geminata* application.

While the commercial impact of *D. geminata* is unknown, studies that investigate the adverse effects of various existing aquatic invasive species (AISs) that are brought to fisheries could be referenced. A report analyzing the economic impact of an AIS on the industries of the Great Lakes states of the U.S. by Rosaen et al. summarized some of the major profit losses by invasive species involving tourism [54]. This nonmarketed loss that is largely due to the rapid spread of AIS can be the result of one of three phenomena: direct operational cost, reduced demand, or decreased productivity. The prevention of AIS spreading requires regular maintenance, which includes but is not limited to water-pipe scraping and chemical treatment. Recreational boating and commercial fishing companies have always relied on the diverse fish species of the local water area. Direct competition within the same habitat due to the occupation of space by AIS can lead to lowered fish stocks and even the extinction of certain species. In addition to the loss of fish population and diversity, the overall demand for tourism can be impacted in several ways. Common examples of reduced customer interest could be that the quality of fish, either its size or health, can no longer meet the expectation of the recreationists, or that beach areas are being occupied with dead fish and fermented rotten algae. Most businesses faced major difficulties opening up, and less-established companies were challenged with permanent shutdown. A reported annual loss of $50 million was due to the reduced demand for businesses and tourism alone [55].

New Zealand, one of the most chronically infected areas, paid an economic cost of $158 million in the first eight years post-*D. geminata* discovery without government intervention [21]. Even with a continuous investment in *D. geminata* control efforts, the preventions themselves could be a nuisance to some of the local businesses. One case study performed in New Zealand focused on the nonmarketed impact that *D. geminata* has on recreational angling industries. AIS management is often extremely complex, with most situations requiring rapid responses to efficiently limit the spread; for one reason, AIS directly competes for the same habitat as some of the most demanded fish species. As reported by Bergey et al., *D. geminata* blooms are observed to thrive in rocky and low-nutrient cold-water streams that are ideal for trout populations [56]. The immediate closure of some of the most demanded trout-fishing streams, regardless of their state of *D. geminata* infection, cost more than it would have to remain open at the time of this study. As a result of the decreased angling accessibility due to mainstem river closure, the fishing effort is thus transferred to smaller stems, threatening the vulnerable fish stock.

It is important to note that the occurrence of *D. geminata* in New Zealand should be considered a special case and not a common situation found in rivers of other regions. In Canada, a study done in 2008 reported that due to the distinct climate in Quebec, salmon rivers are not as impacted as in regions such as the Great Lakes and New Zealand [21]. In addition, studies done on *D. geminata*'s impact on salmon production in British Columbia showed no significance, and some infested rivers of Vancouver were even observed to have huge improvements due to possible natural control mechanisms [48], although *D. geminata* has not yet become a major threat to Canadian fish rivers. However, Canada has endured a reported total of $5.5 billion loss due to the indirect economic burden and the direct management cost from the 16 recognized species [57]. Given the above, the impact of *D. geminata* should still be studied in depth and closely monitored when applying it to the suggested purposes mentioned in this paper.

6. Concluding Remarks

While *D. geminata* can have negative implications on its surrounding environment, due to the specificity of the structural organization of its biosilica-based cell walls as well as polysaccharide stalks it can also offer many positive applications as a unique large-scale source of naturally prestructured functional biological materials. Subsequently, these potential applications extend to both the individual and to the environment. *D. geminata* has the ability to act as a biomaterial for the purposes of cell adhesion, proliferation, drug delivery, and wound healing. Furthermore, it has been demonstrated that the stalks of *D. geminata* have significant adsorption capabilities regarding heavy metals such as lead, nickel, and cadmium, which is particularly promising for heavy-metal remediation in wastewater treatment. To ensure that the extrication of the potential benefits of *D. geminata* does not lead to unwanted spread, it is important that researchers obtain a permit when working with *D. geminata*, that they collaborate with conservation authorities when isolating samples of this species, and that the handling of the samples is only in a certified and contained lab setting. Although past studies have shown that the spread of *D. geminata* has not negatively impacted the salmon population in Canada, proper adherence to protocols is especially important to ensure that Canada is not negatively impacted on an economic level by this species. Overall, *D. geminata* has demonstrated immense potential for use within society under proper guidance.

Author Contributions: Conceptualization, M.A.R.; validation, M.A.R. and H.E.; writing—original draft preparation, M.A.R., E.S., R.S., G.S., N.R. and T.G.; writing—review and editing, M.A.R. and H.E.; supervision, M.A.R.; project administration, M.A.R.; funding acquisition, M.A.R. and H.E. All authors have read and agreed to the published version of the manuscript.

Funding: MAR was supported by the Youth Employment Program (project # 018020008), the Ministry of Employment and Social Development, and Innovation Assistant Program of IRAP-NRC (Project # 967148), Govt. of Canada. HE was partially supported by OPUS 19 grant from the National Science Centre, Poland (2020/37/B/ST5/01909) and by Alexander von Humboldt Polish Honorary Research Scholarship (FNP, Poland).

Institutional Review Board Statement: Not applicable.

Informed Consent Statement: Not applicable.

Data Availability Statement: Not applicable.

Conflicts of Interest: The authors declare no conflict of interest.

References

1. Whitton, B.A.; Ellwood, N.T.W.; Kawecka, B. Biology of the Freshwater Diatom Didymosphenia: A Review. *Hydrobiologia* **2009**, *630*, 1–37. [CrossRef]
2. Hussein, H.A.; Abdullah, M.A. Anticancer Compounds Derived from Marine Diatoms. *Mar. Drugs* **2020**, *18*, 356. [CrossRef]
3. Sundareshwar, P.V.; Upadhayay, S.; Abessa, M.; Honomichl, S.; Berdanier, B.; Spaulding, S.A.; Sandvik, C.; Trennepohl, A. *Didymosphenia geminata*: Algal Blooms in Oligotrophic Streams and Rivers. *Geophys. Res. Lett.* **2011**, *38*. [CrossRef]
4. Wysokowski, M.; Bartczak, P.; Żółtowska-Aksamitowska, S.; Chudzińska, A.; Piasecki, A.; Langer, E.; Bazhenov, V.V.; Petrenko, I.; Noga, T.; Stelling, A.L.; et al. Adhesive Stalks of Diatom *Didymosphenia geminata* as a Novel Biological Adsorbent for Hazardous Metals Removal. *CLEAN-Soil Air Water* **2017**, *45*, 1600678. [CrossRef]
5. Ladrera, R.; Gomà, J.; Prat, N. Effects of *Didymosphenia geminata* Massive Growth on Stream Communities: Smaller Organisms and Simplified Food Web Structure. *PLoS ONE* **2018**, *13*, e0193545. [CrossRef]
6. Ejaz, H.; Somanader, E.; Dave, U.; Ehrlich, H.; Rahman, M.A. Didymo and Its Polysaccharide Stalks: Beneficial to the Environment or Not? *Polysaccharides* **2021**, *2*, 69–79. [CrossRef]
7. Elwell, L.C.; Gillis, C.-A.; Kunza, L.A.; Modley, M.D. Management Challenges of *Didymosphenia geminata*. *Diatom Res.* **2014**, *29*, 303–305. [CrossRef]
8. Taylor, B.W.; Bothwell, M.L. The Origin of Invasive Microorganisms Matters for Science, Policy, and Management: The Case of *Didymosphenia geminata*. *Bioscience* **2014**, *64*, 531–538. [CrossRef]
9. Wustmann, M.; Poulsen, N.; Kröger, N.; van Pée, K.-H. Chitin Synthase Localization in the Diatom Thalassiosira Pseudonana. *BMC Mater.* **2020**, *2*, 10. [CrossRef]
10. Brunner, E.; Richthammer, P.; Ehrlich, H.; Paasch, S.; Simon, P.; Ueberlein, S.; van Pée, K.-H. Chitin-Based Organic Networks: An Integral Part of Cell Wall Biosilica in the Diatom Thalassiosira Pseudonana. *Angew. Chem. Int. Ed.* **2009**, *48*, 9724–9727. [CrossRef]
11. Zgłobicka, I.; Li, Q.; Gluch, J.; Płocińska, M.; Noga, T.; Dobosz, R.; Szoszkiewicz, R.; Witkowski, A.; Zschech, E.; Kurzydłowski, K.J. Visualization of the Internal Structure of *Didymosphenia geminata* Frustules Using Nano X-ray Tomography. *Sci. Rep.* **2017**, *7*, 9086. [CrossRef]
12. Brand, C.; Grech, M. Recent Invasion of *Didymosphenia geminata* (Lyngbye) M. Schmidt in a Patagonian Regulated River Promotes Changes in Composition and Density of Macroinvertebrate Community. *Biol. Invasions* **2020**, *22*, 1903–1915. [CrossRef]
13. Jellyman, P.G.; Harding, J.S. Disentangling the Stream Community Impacts of *Didymosphenia geminata*: How Are Higher Trophic Levels Affected? *Biol. Invasions* **2016**, *18*, 3419–3435. [CrossRef]
14. Jones, L.R.; Manrique, J.M.; Uyua, N.M.; Whitton, B.A. Genetic Analysis of the Invasive Alga *Didymosphenia geminata* in Southern Argentina: Evidence of a Pleistocene Origin of Local Lineages. *Sci. Rep.* **2019**, *9*, 18706. [CrossRef] [PubMed]
15. Bothwell, M.L.; Taylor, B.W.; Kilroy, C. The Didymo Story: The Role of Low Dissolved Phosphorus in the Formation of *Didymosphenia geminata* Blooms. *Diatom Res.* **2014**, *29*, 229–236. [CrossRef]
16. Kilroy, C.; Bothwell, M. Environmental Control of Stalk Length in the Bloom-Forming, Freshwater Benthic Diatom *Didymosphenia geminata* (Bacillariophyceae). *J. Phycol.* **2011**, *47*, 981–989. [CrossRef]
17. Reinoso-Guerra, E.; Aristizabal, J.; Arce, B.; Zurob, E.; Dennett, G.; Fuentes, R.; Suescún, A.V.; Cárdenas, L.; da Cunha, T.H.R.; Cabezas, R.; et al. Nanostructured *Didymosphenia geminata*-Based Membrane for Efficient Lead Adsorption from Aqueous Solution. *J. Environ. Chem. Eng.* **2021**, *9*, 105269. [CrossRef]
18. Wahab, I.F.; Razak, S.I. Polysaccharides as Composite Biomaterials. In *Composites from Renewable and Sustainable Materials*; InteahOpen: London, UK, 2016; pp. 65–84. [CrossRef]
19. Figueroa, F.A.; Abdala-Díaz, R.; Hernández, V.; Pedreros, P.; Aranda, M.; Cabrera-Pardo, J.R.; Pérez, C.; Becerra, J.; Urrutia, R. Invasive Diatom *Didymosphenia geminata* as a Source of Polysaccharides with Antioxidant and Immunomodulatory Effects on Macrophage Cell Lines. *J. Appl. Phycol.* **2020**, *32*, 93–102. [CrossRef]
20. Williams, D.F. On the Nature of Biomaterials. *Biomaterials* **2009**, *30*, 5897–5909. [CrossRef] [PubMed]
21. Bothwell, M.L.; Spaulding, S.A. Synopsis of the 2007 International Workshop on *Didymosphenia geminate*. In *Proceedings of the 2007 International Workshop on Didymosphenia geminate*; Fisheries and Oceans Canada: Ottawa, ON, Canada, 2008; 96p.
22. Zgłobicka, I. Aspects of Structural Biology of *Didymosphenia geminata* (Lyngb.) M. Schmidt (*Bacillariophyta*). *IJA* **2013**, *15*, 291–310. [CrossRef]
23. Zglobicka, I. Frustules of *Didymosphenia geminata* as a Modifier of Resins. *Mater. Eng.* **2018**, *1*, 10–16. [CrossRef]
24. Arora, M.; Arora, E. The Promise of Silicon: Bone Regeneration and Increased Bone Density. *J. Arthrosc. Jt. Surg.* **2017**, *4*, 103–105. [CrossRef]
25. Zgłobicka, I. Exploratory Study of the Use of *Didymosphenia geminata* Stalks as a Functional Biomaterial. Ph.D. Thesis, Division of Materials Design, The Warsaw University of Technology, Warszawa, Poland, 2015.

26. Wang, N.; Cheng, X.; Li, N.; Wang, H.; Chen, H. Nanocarriers and their loading strategies. *Adv. Healthc. Mater.* **2019**, *8*, 1801002. [CrossRef] [PubMed]
27. Przekora, A. The Summary of the Most Important Cell-Biomaterial Interactions That Need to Be Considered during in Vitro Biocompatibility Testing of Bone Scaffolds for Tissue Engineering Applications. *Mater. Sci. Eng. C* **2019**, *97*, 1036–1051. [CrossRef]
28. Andryukov, B.G.; Besednova, N.N.; Kuznetsova, T.A.; Zaporozhets, T.S.; Ermakova, S.P.; Zvyagintseva, T.N.; Chingizova, E.A.; Gazha, A.K.; Smolina, T.P. Sulfated Polysaccharides from Marine Algae as a Basis of Modern Biotechnologies for Creating Wound Dressings: Current Achievements and Future Prospects. *Biomedicines* **2020**, *8*, 301. [CrossRef]
29. Kuznetsova, T.A.; Andryukov, B.G.; Besednova, N.N.; Zaporozhets, T.S.; Kalinin, A.V. Marine Algae Polysaccharides as Basis for Wound Dressings, Drug Delivery, and Tissue Engineering: A Review. *J. Mar. Sci. Eng.* **2020**, *8*, 481. [CrossRef]
30. Medarević, Đ.; Losić, D.; Ibrić, S. Diatoms—Nature materials with great potential for bioapplications. *Hem. Ind.* **2016**, *70*, 613–627. [CrossRef]
31. Marreco, P.R.; Moreira, P.D.L.; Genari, S.C.; Moraes, Â.M. Effects of different sterilization methods on the morphology, mechanical properties, and cytotoxicity of chitosan membranes used as wound dressings. *J. Biomed. Mater. Res.* **2004**, *71*, 268–277. [CrossRef]
32. Ebi, K.L.; Hess, J.J. Health Risks Due to Climate Change: Inequity in Causes and Consequences. *Health Aff.* **2020**, *39*, 2056–2062. [CrossRef]
33. Ehrlich, H.; Motylenko, M.; Sundareshwar, P.V.; Ereskovsky, A.; Zgłobicka, I.; Noga, T.; Płociński, T.; Tsurkan, M.V.; Wyroba, E.; Suski, S.; et al. Multiphase Biomineralization: Enigmatic Invasive Siliceous Diatoms Produce Crystalline Calcite. *Adv. Funct. Mater.* **2016**, *26*, 2503–2510. [CrossRef]
34. Na, Y.; Lee, J.; Lee, S.H.; Kumar, P.; Kim, J.H.; Patel, R. Removal of Heavy Metals by Polysaccharide: A Review. *Polym.-Plast. Technol. Mater.* **2020**, *59*, 1770–1790. [CrossRef]
35. Zeraatkar, A.K.; Ahmadzadeh, H.; Talebi, A.F.; Moheimani, N.R.; McHenry, M.P. Potential Use of Algae for Heavy Metal Bioremediation, a Critical Review. *J. Environ. Manag.* **2016**, *181*, 817–831. [CrossRef]
36. Ubando, A.T.; Africa, A.D.M.; Maniquiz-Redillas, M.C.; Culaba, A.B.; Chen, W.-H.; Chang, J.-S. Microalgal Biosorption of Heavy Metals: A Comprehensive Bibliometric Review. *J. Hazard. Mater.* **2021**, *402*, 123431. [CrossRef] [PubMed]
37. Arbabi, M.; Hemati, S.; Amiri, M. Removal of Lead Ions from Industrial Wastewater: A Review of Removal Methods. *Int. J. Epidemiol. Res.* **2015**, *2*, 105–109.
38. Molazadeh, P.; Khanjani, N.; Rahimi, M.; Nasiri, A. Adsorption of lead by microalgae *Chaetoceros* sp. and *Chlorella* sp. from aqueous solution. *J. Community Health Res.* **2015**, *4*, 114–127.
39. Das, D.; Chakraborty, S.; Bhattacharjee, C.; Chowdhury, R. Biosorption of Lead Ions (Pb^{2+}) from Simulated Wastewater Using Residual Biomass of Microalgae. *Desalination Water Treat.* **2016**, *57*, 4576–4586. [CrossRef]
40. Suganya, S.; Saravanan, A.; Senthil Kumar, P.; Yashwanthraj, M.; Sundar Rajan, P.; Kayalvizhi, K. Sequestration of Pb (II) and Ni (II) Ions from Aqueous Solution Using Microalga *Rhizoclonium hookeri*: Adsorption Thermodynamics, Kinetics, and Equilibrium Studies. *J. Water Reuse Desalination* **2016**, *7*, 214–227. [CrossRef]
41. Gupta, V.K.; Rastogi, A. Biosorption of Lead from Aqueous Solutions by Green Algae Spirogyra Species: Kinetics and Equilibrium Studies. *J. Hazard. Mater.* **2008**, *152*, 407–414. [CrossRef] [PubMed]
42. Deng, L.; Su, Y.; Su, H.; Wang, X.; Zhu, X. Sorption and Desorption of Lead (II) from Wastewater by Green Algae *Cladophora fascicularis*. *J. Hazard. Mater.* **2007**, *143*, 220–225. [CrossRef]
43. Romera, E.; González, F.; Ballester, A.; Blázquez, M.L.; Muñoz, J.A. Comparative Study of Biosorption of Heavy Metals Using Different Types of Algae. *Bioresour. Technol.* **2007**, *98*, 3344–3353. [CrossRef]
44. Rincón, J.; González, F.; Ballester, A.; Blázquez, M.; Muñoz, J. Biosorption of Heavy Metals by Chemically Activated Alga *Fucus vesiculosus*. *J. Chem. Technol. Biotechnol.* **2005**, *80*, 1403–1407. [CrossRef]
45. Pavithra, K.G.; Kumar, P.S.; Jaikumar, V.; Vardhan, K.H.; SundarRajan, P. Microalgae for Biofuel Production and Removal of Heavy Metals: A Review. *Environ. Chem. Lett.* **2020**, *18*, 1905–1923. [CrossRef]
46. Zgłobicka, I.; Chlanda, A.; Woźniak, M.; Łojkowski, M.; Szoszkiewicz, R.; Mazurkiewicz-Pawlicka, M.; Święszkowski, W.; Wyroba, E.; Kurzydłowski, K.J. Microstructure and Nanomechanical Properties of Single Stalks from Diatom *Didymosphenia geminata* and Their Change Due to Adsorption of Selected Metal Ions. *J. Phycol.* **2017**, *53*, 880–888. [CrossRef] [PubMed]
47. Davis, M.A.; Chew, M.K.; Hobbs, R.J.; Lugo, A.E.; Ewel, J.J.; Vermeij, G.J.; Brown, J.H.; Rosenzweig, M.L.; Gardener, M.R.; Carroll, S.P.; et al. Do not Judge Species on Their Origins. *Nature* **2011**, *474*, 153–154. [CrossRef] [PubMed]
48. Ministère du Développement Durable, de L'environnement et des Parcs; Ministère des Ressources Naturelles et de la Faune. *What Is Didymo and How Can We Prevent It from Spreading in Our Rivers?* Ministère du Développement Durable, de l'Environnement et des Parcs: Rouyn-Noranda, QC, Canada, 2008; 13p.
49. Ministry of Primary Industries. Check, Clean, Dry: Preventing Didymo and Other Pests. Available online: https://www.mpi.govt.nz/outdoor-activities/boating-and-watersports-tips-to-prevent-spread-of-pests/check-clean-dry/ (accessed on 27 October 2021).
50. Kilroy, C.; Bothwell, M.L. *Didymosphenia geminata* Growth Rates and Bloom Formation in Relation to Ambient Dissolved Phosphorus Concentration. *Freshw. Biol.* **2012**, *57*, 641–653. [CrossRef]
51. Credit Valley Conservation. CVC Priority Aquatic Invasive Species & Fish Diseases. 2009. Available online: https://cvc.ca/wp-content/uploads/2011/02/CVC_AquaticInvasives_FishDiseases1.pdf (accessed on 15 August 2021).

52. Government of New Brunswick Canada. *Didymosphenia geminate*. Available online: https://www2.gnb.ca/content/gnb/en/departments/erd/natural_resources/content/fish/content/Didymo.html (accessed on 24 October 2021).
53. Government of Saskatchewan. Didymo Rock Snot Fact Sheet. Available online: http://www.environment.gov.sk.ca/Default.aspx?DN=e1c161fd-0a5c-4804-8296-0709df01300d (accessed on 27 October 2021).
54. Rosaen, A.L.; Grover, E.A.; Spencer, C.W. *The Costs of Aquatic Invasive Species to Great Lakes States*; Anderson Economic Group: East Lansing, MI, USA, 2016; 51p.
55. Beville, S.T.; Kerr, G.N.; Hughey, K.F.D. Valuing Impacts of the Invasive Alga *Didymosphenia geminata* on Recreational Angling. *Ecol. Econ.* **2012**, *82*, 1–10. [CrossRef]
56. Bergey, E.A.; Cooper, J.T.; Phillips, B.C. Substrate Characteristics Affect Colonization by the Bloom-Forming Diatom *Didymosphenia geminata*. *Aquat. Ecol.* **2010**, *44*, 33–40. [CrossRef]
57. Government of Canada, Fisheries and Oceans Canada. A Canadian Action Plan to Address the Threat of Aquatic Invasive Species. Available online: https://www.dfo-mpo.gc.ca/species-especes/publications/ais-eae/plan/page01-eng.html (accessed on 24 October 2021).

polysaccharides

MDPI

Article

Rheological Characterization of the Influence of Pomegranate Peel Extract Addition and Concentration in Chitosan and Gelatin Coatings

Mirella Romanelli Vicente Bertolo, Rafael Leme, Virginia da Conceição Amaro Martins, Ana Maria de Guzzi Plepis and Stanislau Bogusz Junior *

Sao Carlos Institute of Chemistry (IQSC), University of Sao Paulo (USP), Avenida do Trabalhador Sao Carlense, 400-Parque Arnold Schimidt, Sao Carlos 13566-590, SP, Brazil; mirella.bertolo@usp.br (M.R.V.B.); rafael2.leme@usp.br (R.L.); virginia@iqsc.usp.br (V.d.C.A.M.); amplepis@iqsc.usp.br (A.M.d.G.P.)
* Correspondence: stanislau@iqsc.usp.br

Abstract: In this study, the effects of an agro-industrial residue with active properties, pomegranate peel extract (PPE), were evaluated on the rheological properties of potential coatings based on chitosan (C) and gelatin (G). For this, rheological properties of the polymeric solutions were investigated in relation to PPE concentration (2 or 4 mg PPE g^{-1} solution), and to its incorporation order into the system (in C or in CG mixture). All solutions were more viscous than elastic ($G'' > G'$), and the change in PPE concentration had a greater influence accentuating the viscous character of the samples in which PPE was added to the CG mixture (CGPPE2 and CGPPE4). PPE addition to the CG mixture increased the angular frequency at the moduli crossover, indicating the formation of a more resistant polymeric network. This tendency was also observed in flow results, in which PPE addition decreased the pseudoplastic behavior of the solutions, due to a greater cross-linking between the polymers and the phenolic compounds. In general, all the studied solutions showed viscosities suitable for the proposed application, and it was possible to state the importance of standardizing the addition order of the components during the preparation of a coating.

Keywords: chitosan; gelatin; coatings; pomegranate peel extract; rheology

Citation: Romanelli Vicente Bertolo, M.; Leme, R.; da Conceição Amaro Martins, V.; de Guzzi Plepis, A.M.; Bogusz Junior, S. Rheological Characterization of the Influence of Pomegranate Peel Extract Addition and Concentration in Chitosan and Gelatin Coatings. *Polysaccharides* **2021**, 2, 648–660. https://doi.org/10.3390/polysaccharides2030039

Academic Editor: Azizur Rahman

Received: 1 July 2021
Accepted: 2 August 2021
Published: 7 August 2021

1. Introduction

The search for edible food coatings that are biodegradable, non-toxic, and that have active properties has been increasing in recent years, aiming to extend the shelf life of foods while preserving their sensory characteristics and improving consumer safety [1,2]. In the literature, there is an increasing number of studies that evaluate the most diverse polymers, natural or synthetic, and their physical-chemical, structural, and active properties in foods. In these studies, polysaccharides stand out as one of the groups of polymers that are most applied for coating purposes, due to their attractive properties of brightness, transparency, flexibility, barrier to the passage of water and gases, antimicrobial activity, among others [3].

Chitosan is a polysaccharide derived from the partial deacetylation of the chitin found in the exoskeleton of crustaceans and insects, as well as in the structure of mollusks and fungi [4]. Its cationicity, biodegradability, non-toxicity, and its filmogenic and antimicrobial properties make chitosan one of the most studied polysaccharides for the development of edible coatings [5,6]. To improve the mechanical and barrier properties of chitosan-based coatings, the association of this polysaccharide with other natural polymers, such as gelatin, is commonly evaluated [7]. Obtained by the partial hydrolysis of collagen, this protein is well-used in the food industry as a thickening and emulsifying agent, and several studies already successfully applied a combination of chitosan and gelatin for edible coating development [8–11].

Furthermore, studies on food coatings aim to develop materials with active properties, which can also protect the food to be coated from oxidation by light, or which improve/provide antimicrobial activity of coatings. In this context, polyphenols, such as vegetable extracts (grape seed extract, jaboticaba peel extract, etc.), catechins and epicatechins, gallic acid, spent coffee grounds, among others, have been extensively used as active compounds [1,2,10,12]. Pomegranate peel extract (PPE), in addition to being a compound rich in polyphenols, is also considered an agro-industrial residue, and its application in edible coatings can add value to this waste from the food and cosmetics industry [9].

One of the areas of study in the development of coatings that has been explored more recently is rheology; the rheological behavior of the candidate solutions for food coatings can affect several properties that will be vital for the desired application, such as the uniformity of the solution, its thickness, the mechanical properties of the films formed by it, among others [2,3]. Understanding the behavior of polymeric chains in the developed coating, their degree of entanglement, their flow and viscoelastic properties, and how they are affected by the addition and concentration of the active agents mentioned above, is a necessary step in the study of coatings. Regarding chitosan, gelatin, and pomegranate peel extract materials, Bertolo et al. (2020) [9] carried out a rheological study of how the concentration of the extract can interfere in the rheological properties of a solution similar to this study. They utilized a chitosan of different molar mass and degree of acetylation, solubilized in acetic acid, and mixed with gelatin in another proportion (2:1); however, in order to scale-up the coating process, a deeper study of the rheological properties of this system is necessary, to ensure that all the needs of processes involving the flow of a fluid (extraction, spray, filtration, purification, immersion) are met by the developed material.

Thus, the aim of this study is to complete a complete rheological assessment of viscoelastic and flow parameters of materials, based on chitosan, gelatin, and pomegranate peel extract. For this, we sought to evaluate how the order of addition of the extract to the polymeric system could affect its rheological properties, an important advance in process design, in order to standardize the steps in the coating production [13,14]. Furthermore, it was evaluated how the extract concentration can differently affect the rheological properties of the materials, depending on whether it was initially added to chitosan, or whether it was added to the mixture of chitosan and gelatin already formed. The results showed that PPE phenolics act by decreasing the pseudoplastic behavior of the coatings, since they mediate stronger interactions between chitosan and gelatin, making them more resistant to the applied shear. It was also analyzed that the order of addition of the extract has an influence on the coatings' rheological properties, leading to a possible saturation of the active polymeric sites when first added to chitosan.

2. Experimental

2.1. Materials

All solvents and reagents used were PA grade and used as such. Chitosan was obtained from the partial deacetylation of β-chitin, found in the squid pens of *Doryteuthis* spp., obtained from Miami Comércio e Exportação de Pescados Ltd.a in Cananéia city–Sao Paulo State, Brazil. Gelatin was commercial (Sigma-Aldrich®, St. Louis, MO, USA), type A, swine, with approximately 300 bloom. The red pomegranates (*Punica granatum* L.) used (Peruvian variety) were purchased at the Supply Centers of Campinas-S.A (CEASA, Campinas city-SP, Brazil) in February, 2019. Before extraction, the fruits were peeled, and their peels were sanitized with 2% (*v/v*) sodium hypochlorite, dried at room temperature under air flow for 2 h, frozen in the freezer and lyophilized for 16 h (Edwards equipment, model Freeze Dryer Modulyo, Burgess Hill, West Sussex, United Kingdom). Finally, they were crushed to obtain a thin powder, which was stored in polypropylene flasks under refrigeration and protected from light until use. The time between extraction and experiments was not longer than one month.

2.2. Methods

2.2.1. Chitosan Obtention and Characterization

Chitosan was obtained by the partial deacetylation process, to which the β-chitin extracted from squid pens was submitted, a procedure described by Horn et al. (2009), which involves the conversion of the acetoamide groups from the (1,4)-*N*-acetyl-D-glucos-2-amine chains of β-chitin to amino groups, in a basic medium (40% NaOH (*v/v*), constant stirring for 3 h at 80 °C, under N_2 flow) [15]. The obtained powder was characterized by its viscosimetric molar mass, being classified as chitosan of high molar mass (300 kDa); its degree of acetylation of 11.1% was determined by nuclear magnetic resonance spectroscopy (^{1}H NMR) [16,17].

2.2.2. Pomegranate Peel Extraction and Characterization

The pomegranate peel extract (PPE) was obtained according to the procedure described by Bertolo et al. (2020) [9]. The powder obtained after crushing the dried peels was added to a 60% (*v/v*) hydroethanolic solution, in the proportion of 1 g of pomegranate peel powder to 30 g of hydroethanolic solution. The extraction took place at 50 °C, for 1 h, under slow and constant stirring. Then, the solution was filtered through a filter paper (Whatman, n. 1), and the ethanol was removed slowly by evaporation. Finally, the extract was lyophilized to obtain a dry powder.

PPE was characterized in terms of its total phenolic content. For this, the Folin–Ciocalteu colorimetric method was adopted, with a procedure adapted for a 96-well microplate [18]. Gallic acid (Sigma-Aldrich®, St. Louis, MO, USA) was used as the standard molecule for the construction of the method's calibration curve, from eight aqueous solutions with concentrations of 4, 8, 12, 16, 20, 24, 28, and 32 μg mL^{-1}. In the microplate wells, 50 μL of PPE solution (0.1 mg mL^{-1}) and 50 μL of Folin's reagent were placed. After stirring and 5 min of reaction, 200 μL of a 20% sodium carbonate solution (*w/w*) was added to each well, followed by further stirring. All samples were taken in triplicate. After 15 min, the absorbance of the samples was read at 725 nm on a Thermo Scientific ™ UV-Vis spectrophotometer VL0L00D0 (Waltham, MA, USA). The 60% (*v/v*) hydroethanolic solution was used as the blank extract.

2.2.3. Coating Preparation

The 1% (*w*/*w*) initial solutions of chitosan (C) and gelatin (G) were prepared as follows: chitosan was dissolved in a 1% lactic acid solution (*w/w*), by slow and constant stirring for 24 h; gelatin was dissolved in water at 60 °C for 30 min, and its gelation was carried out under refrigeration for 2 h. The CG control sample (that is, without PPE) was prepared by mixing the polymers in the proportion of 1:1 at 45 °C for 2 h, with the addition of 1 mL of a 50% hydroethanolic solution (*v/v*), in order to maintain the concentration in relation to the other samples to be prepared.

Then, the incorporation of PPE into chitosan/gelatin solutions took place. For this, 2 different concentrations of a hydroethanolic solution (50%, *v/v*) of extract, 100 and 200 mg mL^{-1}, were prepared. To incorporate these solutions, two additional orders were tested:

(1) Addition of 1 mL of PPE to C, followed by the mixture of G (CPPE2G and CPPE4G samples);

(2) Mixture of C and G, followed by the addition of 1 mL of PPE (CGPPE2 and CGPPE4 samples).

The PPE addition order to the polymeric system was changed to evaluate how the phenolic hydroxyls of the PPE components would interact with the protonated amino groups (NH_3^+) of chitosan, in the presence (CG mixture) or in the absence (C solution) of gelatin. This protein has carboxylic groups of its amino acids responsible for the formation of the polymeric network by electrostatic interaction with chitosan (please see the GA of our previous work [7]). Thus, different orders of PPE addition to the system can lead to different rheological responses, due to the greater or lesser degree of interaction between the polymers, as well as between the polymers and the extract. Regardless of the PPE order of addition, the proportion adopted was that of 1 mL of extract for each 50 g of mixture of chitosan and gelatin. Thus, the final concentrations of PPE in the polymeric mixture were 2 mg PPE g^{-1} mixture (for CPPE2G and CGPPE2 samples) and 4 mg PPE g^{-1} mixture (for CPPE4G and CGPPE4 samples).

All the prepared solutions (the controls C, G, and CG, as well as those containing pomegranate peel extract, PPE) presented a homogeneous aspect, without the presence of precipitates, and with varied yellow coloration, according to the extract concentration and its order of addition to the system.

2.2.4. Rheological Measurements

The rheological study was carried out with the samples CPPE2G, CPPE4G, CGPPE2, CGPPE4, CG, and with the 1% chitosan solution (C). An AR-1000 N controlled stress rheometer (TA Instruments, New Castle, DE, USA) was used, with a cone/plate geometry of stainless-steel of 20 mm in diameter, at 2° angle, and a fixed gap of 69 µm. The temperature was controlled with a Peltier system, and all the measurements described below were performed in triplicate. For all measurements (strain, frequency, temperature sweep, and steady shear), the prepared solutions (in the final state of a gel) were stored under refrigeration in the dark until analysis. After stabilizing the temperature of the solutions to room temperature, they were placed on the Peltier plate of the rheometer, and their excess was removed after setting the zero gap of the equipment.

2.2.5. Strain Sweep Measurements

Fixed values of temperature (25 °C) and frequency (1.0 Hz) were adopted in an amplitude range of 0.05 to 500 Pa, to determine the linear viscoelastic region (LVR) of the coating solutions, that is, the strain range in which their elastic (G′) and viscous (G″) moduli do not change. The software Rheology Advantage Data Analysis, version V5.7.0 (TA Instruments Ltd., New Castle, DE, USA) was employed to analyze the parameters obtained from strain sweep measurements, which were: G'_{LVR}, the elastic modulus at the end of LVR; γ_L, the maximum strain value to which the solution can be submitted before their moduli start to decrease; tanδ, the ratio between G″ and G′; and G″-G′, the difference between the moduli at 10% of strain.

2.2.6. Frequency Sweep Measurements

For frequency sweep measurements, a frequency range from 0.1 to 100 rad s^{-1} was adopted, at fixed values of temperature (25 °C) and strain (10% in the LVR). The behavior of G′ and G″, according to the increase in frequency, was analyzed in terms of $G'_{crossover}$ and $\omega_{crossover}$, that is, G′ and frequency values where G′ = G″.

2.2.7. Temperature Sweep Measurements

G′ and G″ behavior was also evaluated, according to temperature sweep tests, performed from 25 to 75 °C, at 5 °C min ^{1}, 1.0 Hz of frequency, and 10% of strain.

2.2.8. Steady Shear Measurements

Finally, flow measurements were conducted at 25 °C, with a range and shear rate of 0.1 to 1000 s^{-1}. The viscosity behavior of the solutions was characterized as Newtonian or pseudoplastic, according to PPE addition concentration. The experimental curves of viscosity were fitted according to the Cross model (Equation (1).

$$\frac{\eta - \eta_\infty}{\eta_0 - \eta_\infty} = \frac{1}{(1 + (k\gamma)^n)} \quad (1)$$

In Equation (1), η_0 is the zero-shear viscosity (Pa s), η_∞ is the viscosity limit at infinite shear (Pa s), γ is the shear rate (s^{-1}), k is the so-called consistency index (s), and n the rate index (dimensionless).

3. Statistical Analysis

The Software Action (Estatcamp Team, 2014,São Carlos–SP, Brazil) was used for the statistical analysis of the data. The Shapiro–Wilk test was applied to verify data distribution. Parametric rheological results were examined using analysis of variance (ANOVA), followed by Tukey's test. Non-parametric rheological results were examined using Kruskal–Wallis test. Significance level was set at equal or higher than 5% in all cases.

4. Results and Discussion

4.1. Pomegranate Peel Extraction and Characterization

After the extraction process to which the pomegranate peels were subjected, and the lyophilization of the extract, a thin, dry, and yellowish powder was obtained, with a yield of 54.5%. Its total phenolic content (TPC), determined by the Folin–Ciocalteu method, was 213 $\pm$ 6 mg gallic acid equivalent (GAE) g^{-1} extract. This result is consistent with that found by Derakshan et al. (2018), who analyzed the TPC of extracts from the peel and the seed of three different types of pomegranates, finding a range of 276–413 mg EAG g^{-1} for the peel extracts [19]. Bertolo et al. (2020), in a previous study, determined 492 $\pm$ 82 mg GAE g^{-1} extract, using a yellow pomegranate of Brazilian variety [9]. It is worth mentioning that the variations observed in the TPC of different phenolic extracts are related to the most diverse factors: from intrinsic ones, such as the origin of the fruit and its harvest time, to extrinsic factors, such as the method of extraction adopted, the time of extraction, and the solvents used.

4.2. Rheological Measurements

Strain Sweep Measurements

Initially, the sweep tests of the elastic (G') and viscous (G'') moduli of the material were performed as a function of the percentage of deformation, essential for determining the linear viscoelastic region of the solutions (LVR) where loss and storage moduli (G'' and G', respectively) are practically constant, regardless of the applied deformation. [14] The extent of the LVR can be directly related to the structural strength of the material: solutions that are more resistant to the applied deformation tend to have a more extensive LVR, requiring greater values of deformation for the moduli to cease stability and start to decrease [7,14].

As can be seen in the graphs of Figure 1, divided into A and B according to the order of addition of PPE in the polymeric solutions, all samples showed $G'' > G'$, which indicates that their viscous behavior exceeded the elastic behavior, regardless of the addition of the extract or its concentration, a typical liquid-like behavior [7,9,10,20]. From the curves in Figure 1, it was possible to determine the parameters presented in Table 1: the first, γ_L, represents the highest strain value to which the solution could be subjected, before leaving the LVR; the higher this value, the greater the resistance offered by the sample to the applied deformation. It can be said that the addition of gelatin to sample C led to a significant decrease in γ_L, from 48.87 $\pm$ 3.24% in C to 40.78 $\pm$ 0.32% in CG, an indication that the incorporation of gelatin and its interaction with the polymer chains of chitosan can make the polymeric system less resistant to the applied deformation.

However, the addition of PPE in CGPPE2 and CGPPE4 solutions led to an increase in their critical deformation, to values significantly equal to the original chitosan solution, not influenced by the concentration of the extract. Similar results were found by Bertolo et al. (2020) when analyzing chitosan and gelatin systems incorporated with grape seed extracts. The addition of gelatin led to a decay of the critical deformation of chitosan, and the concentration of the added phenolics changed this parameter significantly. Lower concentrations promoted greater stability for the solutions [7]. Finally, for the samples in which PPE was initially added to C, the incorporation of the extract did not significantly alter γ_L, and the subsequent addition of gelatin did not promote the effect observed in the other samples, by decreasing the critical strain.

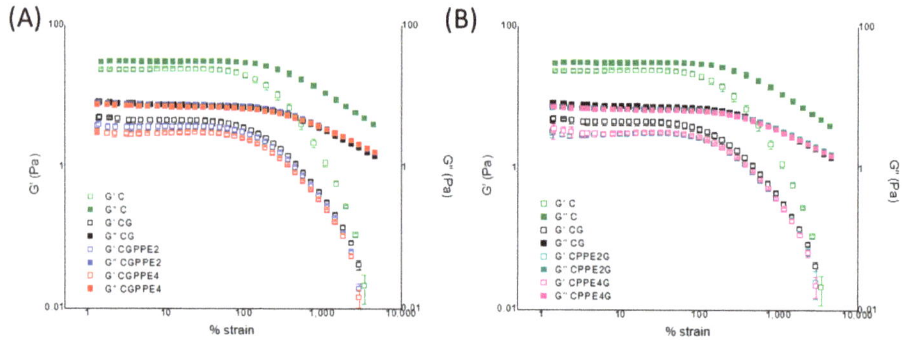

Figure 1. G″ and G′ moduli as a function of % strain, at 1 Hz frequency and 25 °C, for: (**A**) C, CG, CGPPE2, and CGPPE4; and (**B**) C, CG, CPPE2G, and CPPE4G.

Table 1. Parameters determined by strain sweep measurements: critical strain (γ_L), G′ at the LVR limit (G'_{LVR}), loss tangent value (tanδ) and G″–G′ moduli at 10% strain. Measurements were performed at 1 Hz and 25 °C.

Sample	γ_L (%)	G'_{LVR} (Pa)	tanδ	G″–G′ (Pa)
C	48.87 ± 3.24 [a]	22.80 ± 1.49 [a]	1.32 ± 0.02 [e]	6.51 ± 0.27 [a]
CG	40.78 ± 0.32 [b]	4.27 ± 0.45 [b]	1.68 ± 0.06 [d]	2.94 ± 0.06 [d]
CGPPE2	48.93 ± 0.30 [a]	3.56 ± 0.28 [b]	2.05 ± 0.11 [c]	3.74 ± 0.07 [b, c]
CGPPE4	48.26 ± 0.40 [a]	2.92 ± 0.16 [b]	2.33 ± 0.06 [a]	4.04 ± 0.14 [b]
CPPE2 G	49.01 ± 0.47 [a]	2.97 ± 0.12 [b]	2.23 ± 0.01 [a, b]	3.77 ± 0.18 [b, c]
CPPE4 G	49.76 ± 0.33 [a]	3.04 ± 0.11 [b]	2.12 ± 0.02 [b, c]	3.57 ± 0.10 [c]

Different lowercase letters on the same column indicate significant differences by the ANOVA and Tukey test with $p \leq 0.05$.

Even though γ_L did not undergo significant changes with the incorporation of PPE into the system, the second parameter in Table 1, G'_{LVR}, showed more pronounced variations: the value of the elastic modulus of the solutions at the limit of LVR suffered an abrupt decrease from 22.80 ± 1.49 Pa to 4.27 ± 0.45 Pa from C to CG. The decay trend continued with the incorporation of PPE into the system, regardless of its concentration or order of addition. The third parameter in Table 1, tanδ, is the ratio between the viscous and elastic moduli at the LVR limit, and allows particularly important rheological classifications of the material: if tanδ > 1, G″ > G′ and the samples are classified as viscous; the opposite is also valid, and if tanδ < 1, the elastic character predominates. However, if tanδ > 0.1, the behavior of the samples is situated between that of a highly concentrated polymeric solution and that of a real gel [20]. In the samples of this study, tanδ varied between 1.32 ± 0.02 (C) and 2.33 ± 0.06 (CGPPE4), which classifies them as being highly concentrated polymeric solutions, with a predominant viscous character, as had already been observed in the curves of Figure 1.

The incorporation of gelatin in CG increased tanδ value from 1.32 ± 0.02 to 1.68 ± 0.06, reflecting the decrease observed in G'_{LVR} for that same sample; the incorporation of PPE into CG mixture led to values greater than two for tanδ, indicating that the incorporation of the extract accentuated the viscous character of the solutions; when doubling the PPE concentration from CGPPE2 to CGPPE4, tanδ increased significantly from 2.05 ± 0.11 to 2.33 ± 0.06. The same effect of the incorporation of PPE is valid for the samples CPPE2G and CPPE4G, with tanδ values also greater than two, but without significant changes observed when the concentration of the extract was doubled. Despite the proportion between the polymers prepared by Bertolo et al. (2020) being different (2:1), they obtained similar tanδ results for their solutions, ranging from 1.25 ± 0.16 for the sample without

PPE, to 2.59 ± 0.12 for the sample with one of the highest concentrations of PPE [9]. Their results agree with the results of this study, reinforcing the tendency of phenolic compounds to increase the viscous character of the solutions.

Finally, to confirm the effects of the incorporation and concentration of PPE in the polymeric system, the difference between G″ and G′ at 10% of deformation was calculated (G″–G′ parameter from Table 1); this difference indicates whether the moduli are approaching or moving away, according to the incorporation of the phenolics, and is a new indication of which character is more pronounced. The incorporation of PPE into the CG mixture led to a significant increase in G″–G′ difference from 2.94 ± 0.06 Pa to 3.74 ± 0.07 Pa in CGPPE2, and doubling the concentration in CGPPE4, the difference between the moduli increased to 4.04 ± 0.14 Pa; this result reinforces the effect that had already been observed, that is, that the PPE addition accentuates the viscous character of the solutions, making its difference with the elastic modulus even greater. However, when the order of PPE addition was reversed, this tendency to increase the G″–G′ difference with twice the extract concentration was no longer observed. This result also meets the other parameters analyzed for CPPE2G and CPPE4G samples, by not showing significant differences between them, and by not following the trends observed for CGPPE2 and CGPPE4. Furthermore, these samples indicate that the addition of the extract in high concentrations to chitosan, for later mixing with gelatin, may be saturating the binding sites and making the formation of the polymeric network more difficult. Thus, even if the PPE concentration in the system is doubled, the changes observed are not as significant as those obtained in the other order of addition (Table 1).

4.3. Frequency Sweep Measurements

The behavior of G′ and G″ moduli was also evaluated in relation to the angular frequency (Figure 2A,B). In all cases, G″ was predominant over G′ during a wide sweeping range, occurring in the crossing of the moduli and their inversion in variable angular frequency values, according to the PPE concentration and addition. This fact reinforces the classification of the samples of this study as being concentrated polymeric solutions/weak gels, since G″ > G′ at the beginning and the crossover occurred within the frequency range adopted; real gels or diluted polymeric solutions would not have the same profile [12,14]. The values of $G'_{crossover}$ and $\omega_{crossover}$, that is, the elastic modulus and the angular frequency at G″ = G′, are reported in Table 2.

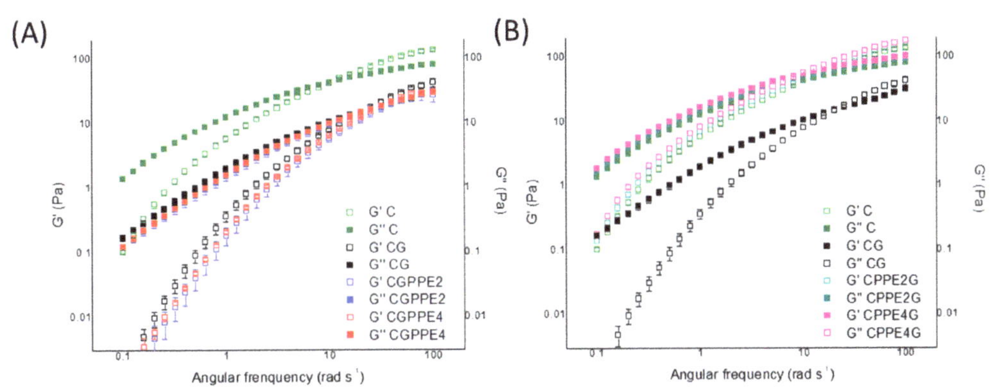

Figure 2. Frequency dependency of G′ and G″ moduli at 10% of strain and 25 °C for: (**A**) C, CG, CGPPE2, and CGPPE4; and (**B**) C, CG, CPPE2G, and CPPE4G.

Table 2. Parameters determined by frequency sweep measurements: angular frequency at the crossover point ($\omega_{crossover}$) and G′ modulus at the crossover point ($G'_{crossover}$). Measurements performed at γ = 10% and 25 °C.

Solution	$\omega_{crossover}$ (rad s^{-1})	$G'_{crossover}$ (Pa)
C	9.99 ± 0.01 [d]	41.54 ± 2.97 [a]
CG	19.98 ± 0.01 [c, d]	14.65 ± 0.78 [d]
CGPPE2	25.13 ± 0.01 [b, c]	18.81 ± 1.33 [c]
CGPPE4	37.04 ± 4.64 [a]	22.01 ± 1.51 [a, b]
CPPE2 G	37.04 ± 4.64 [a]	19.63 ± 0.75 [b, c]
CPPE4G	31.69 ± 0.01 [a, b]	17.85 ± 0.26 [c, d]

Different lowercase letters in the same column indicate significant differences by the ANOVA and Tukey test with $p \leq 0.05$.

Chitosan solution (C) was the sample with the lowest $\omega_{crossover}$ value (9.99 ± 0.01 rad s^{-1}), and the addition of gelatin promoted an increase in that value to 19.98 ± 0.01 rad s^{-1} in CG, reflecting the effect of gelatin in delaying the inversion between moduli, probably due to the higher number of interactions formed between the polymers in the mixture. The addition of PPE in the CG mixture further increased $\omega_{crossover}$ values, reaching 37.04 ± 4.64 rad s^{-1} in CGPPE4. It is worth mentioning that half of the PPE concentration, in the inverse order of addition (CPPE2G), presented the same crossing frequency value as CGPPE4. Again, the trend observed for CPPE2G and CPPE4G was different from the other samples: in these two cases, twice the extract concentration caused a decline in the angular frequency value to 31.69 ± 0.01 rad s^{-1} in CPPE4G, which can be another indication that the addition of PPE to chitosan for subsequent mixing of gelatin may make the polymeric network more susceptible to changes.

In relation to the $G'_{crossover}$ values, there was a sharp decrease from 41.54 ± 2.97 Pa from C to 14.65 ± 0.78 Pa in CG; the concentration of PPE did not promote significant changes in this parameter, but the samples in which the extract was added to the CG mixture showed, in general, $G'_{crossover}$ values slightly higher than the others. Regarding these frequency results, Sun et al. (2020) also evaluated the behavior of G′ and G″ as a function of angular frequency for solutions based on glucomannan/carboxymethyl chitosan incorporated with epigallocatechin gallate (EGCG), one of the major phenolic compounds in tea leaves. They concluded that the addition of EGCG in high concentrations increased the values of the G″ and G′ moduli at high frequencies, implying the formation of a close non-covalent entangled network between the polymers and the phenolic [2].

Hosseini et al. (2021) reached similar findings when analyzing the dynamic behavior of film forming solutions (FFSs) based on chitosan, polyvinyl alcohol, and fish gelatin, incorporated with cinnamaldehyde. While the FFSs without the phenolic compound showed a liquid-like behavior (G″ > G′) at low angular frequencies, the elastic behavior (solid-like) was predominant at higher frequency values; however, once cinnamaldehyde was added to the system, the high number of interactions between the oil droplets and the polymers led to an elastic behavior that was 100%prevalent throughout the entire frequency range adopted, given the formation of an elastic solid-state [3].

4.4. Temperature Sweep Measurements

In rheological studies involving materials with possible applications such as food coatings, temperature is one of the main factors influencing the viscoelastic behavior that should be studied. In general, it is necessary to predict the behavior of the material (whether it will be more viscous or more elastic, and in what proportion, for example) over a wide range of temperatures, which simulate possible situations of transport, storage, and even cooking of the food to be coated. The graphs in Figure 3A,B represent the variation of G′ and G″ moduli for the samples of this study.

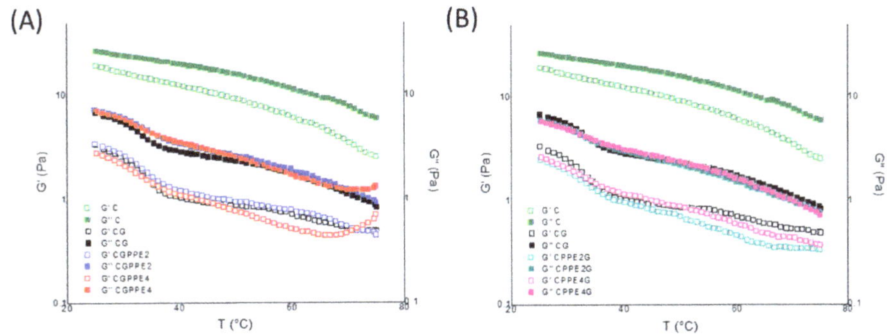

Figure 3. G′ and G″ moduli as a function of temperature (from 25 to 75 °C) for: (**A**) C, CG, CGPPE2, and CGPPE4; and (**B**) C, CG, CPPE2G, and CPPE4G.

As observed in the previous tests, G″ > G′, confirming once again the predominant viscous character of the samples. With the increase in temperature, both moduli of all samples started to decrease, a typical polymeric behavior. For the chitosan solution (C), it was observed that G″ modulus decreased in a greater proportion than G′, going from 26.58 Pa at 25 °C to 5.93 Pa at 75 °C (more than 20 Pa of difference); for G′, this difference between the final and the initial modulus values was not greater than 17 Pa (values obtained from the curves in Figure 3).

With the addition of gelatin in CG, although the initial values of both moduli were significantly lower compared to C, the observed trend did not change; the viscous modulus continued to decrease with greater intensity as the temperature rose, reaching values lower than 1 Pa at the end of the test for almost all samples, except for CGPPE4. In this case, from 65 °C onwards, a slight increase was observed for both moduli, indicating a tendency to cross at temperatures higher than 75 °C.

This tendency was not observed for any other sample, regardless of the order of addition or concentration of PPE. This may be related to a greater ease of inversion for elastic behavior at high temperatures, due to the elimination of energized water molecules and the consequent association between polymeric molecules [7,13]. Apparently, the addition of PPE in high concentrations to the polymeric network already formed in CGPPE4 may have facilitated the elimination of water molecules at high temperatures, leading to a rearrangement of chains with a greater increasing tendency of the moduli, which could result in an inversion between them at even higher temperatures. Luciano et al. (2021) analyzed the behavior of the viscous and elastic moduli of polymeric solutions based on gelatin with different concentrations of nisin, an antibacterial peptide, as a function of temperature. Their film-forming solutions showed G′ > G″, and the addition of nisin was able to increase the sol-gel and gel-sol transition temperatures by 4 °C, which occurred around 14–18 °C and 25–29 °C, respectively [21]. In our study, similar transitions were not observed for the solutions containing gelatin, probably due to the high concentration of chitosan and the formation of an intricate polymeric network between the polymers.

In general, we can conclude the discussion of oscillatory rheology results by saying that both factors of concentration and order of addition of PPE contributed to the differences observed in the results of deformation, frequency, and temperature. The viscous and elastic moduli of the samples can be influenced to a greater or lesser extent, according to the highest or lowest concentration of extract, as well as if it was added to the polymeric network already formed or to the chitosan solution first. For CGPPE4 sample, the effects of the highest concentration of PPE were noticeably clear, leading to a sample with a higher viscous character, with inversion between the moduli in higher angular frequency values, and with a possible crossing between them in temperature values lower than the others.

4.5. Steady Shear Measurements

Figure 4A,B shows the viscosity curves of the samples as a function of the shear rate; in all cases, a decrease in viscosity was observed according to the increase in shear rate, a typical polymeric behavior known as shear-thinning or pseudoplastic. This behavior is associated with the fact that the polymeric chains orient themselves towards the applied stress, increasing their ordering and, consequently, decreasing viscosity [22].

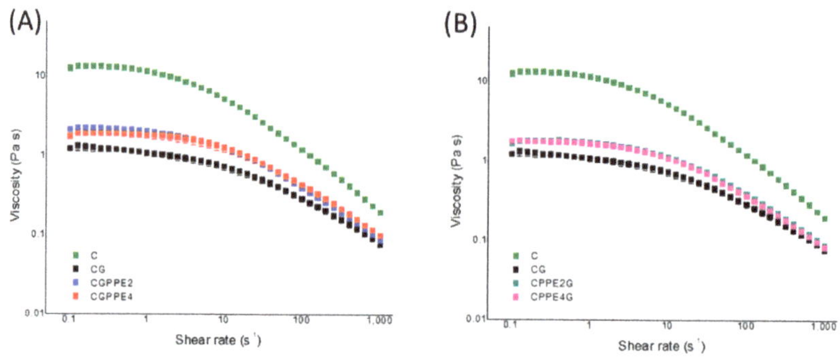

Figure 4. Viscosity curves as a function of the shear rate (0.1 to 1000 s^{-1}) at 25 °C for: (**A**) C, CG, CGPPE2, and CGPPE4; and (**B**) C, CG, CPPE2G, and CPPE4G.

Knowing the viscosity of the polymeric solution that will act as a coating is extremely important to predict its behavior. In general, we look for samples with intermediate viscosities, which escape from the extremes of high viscosity (which would be an obstacle to coat food by immersion or spray, for example) and low viscosity (which would make it difficult to spread the solution over the food surface, due to its dripping effect). The literature points to viscosities in the range of 1–10 Pa s as being suitable for experimental coating processing conditions [23].

As can be seen in Table 3, the adjustment of the viscosity data with the Cross model allowed for the identification of some important parameters. The first one is η_0, the initial viscosity of the solutions at zero shear rate, when the polymeric molecules are still randomly oriented, except for sample C, which had the highest initial viscosity value ($\eta_0 = 14.90 \pm 1.27$ Pa·s). All samples containing gelatin and/or PPE showed viscosities between 1.92 and 2.85 Pa s, an indication that all of them are suitable for the proposed application. Although no significant differences were observed ($p < 0.05$), according to the addition of PPE or its concentration, some trends can be elucidated. The addition of PPE to the polymeric network already formed in CG tended to decrease its viscosity, probably due to the interaction of the extract in the active sites that had not yet been occupied by the polymers. The increase in its concentration in CGPPE4 followed the same decreasing tendency, with a viscosity of 2.05 ± 0.18 Pa s. In the samples in which PPE was initially added to chitosan, η_0 values were slightly lower than 2 Pa s, and CPPE4G showed the lowest viscosity value among all samples, of 1.92 ± 0.14 Pa s.

The second parameter presented in Table 3 is k, the so-called consistency index: the lower k, the greater the Newtonian plateau of the sample before its pseudoplastic behavior begins, that is, the greater the value of the critical shear rate necessary for the sample viscosity starts to decrease [14]. Sample C was the one with the highest k value (0.212 ± 0.020 s), and the addition of gelatin in CG decreased this value to 0.119 ± 0.004 s. This indicated that the formation of the polymeric network tended to better stabilize the polymeric chains, making them most resistant to applied shear and slowing the decline in viscosity. The k index continued to decrease with the addition of PPE and the increase in its concentration, with CGPPE4 sample having the lowest k (0.056 ± 0.001 s) and, therefore, being the most resistant to shear. For CPPE2G and CPPE4G samples, twice the

PPE concentration did not significantly change k, and the values remained between 0.066 and 0.069 s.

Table 3. Parameters obtained by modeling the viscosity curves as a function of the shear rate by the Cross model: zero-shear viscosity (η_0), consistency index (k), and rate index (n). Measurements performed from 0.1 to 1000 s^{-1} at 25 °C.

Solution	η_0 (Pa s)	k (s)	n
C	14.90 ± 1.27^{a}	0.212 ± 0.020^{a}	0.800 ± 0.002^{a}
CG	2.85 ± 0.14^{b}	0.119 ± 0.004^{b}	0.720 ± 0.006^{c}
CGPPE2	2.37 ± 0.01^{b}	0.086 ± 0.004^{c}	$0.744 \pm 0.011^{b,c}$
CGPPE4	2.05 ± 0.18^{b}	0.056 ± 0.001^{d}	$0.771 \pm 0.031^{a,b}$
CPPE2G	1.97 ± 0.11^{b}	$0.066 \pm 0.002^{c,d}$	$0.744 \pm 0.006^{b,c}$
CPPE4G	1.92 ± 0.14^{b}	$0.069 \pm 0.003^{c,d}$	$0.759 \pm 0.009^{b,c}$

Different lowercase letters in the same column indicate significant differences by the ANOVA and Tukey test with $p \leq 0.05$.

Finally, the last parameter, n, the rate index, represents the dependence of viscosity in relation to the shear rate: samples with values of n between zero and one are considered pseudoplastic, while values of n greater than one characterize Newtonian samples [24]. For the samples in this study, n varied from 0.744 (CGPPE2 and CPPE2G) to 0.800 (C), which reinforces the pseudoplastic behavior of all of them, and indicates that the addition of PPE promotes a slight decline in it. The results found here for k and n parameters agree with that observed by Tudorache & Tordenave (2019). They analyzed the pseudoplastic behavior of different polysaccharides (β-glucan, xanthan gum, and guar gum) complexed with phenolic compounds (vanillin, ferulic acid, acid caffeine, among others); in all cases, the addition of phenolics also caused a decline in the pseudoplastic behavior of the samples. At a molecular level, these results can be interpreted in view of the weak associations existing between the polymeric chains, which flow easily upon shearing, presenting pseudoplastic behavior. Once the phenolic compounds are added, they mediate a stronger cross-linking between the polymeric molecules, making them more resistant to the applied shear, which explains the accentuation of the Newtonian behavior for samples containing PPE [25]. Rodrigues et al. (2020) also evaluated the rate index of solutions based on chitosan and gelatin incorporated with extracts of grape seed and jaboticaba peel; in their case, it was possible to observe a decline in the values of n (and a consequent decrease in the pseudoplastic behavior) only for the highest concentrations of extract (5 mg g^{-1} mixture) when added to the mixture of chitosan and gelatin already formed. In this sense, PPE was able to promote more marked changes in the behavior of the polymeric system, even in lower concentrations [10].

Thus, it can be said that the rheological results of flow agree with the oscillatory results previously discussed for the samples of chitosan and gelatin with pomegranate peel extract. It was observed that the order of addition of PPE to the polymeric matrix is a parameter to be standardized during the formulation of new coatings, since it is able to change their rheological characteristics. The concentration of PPE was again a factor of great influence, especially for the samples in which the extract was added to the polymeric network already formed by CG.

5. Conclusions

PPE incorporation into chitosan and gelatin coatings affected both oscillatory and flow properties. The effects of PPE concentration were more accentuated when it was added to the CG mixture, indicating that the other addition order evaluated may have led to a possible saturation of chitosan binding sites, making it difficult for this polysaccharide to form a polymeric network with the subsequent incorporation of gelatin. The viscous character of the solutions was accentuated with higher concentrations of PPE. CGPPE4 solution showed the highest tanδ value, as well as higher angular frequency and G′ values at the moduli crossover, indicating the formation of a more intricate polymeric network. Temperature also affected the viscoelastic behavior of the samples, with CGPPE4 being the only solution with a different profile of increasing the moduli at temperatures above 65 °C. Finally, the flow tests evidenced the influence of PPE in decreasing the pseudoplastic behavior of the polymeric solutions, making them more resistant to the applied shear. The results of this study contribute to the in-depth understanding of the molecular relationships between a polysaccharide, a protein, and phenolic compounds in the prepared coating solutions, and how these relationships can be related to the intended application for the materials.

Author Contributions: M.R.V.B.: conceptualization, methodology, data curation, writing (original draft manuscript). R.L.: methodology, data curation. V.d.C.A.M.: conceptualization, writing (review and editing), supervision. A.M.d.G.P.: writing (review and editing), supervision, project administration. S.B.J.: writing (review and editing), supervision, project administration. All authors have read and agreed to the published version of the manuscript.

Funding: This work was supported by the São Paulo Research Foundation (FAPESP) (grant number 2019/18748-8 for MRVB).

Institutional Review Board Statement: Not applicable.

Informed Consent Statement: Not applicable.

Data Availability Statement: Data sharing not applicable.

Acknowledgments: The authors would like to thank Júlio César Borges for the use of the Thermo Scientific ™ UV-Vis Spectrophotometer VL0L00D0, belonging to the Protein Biochemistry and Biophysics Group at IQSC/USP, under his responsibility.

Conflicts of Interest: The authors declare they have no competing interests.

References

1. Lei, Y.; Wu, H.; Jiao, C.; Jiang, Y.; Liu, R.; Xiao, D.; Lu, J.; Zhang, Z.; Shen, G.; Li, S. Investigation of the Structural and Physical Properties, Antioxidant and Antimicrobial Activity of Pectin-Konjac Glucomannan Composite Edible Films Incorporated With Tea Polyphenol. *Food Hydrocoll.* **2019**, *94*, 128–135. [CrossRef]
2. Sun, J.; Jiang, H.; Li, M.; Lu, Y.; Du, Y.; Tong, C.; Pang, J.; Wu, C. Preparation and Characterization of Multifunctional Konjac Glucoman-nan/Carboxymethyl Chitosan Biocomposite Films Incorporated With Epigallocatechin Gallate. *Food Hydrocoll.* **2020**, *105*, 105756. [CrossRef]
3. Hosseini, S.F.; Ghaderi, J.; Gómez-Guillén, M.C. Trans-Cinnamaldehyde-Doped Quadripartite Biopolymeric Films: Rheological Behavior of Film-Forming Solutions and Biofunctional Performance of Films. *Food Hydrocoll.* **2021**, *112*, 106339. [CrossRef]
4. Nair, M.S.; Saxena, A.; Kaur, C. Effect of Chitosan and Alginate Based Coatings Enriched With Pomegranate Peel Extract to Extend the Postharvest Quality of Guava (*Psidium Guajava* L.). *Food Chem.* **2018**, *240*, 245–252. [CrossRef]
5. Ngo, D.-H.; Vo, T.-S.; Kang, K.-H.; Je, J.-Y.; Pham, H.N.-D.; Byun, H.-G.; Kim, S.-K. Biological Effects of Chitosan and Its Derivatives. *Food Hydrocoll.* **2015**, *51*, 200–216. [CrossRef]
6. Bonilla, J.; Sobral, P.J. Investigation of the Physicochemical, Antimicrobial and Antioxidant Properties of Gelatin-Chitosan Edible Film Mixed With Plant Ethanolic Extracts. *Food Biosci.* **2016**, *16*, 17–25. [CrossRef]
7. Bertolo, M.R.V.; Martins, V.D.C.A.; Plepis, A.M.D.G.; Junior, S.B. Rheological Study of the Incorporation of Grape Seed Extract in Chitosan and Gelatin Coatings. *J. Appl. Polym. Sci.* **2021**, *138*, 50052. [CrossRef]
8. Nowzari, F.; Shábanpour, B.; Ojagh, S.M. Comparison of chitosan–gelatin Composite and Bilayer Coating and Film Effect on the Quality of Refrigerated Rainbow Trout. *Food Chem.* **2013**, *141*, 1667–1672. [CrossRef] [PubMed]
9. Bertolo, M.R.; Martins, V.; Horn, M.M.; Brenelli, L.B.; Plepis, A. Rheological and Antioxidant Properties of chitosan/Gelatin-Based Materials Functionalized by Pomegranate Peel Extract. *Carbohydr. Polym.* **2020**, *228*, 115386. [CrossRef]
10. Rodrigues, M.; Álison, V.; Bertolo, M.R.V.; Marangon, C.; Martins, V.; Plepis, A.M.D.G. Chitosan and Gelatin Materials Incorporated With Phenolic Extracts of Grape Seed and Jabuticaba Peel: Rheological, Physicochemical, Antioxidant, Antimicrobial and Barrier Properties. *Int. J. Biol. Macromol.* **2020**, *160*, 769–779. [CrossRef] [PubMed]
11. Cardoso, G.P.; Dutra, M.P.; Fontes, P.R.; Ramos, A.D.L.S.; Gomide, L.A.D.M.; Ramos, E.M. Selection of a Chitosan Gelatin-Based Edible Coating for Color Preservation of Beef in Retail Display. *Meat Sci.* **2016**, *114*, 85–94. [CrossRef] [PubMed]
12. Wu, C.H.; Li, Y.Z.; Yu, D.; Lin, W.; Cailing, T.; Hu, Y.; Pang, J.; Yan, Z. Preparation and Characterization of Konjac Glucomannan-Based Bionanocompo-Site Film for Active Food Packaging. *Food Hydrocoll.* **2019**, *89*, 682–690. [CrossRef]

13. Silva-Weiss, A.; Bifani, V.; Ihl, M.; Sobral, P.; Gómez-Guillén, M.C. Structural Properties of Films and Rheology of Film-Forming Solutions Based on Chitosan and Chitosan-Starch Blend Enriched With Murta Leaf Extract. *Food Hydrocoll.* **2013**, *31*, 458–466. [CrossRef]
14. Steffe, J.F. *Rheological Methods in Food Process Engineering*; Freeman Press: East Lansing, MI, USA, 1996.
15. Horn, M.M.; Martins, V.; Plepis, A.M.D.G. Interaction of Anionic Collagen With Chitosan: Effect on Thermal and Morphological Characteristics. *Carbohydr. Polym.* **2009**, *77*, 239–243. [CrossRef]
16. Rinaudo, M. Chitin and Chitosan: Properties and Applications. *Prog. Polym. Sci.* **2006**, *31*, 603–632. [CrossRef]
17. Lavertu, M.; Xia, Z.; Serreqi, A.; Berrada, M.; Rodrigues, A.G.; Wang, D.; Buschmann, M.; Gupta, A. A Validated 1H NMR Method for the Determination of the Degree of Deacetylation of Chitosan. *J. Pharm. Biomed. Anal.* **2003**, *32*, 1149–1158. [CrossRef]
18. Singleton, V.L.; Orthofer, R.; Lamuela-Raventos, R.M. [14] Analysis of Total Phenols and Other Oxidation Substrates and Antioxidants by Means of Folin-Ciocalteu Reagent. *Methods Enzymol.* **1999**, *299*, 152–178. [CrossRef]
19. Derakhshan, Z.; Ferrante, M.; Tadi, M.; Ansari, F.; Heydari, A.; Hosseini, M.S.; Conti, G.O.; Sadrabad, E.K. Antioxidant Activity and Total Phenolic Content of Ethanolic Extract of Pomegranate Peels, Juice and Seeds. *Food Chem. Toxicol.* **2018**, *114*, 108–111. [CrossRef]
20. Razmkhah, S.; Mohammad, S.; Razavi, A. Dilute Solution, Flow Behavior, Thixotropy and Viscoelastic Characterization of Cress Seed (*Lepidium Sativum*) Gum Fractions. *Food Hydrocoll.* **2017**, *63*, 404–413. [CrossRef]
21. Luciano, C.G.; Tessaro, L.; Lourenço, R.V.; Barbosa Bittante, A.M.Q.; Fernandes, A.M.; Freitas Moraes, I.C.; do Amaral Sobral, P.J. Effects of Nisin Concentration on Properties of Gelatin Film Forming so-Lutions and Their Films. *Int. J. Food Sci. Technol.* **2021**, *56*, 587–599. [CrossRef]
22. Mandala, I.; Savvas, T.; Kostaropoulos, A. Xanthan and Locust Bean Gum Influence on the Rheology and Structure of a White Model-Sauce. *J. Food Eng.* **2004**, *64*, 335–342. [CrossRef]
23. Nair, S.B.; Jyothi, A.N.; Sajeev, M.S.; Misra, R. Rheological, Mechanical and Moisture Sorption Charac-Teristics of Cassava starch-konjac Glucomannan Blend Films. *Starch* **2011**, *63*, 728. [CrossRef]
24. Tudorache, M.; Bordenave, N. Phenolic Compounds Mediate Aggregation of Water-Soluble Polysaccha-Rides and Change Their Rheological Properties: Effect of Different Phenolic Compounds. *Food Hydrocoll.* **2019**, *97*, 105193. [CrossRef]
25. Phillips, G.O.; Williams, P.A. *Handbook of Hydrocolloids*; Elsevier: Amsterdam, The Netherlands, 2009. [CrossRef]

polysaccharides

Article

Control of Surface Properties of Hyaluronan/Chitosan Multilayered Coatings for Tumor Cell Capture

Giulia G. Lima [1], João B. M. Rocha Neto [1], Hernandes Faustino de Carvalho [2] and Marisa Masumi Beppu [1,*]

[1] School of Chemical Engineering, University of Campinas, Campinas, São Paulo 13083-852, Brazil; giuliagrassalima@gmail.com (G.G.L.); jbmrneto@gmail.com (J.B.M.R.N.)
[2] Institute of Biology, University of Campinas, Campinas, São Paulo 13083-862, Brazil; hern@unicamp.br
* Correspondence: beppu@unicamp.br

Abstract: Prostate cancer (PCa) is a slow-growing neoplasm that has, when diagnosed in its early stages, great chances of cure. During initial tumor development, current diagnostic methods fail to have the desired accuracy, thus, it is necessary to develop or improve current detection methods and prognostic markers for PCa. In this scenario, films composed of hyaluronic acid (HA) and chitosan (CHI) have demonstrated significant capture potential of prostate tumor cells (PC3 line), exploring HA as a CD44 receptor ligand and direct mediator in cell-film adhesion. Here, we present a strategy to control structural and cell adhesion properties of HA/CHI films based on film assembly conditions. Films were built via Layer-by-layer (LbL) deposition, where the pH conditions (3.0 and 5.0) and number of bilayers (3.5, 10.5, and 20.5) were controlled. The characterization of these films was carried out using profilometry, ultraviolet-visible (UV-VIS), atomic force microscopy (AFM) and contact angle measurements. Multilayer HA/CHI films produced at pH 3.0 gave optimum surface wettability and availability of free carboxyl groups. In turn, at pH 5.0, the coverings were thinner and presented a smoother surface. Films prepared with 3.5 bilayers showed greater tumor cell capture regardless of the pH condition, while films containing 10.5 and 20.5 bilayers presented a significant swelling process, which compromised their cell adhesion potential. This study shows that surface chemistry and morphology are critical factors for the development of biomaterials designed for several cell adhesion applications, such as rapid diagnostic, cell signaling, and biosensing mechanisms.

Keywords: layer-by-layer; adsorption; hyaluronic acid; cell adhesion; multilayer films

check for updates

Citation: Lima, G.G.; Rocha Neto, J.B.M.; Carvalho, H.F.d.; Beppu, M.M. Control of Surface Properties of Hyaluronan/Chitosan Multilayered Coatings for Tumor Cell Capture. *Polysaccharides* **2021**, *2*, 387–399. https://doi.org/10.3390/polysaccharides2020025

Academic Editor: Azizur Rahman

Received: 30 April 2021
Accepted: 25 May 2021
Published: 30 May 2021

1. Introduction

Prostate cancer (PCa) is a slow-growing neoplasm that causes second highest mortality among men, though presenting good chances of cure when diagnosed in its early stages [1]. Currently, the diagnosis is performed through prostate specific antigens (PSA) detection and rectal examination, which do not present the desired accuracy and thus result in many false negative results [2]. Therefore, in this context, the development and improvement of rapid detection methods and prognostic markers for prostate cancer are of great importance in order to contribute to more effective treatments [3].

Towards that end, previous studies have shown great potential for multilayer polymeric films composed of hyaluronic acid (HA) and chitosan (CHI) in the selective capture of tumor cells [4,5]. This interaction is driven by the high affinity of the hyaluronate, present in the HA molecule, with CD44 receptors. CD44 is an extracellular glycoprotein that is particularly upregulated in immune cells, such as B lymphocytes, macrophages, and tumor cells [6,7]. When bonded to HA, one of its main ligands [8], CD44 is responsible for a range of cellular processes, such as cell proliferation, migration, and adhesion [9]. Furthermore, current discussions point to a possible correlation between CD44 expression and tumor cell metastasis [10–12], which supports the potential use of HA-based devices as diagnostic tools.

Hyaluronic acid and chitosan are natural polymers that offer a number of advantages for use in biomedical materials, such as biodegradability, biocompatibility, and low toxicity [13]. In this way, the risk of triggering immune responses when interacting with living organisms becomes very low. CHI is a polymer obtained from the deacetylation of chitin [14], which can be found in a wide variety of sources, such as insects [15–17], fungi [18], and crustaceans [19]. Chitosan can vary in terms of molecular weight, degree of deacetylation, and dispersity, which are key properties in determining the molecule's physicochemical characteristics and biological behavior [20]. Moreover, chitosan can be synthesized in a range of derivative types in order to enhance a particular property [21]. For instance, molecular weight and degree of deacetylation, which is determined by D-glucosamine and *N*-acetyl-*D*-glucosamine groups in the polymer backbone, are determinant factors on modulating CHI antibacterial activity [22]. This particular application of chitosan CHI has been extensively investigated due to the electrostatic interaction between the positive charges of the amine groups of CHI and the negative charges in bacterial membranes, thus disrupting bacterial cell walls [23,24]. Furthermore, the use of chitosan as a building block for polyelectrolytic-based materials extends CHI applications to a variety of biomedical purposes, such as the development of drug delivery systems [25,26], biomarkers [27,28], wound dressings [29], tissue engineering [30,31], and as both treatment and prevention of microbial infections [32–35]. In turn, HA is a copolymer that occurs naturally in the human body as one of the main components of the extracellular matrix, being found in larger amounts in the eyes, skin, and connective tissues [36–38]. As well as CHI, hyaluronic acid is widely explored in the medical field, with applications in joint disorder therapies [39,40], implants [41], regenerative medicine [42,43], and drug delivery [44].

Moreover, HA and CHI are polyelectrolytes that have ionizable functional groups, being classified as anionic (negative) and cationic (positive), respectively. Thus, by using oppositely charged polyelectrolytes, it is possible to build multilayer films via Layer-by-layer (LbL) deposition, in such a way that the cohesion between the layers is maintained by electrostatic forces. Films based on HA and CHI have been used in several applications, such as antimicrobial surfaces [45,46], cell adhesion [47], drug delivery [48], biosensing [5], wound dressings [29,49], gene delivery systems [50,51], and tissue engineering [52].

The Layer-by-layer assembly technique, developed by Iler in 1996, consists of alternating immersions of a substrate in solutions of opposite charges [53], so that the deposition of the coatings occurs by adsorption [54]. It is a self-assembly process with no restrictions regarding the type of geometry and substrate [55], which allows the incorporation of a multitude of materials to the multilayer.

In this process, there are several possible formation conditions that can directly interfere on the functionality of the multilayer films, since they can lead to different topography conditions and physicochemical properties of the coatings [56,57]. Such parameters include pH, ionic strength, and concentration of the polyelectrolytic solutions [58–60] as well as the number of bilayers deposited on the substrate [61]. For instance, previous studies have shown that it is possible to obtain tunable topography [62], hydrophilicity [63], drug delivery parameters [64], and cell adhesion conditions [65] by adjustment of polymeric solutions in Layer-by-layer assembly.

Given this context, this project proposes investigating the influence of the pH of the polyelectrolytic solutions and the number of bilayers deposited on the morphology and composition of HA and CHI films via LbL. In addition, the selective capture potential of the coatings in regard to tumor cells of the PC3 strain, as well as the quality (form factor) and kinetics of adhesion, were then explored.

2. Materials and Methods

2.1. Materials

Chitosan (CHI), hyaluronic acid sodium salt (HA, from *Streptococcus equi* sp.), Polyethylenimine (PEI), Alcian blue (AB), and Rose Bengal (RB) were purchased from Sigma-Aldrich, St. Louis, MO, USA. All polyelectrolytes were used without further purification. Sodium hydroxide (NaOH), sodium chloride (NaCl), and hydrochloric acid (HCl) were purchased from Synth (São Paulo, Brazil). Glass slides were used as substrates and purchased from Kasvi (São José dos Pinhais, Brazil). HAM-F12K cell culture media and streptomycin/penicillin (S/P, 5000 U.I./mL) were purchased from Lonza (Basel, Switzerland). Fetal bovine serum (FBS) and Dulbecco's Phosphate Buffered Saline (DPBS), DAPI, and Phalloidin were purchased from Sigma-Aldrich (St. Louis, MO, USA). The PC3 tumor line was obtained from the American Type Culture Collection (ATCC, Manassas, VA, USA).

2.2. *Methods*

2.2.1. Polyelectrolyte Solutions

Chitosan, hyaluronic acid, and polyethyleneimine (PEI) solutions were prepared at a concentration of 1% (w/v) with an ionic strength of 0.1 M NaCl. CHI solution was solubilized in a 0.1 M glacial acetic acid (HAc) solution, while HA and PEI were solubilized in Milli-Q water. The solutions were stirred for 24 h in the presence of NaCl. The solutions were then divided into two portions each and adjusted to pH levels of 3.00 $\pm$ 0.05 or 5.00 $\pm$ 0.05, respectively.

2.2.2. Preparation of Substrates

Glass slides (24mm $\times$ 24 mm) were cleaned by ultrasonication with an aqueous solution of commercial detergent and Milli-Q water for 10 min each, respectively. The substrates were then dried at room temperature. Subsequently, they were treated with O_2 plasma at 100 mTorr for 15 min (720 V DC, 25 mA DC, 18 W; Harrick Plasma Cleaner, PDC-32G) and covered with a PEI precoating [4].

2.2.3. Film Assembly via Layer-by-Layer Deposition

Multilayer films were assembled by alternating immersions of the PEI precoated glass slides in HA and CHI solutions for 10 min each, followed by three rinsing steps with ultrapure water (2 min, 1 min, and 1 min, respectively) between each polyelectrolyte immersion. The LbL procedure was repeated for 3.5, 10.5, and 20.5 cycles for all films, setting HA as the outermost layer.

2.2.4. Cell Adhesion Assays

Prostate tumor cells of the PC3 strain were cultured in HAM F12 K medium containing 10% fetal bovine serum and 1% penicillin/streptomycin. The culture was kept in a humidified incubator at 37 °C and 5% CO_2, with periodic changes of the culture medium.

To highlight the cells adhered to the culture flasks surface, 1 mL of 0.25% (v/v) trypsin was used. In this way, the cells in suspension could be counted using a Neubauer Chamber. Then, the HA/CHI films were placed in culture wells, pipetted with the suspension, and left for 1 h in an oven for the incubation step.

Finally, the wells were rinsed with a DPBS Flush buffer at pH 7.4 in order to remove the nonadhered cells. Micrographs were acquired using the Axio Observer.Z1 Zeiss inverted confocal L510 microscope (Carl Zeiss AG, Oberkochen, Germany).

2.2.5. Physicochemical Characterization

Film thickness was determined with a Dektak 150 stylus profilometer (Veeco, Plainview, NY, USA) at a force of 1.0 mg and scan speed of 17 μm/s. Measurements were taken in quintuplicate. To evaluate the availability of carboxyl groups in HA, the films were stained with the polycationic dye Alcian Blue (AB), whose absorbance reading is performed at a wavelength of 617 nm. The analysis was carried out using an ultraviolet-visible (UV-VIS) spectrophotometer (Schimadzu, Model 1800). The hydrophilic character of the films was assessed through contact angle measurements, which were carried out in a goniometer Krüss Easy DropDSA-150 (Hamburg, Germany) in the static sessile drop mode. Measurements were taken in sextuplicate. Film topography data was obtained with atomic force microscopy (AFM) images, which were taken with a Park NX-10 Atomic Force

Microscope (Suwon, South Korea) in tapping mode at room temperature and a humidity of approximately 5%. The AFM images and roughness data were analyzed using Gwyddion open software.

3. Results and Discussion

3.1. Film Characterization

3.1.1. Profilometry

Film thickness was assessed through profilometry. The results presented in Figure 1 indicate that the increase in pH range led to thinner films. This result is a consequence of the charge density assumed by both HA and CHI in the studied pH range.

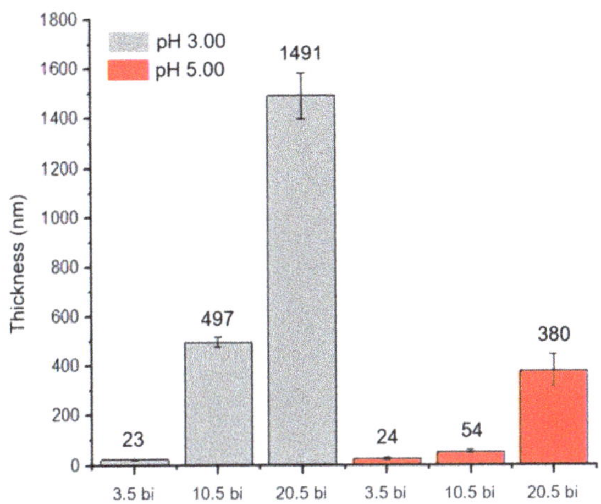

Figure 1. Average thickness of films prepared under different pH conditions and number of bilayers.

At both pH 3.0 and pH 5.0, chitosan (pKa = 6.0) is highly charged, with about 99.90% and 90.91% of its functional groups ionized respectively. In turn, hyaluronic acid (pKa = 2.9) is partially charged at pH 3.0, with 55.73% ionized functional groups, while it is highly charged at pH 5.0, where ionization levels reach 92.64% [37,66].

The charge density of the polymers in solution are determining factors for the conformation of their chains, in a way that the higher the charge density, the more linear the molecule is, due to the phenomenon of charge repulsion [67]. In this scenario, CHI adopts a stiff conformation regardless of the pH condition, while HA chains assume a random coil conformation at pH 3.0 and a stiff conformation at pH 5.0 [67]. Thus, the result of thicker films at pH 3.0 is explained.

In addition, the coiled conformation of HA at pH 3.0 provides a smaller surface area for approximation with other molecules, which means that a smaller portion of the negative charges of HA interact with the positive ones of CHI. For this reason, a greater amount of HA molecules is required in order to compensate charges with CHI [66]. We suggest that a greater number of HA molecules is required in order to compensate charges with CHI for the film assembly [55]. Thus, there is a greater deposition of hyaluronic acid at pH 3.0, which contributes to an increase in the overall film thickness. Similar results were described by Montelongo et al. (2016) on the assessment of HA/CHI films under different pH conditions, where films assembled at pH 3.0 presented the highest thickness measurements among the analyzed samples [46].

3.1.2. UV-Vis

The UV-Vis technique was employed to quantify the free carboxylic groups in the films using the Alcian Blue dye. The AB binds to the free carboxylic groups of hyaluronic acid, that is, those that do not participate in any interaction with the amino groups of chitosan [61]. Therefore, the higher the absorbance value, the more dye was incorporated by the film. AB absorbance measurements are shown in Figure 2 and are relative to both sides of the films.

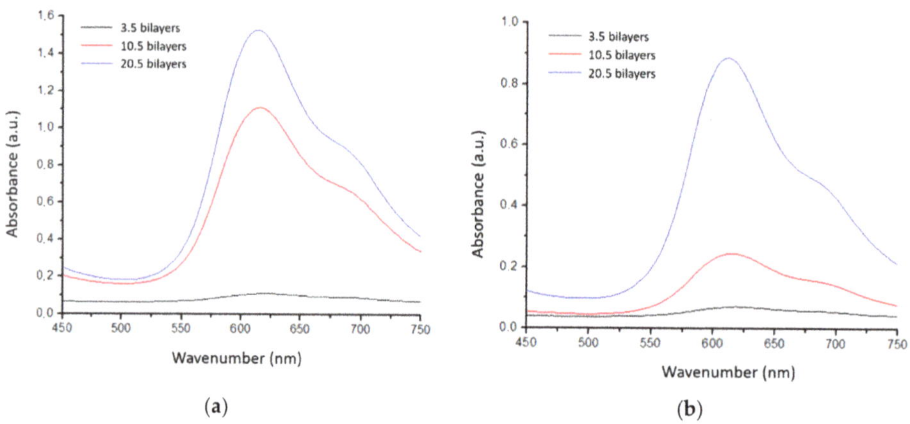

Figure 2. Absorbance of the films for 3.5, 10.5, and 20.5 bilayers under the conditions of (**a**) pH 3.0 and (**b**) pH 5.0.

It is observed that the increase in the number of bilayers, regardless of pH, increases the absorbance of the dye in the films, which is expected since a greater number of deposited cycles promotes a greater adsorption of HA to multilayers. When comparing the films according to the pH conditions, the influence of charge density on the polyelectrolytes is again noted. At pH 3.0, the absorbance peaks are higher than at pH 5.0, which occurs for two reasons. The first is the higher concentration of hyaluronic acid in films prepared at pH 3.0, due to the need of more molecules to compensate charges with chitosan. The second is the fact that there is a greater number of free carboxylic groups to interact with the dye, since the coiled conformation of the HA chains creates a spatial impediment, allowing only a part of these groups to come into contact with the CHI molecules. In contrast, at pH 5.0, the carboxylic groups of HA are committed to the electrostatic interaction responsible for the formation of multilayers, since CHI is also highly charged and available. Nascimento et al. (2018) observed a similar AB incorporation trend regarding the number of bilayers and pH on HA/CHI films [68].

3.1.3. Contact Angle

The hydrophilic character of the films was evaluated through contact angle measurements. With the aid of a goniometer coupled with software, values of the angles formed between drops of water and the surface of the coverings were obtained over time, as shown in Figure 3. The affinity for aqueous environments determines the permeation of the culture medium in the samples so that more hydrophilic surfaces provide a greater area of contact with the cellular environment. In this way, films with greater wettability tend to favor cell contact and subsequent adhesion [69].

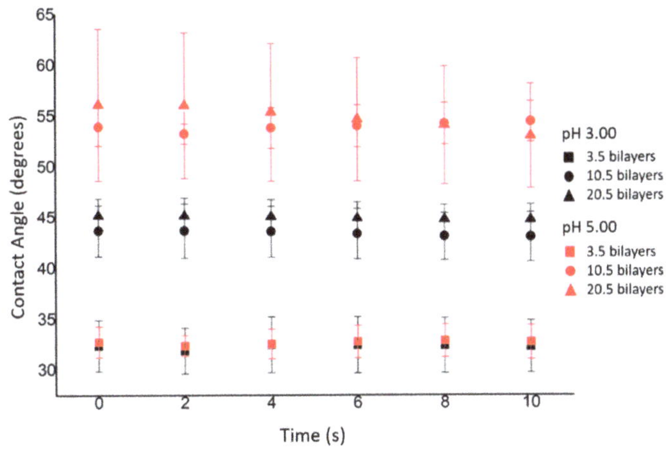

Figure 3. Measurement of contact angle over time for films built under different pH conditions and number of bilayers.

According to the results, there is no statistical difference in contact angle measurements over time, which shows that there is no considerable spread of the drop. Thus, it can be pointed out that the films offer good stability for the drop.

Moreover, a decrease in the contact angle was observed with the decrease of the pH condition from 5.0 to 3.0 for films assembled with 10.5 and 20.5 bilayers. As discussed before, due to the lower density of charges at pH 3.0, there is a higher amount of HA chains in the films prepared in this condition. Hyaluronic acid is one of the most hydrophilic molecules found in nature [37], which corroborates to the greater surface wettability found in HA/CHI films assembled at pH 3,0, since they contain a higher amount of HA.

Regarding the films assembled with 3.5 bilayers, no significant changes were observed in contact angle results with an increase in pH levels. The same effect was found for films with 10.5 and 20.5 bilayers within the same pH condition. Therefore, we suggest that up to 10.5 bilayers, the number of deposited layers is the leading factor for modulating the hydrophilic character of the coverings, whereas for films built with over 10.5 bilayers, the increase in pH levels then becomes the key factor for controlling this property.

3.1.4. Atomic Force Microscopy (AFM)

Through atomic force microscopy analysis, the mean square roughness of the nanometric films was determined, the values of which are shown in Table 1. Figure 4 shows the AFM images of the coatings.

Table 1. Roughness values of nanometric coatings for different pH conditions and number of bilayers.

pH Condition	Number of Bilayers	Roughness (nm)
3.00	3.5	13 ± 3
	10.5	44 ± 12
	20.5	42 ± 6
5.00	3.5	8 ± 1
	10.5	23 ± 4
	20.5	21 ± 6

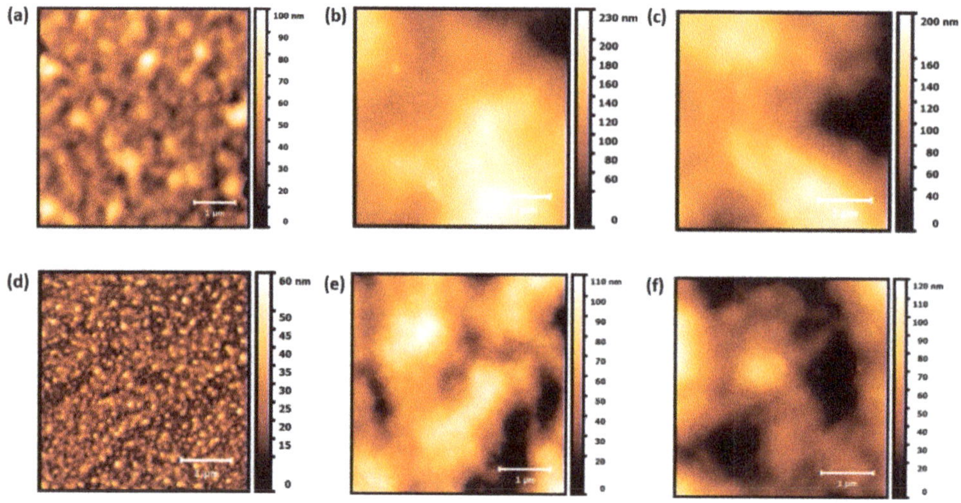

Figure 4. AFM images for films assembled under the conditions of pH 3.0 and (**a**) 3.5 bilayers, (**b**) 10.5 bilayers, and (**c**) 20.5 bilayers, respectively; and of pH 5.0 and (**d**) 3.5 bilayers, (**e**) 10.5 bilayers, and (**f**) 20.5 bilayers, respectively.

AFM results revealed that the increase in pH promoted the formation of more regular films with smoother surfaces. This result corroborates with the discussion on the conformation of the hyaluronic acid molecules within the studied pH range. The adsorption of a more coiled-shaped HA at pH 3.0 results in a rougher surface for films assembled in this pH condition.

On the other hand, the stiff conformation of both HA and CHI at pH 5.0 leads to a smoother surface on films built in this condition. It is also important to emphasize that the outermost layer of the films is composed of HA, which reinforces the role of charge density and assumed conformation of this polyelectrolyte not only in the film topography, but also in all surface properties. The increase in the number of bilayers, in turn, reveals a kind of roughness saturation in the coatings, which have a growth profile based on the construction of polymeric islands [70], which was also described in previous literature reports [48,71].

3.1.5. Capacitance

Through atomic force microscopy analysis, the films were also characterized as to their capacitance. Figure 5 shows the obtained AFM images, while Table 2 contains the capacitance measurement values for some of the coatings produced in this project.

Table 2. Capacitance measurements of the nanometric coatings for different pH conditions and number of bilayers.

pH Condition	Number of Bilayers	Capacitance (mV)
3.00	3.5	119 ± 23
	10.5	122 ± 29
	20.5	150 ± 10
5.00	3.5	155 ± 52
	10.5	152 ± 47
	20.5	146 ± 10

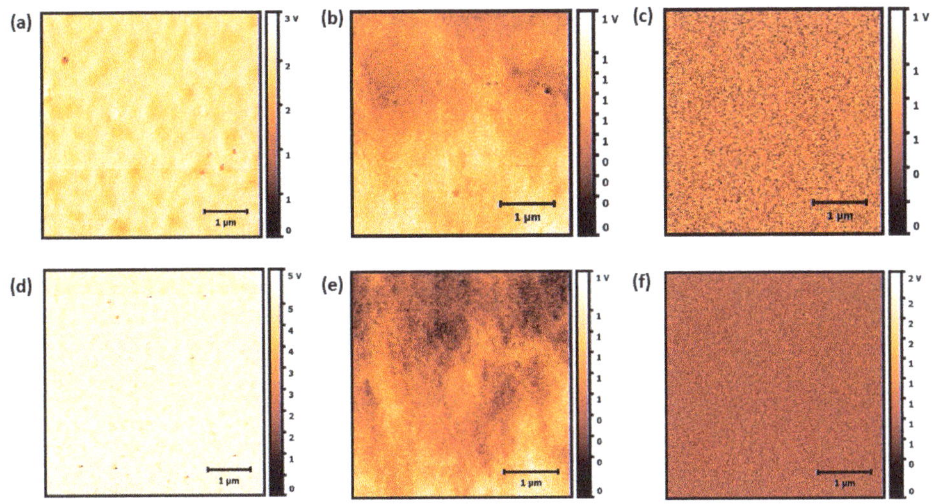

Figure 5. Distribution of loads in coatings prepared under the conditions of pH 3.0 and (**a**) 3.5 bilayers, (**b**) 10.5 bilayers, and (**c**) 20.5 bilayers, respectively; and of pH 5.0 and (**d**) 3.5 bilayers, (**e**) 10.5 bilayers, and (**f**) 20.5 bilayers, respectively.

The capacitance results indicate that the films are electrically similar, though the films prepared at pH 5.0 showed higher average capacitance and larger variability between trials.. Again, this result is a consequence of the greater electrostatic character assumed by the polyelectrolytes in this pH range. Moreover, previous work by our group has shown an association between smoother surfaces and higher charge mobility, which is in accordance with the results presented in this paper [4].

The surface charge of a substrate is known to have a significant influence on the cell adhesion process of several strains [56]. However, this property of HA/CHI coatings still has an exploratory character, with the aim to investigate the correlation between charge mobility on the surface and the selectivity of films in the adhesion of circulating tumor cells. Nevertheless, recent studies on HA/CHI films have pointed out to an inversely proportional relation between charge mobility and the average number of PC3 cells adhered to the films [72].

3.2. Cell Adhesion Assays

Selective Potential of the Multilayer Films

The images of the tumor cells adhered to the films are shown in Figure 6 for different pH conditions and number of bilayers. For the 3.5 bilayers films, it was possible to verify the cellular adhesion in a clear way, with visually similar numbers of adhered cells in both pH conditions. These results corroborate the findings of previous works by Rocha Neto et al. (2020), which revealed an increase in the number of captured PC3 cells on HA/CHI films with a decrease in pH levels, suggesting pH as a key factor for modulating cell adhesion [72]. As for the films with a higher number of bilayers, swelling of the films was observed in several regions, which made quantitative analyses of cell adhesion impossible. This event is evident in Figure 6d, where the upper part of the film reveals a very different structure to the lower part.

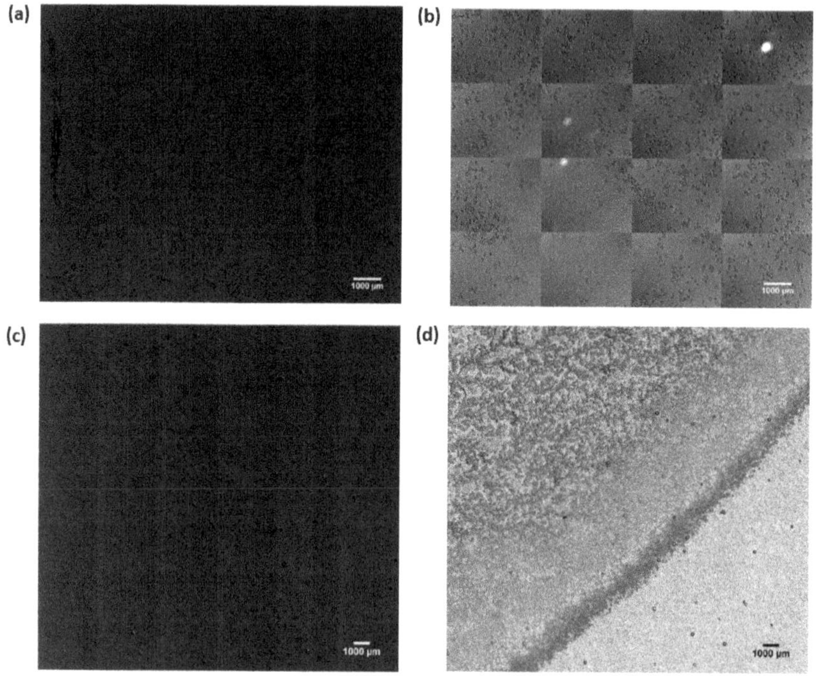

Figure 6. Microscope images of cell adhesion in the conditions of (**a**) pH 3.0 and 3.5 bilayers, (**b**) pH 5.0 and 3.5 bilayers, (**c**) pH 3.0 and 10.5 bilayers, and (**d**) pH 3.0 and 20.5 bilayers.

By adjusting experimental variables such as pH of the polyelectrolytic solutions and number of bilayers, it was possible to promote changes in the physical and chemical properties of HA/CHI multilayer films, changing the topographic profile, capacitance, thickness, availability of free functional groups, and hydrophilicity. For 3.5 bilayers, the films showed similar thicknesses across the entire pH range. Thus, according to the adhesion images obtained, it is suggested that thinner films have a high selective potential for prostate tumor cells.

However, it was not possible to verify the relationship between thicker films and the number of cells adhered due to the swelling of films prepared with more than 10.5 bilayers. One of the hypotheses raised for this issue is the dragging of the film during rinsing steps to remove the nonadherent cells.

Another possibility that can be considered is a change on film structure due to pH variations. The films were assembled at specific pH conditions, where the ionization of the polyelectrolytes is known. However, the cell adhesion assays are performed in a culture medium at pH 7.4. Thus, it is suggested that the change in pH condition during 1 h could have altered the electrostatic forces that maintain the cohesion between layers for thicker films. Kumorek et al. described the effects of medium pH on the disassembly of tannic acid (TA) and chitosan films. It was suggested that under significantly different pH conditions than those of film assembly, a change in the ionization profile of CHI and TA occurs, which compromises the electrostatic interaction between the molecules and, thus, leads to the disintegration of the multilayer films [73]. Moreover, recent investigations carried out by our group on the use of chitosan molecules with different degrees of deacetylation pointed to the control of CHI properties as a promising strategy to promote a higher stability of HA/CHI films under pH levels close to physiological conditions [48]. Considering that, in this study, films developed under the same pH conditions were exposed in the same manner to the cell culture medium at physiological pH, we also suggest that the disintegration of the films can be aggravated by an increase in the number of bilayers, therefore corroborating with the results observed for films of 10.5 and 20.5 bilayers in cell adhesion assays.

The increase in the pH of the polyelectrolytic solutions promoted the formation of films with smoother surfaces. As for the number of bilayers, roughness saturation was observed from 10.5 bilayers on. However, the roughness values have an order of magnitude of nanometers, while the cell size is in the order of micrometers. Thus, it is proposed that the tumor cells are not sensitive to the differences in roughness expressed by the films.

The capacitance of the coatings varied from 0.090 V to 0.163 V. The assembly conditions explored did not promote large electrical differences between the films, disfavoring any comprehension of this property on the cell adhesion mechanism.

Regarding the hydrophilicity of nanometric films, it was found that the pH 3.0 condition led to more hydrophilic films. However, the contact angle tests were performed over 10 s, where the spread of water droplets on the surfaces of the coverings was monitored. In this period, very similar values were observed in the angles obtained for the 3.5-layer films in the studied pH range. In contrast, the cell adhesion tests lasted for 1 h, which is assumed to be enough time for the culture medium to spread the same way in both 3.5-layer films, regardless of the pH condition.

4. Conclusions

The HA/CHI coatings developed via LbL exhibited significant selective potential in capturing the prostatic tumor line PC3, exploring the CD44-HA interaction. In addition to the presence of HA to promote cell adhesion into the films, we suggest that film thickness plays an important role in this mechanism. Thicker films tend to be unstable for applications that require contact with physiological environments. Therefore, our findings suggest that thinner films are more suitable to induce tumor cell capture. Moreover, it was found that HA/CHI films with 3.5 bilayers at pH 3.0 provided the optimum condition for cell adhesion in this study.

Since cell adhesion on substrates is a surface phenomenon, this study suggests that the topography and availability of functional groups act as key factors for the development of biomaterials suitable for this application. By understanding the PC3 lineage as a tumor cell model, there is a potential application of HA/CHI coatings as platforms for the selective capture of tumor lineages, as the CD44-HA interaction is susceptible to applications in rapid diagnostic devices, cascades of cell signaling, and biosensor mechanisms.

Author Contributions: Conceptualization, G.G.L., J.B.M.R.N., and M.M.B.; methodology, G.G.L. and J.B.M.R.N.; software, J.B.M.R.N.; validation, G.G.L., H.F.d.C., and J.B.M.R.N.; formal analysis, G.G.L. and J.B.M.R.N.; investigation, G.G.L. and J.B.M.R.N.; resources, H.F.d.C. and M.M.B.; data curation, G.G.L. and J.B.M.R.N.; writing—original draft preparation, G.G.L. and J.B.M.R.N.; writing—review and editing, all authors; visualization, all authors; supervision, M.M.B. and H.F.d.C.; project administration, M.M.B.; funding acquisition, M.M.B. All authors have read and agreed to the published version of the manuscript.

Funding: This research was funded by São Paulo Research Foundation (FAPESP, grant No. 2018/20560-4) and National Council for Scientific and Technological Development (CNPq, grant No.147536/2016-2).

Institutional Review Board Statement: Not applicable.

Informed Consent Statement: Not applicable.

Acknowledgments: We thank Analytical Resources and Calibration Laboratory (LRAC) from School of Chemical Engineering (Unicamp) and the Brazilian Nanotechnology National Laboratory (LNNano, CNPEM) for the analytical facilities. We acknowledge CNPq (grant n.147536/2016-2) and São Paulo Research Foundation (FAPESP, grant n. 2018/20560-4) for the financial support to conduct this project.

Conflicts of Interest: The authors declare no conflict of interest.

References

1. Dahm, P.; Neuberger, M.; Ilic, D. Screening for prostate cancer: Shaping the debate on benefits and harms. *Cochrane Database Syst. Rev.* **2013**, ED000067. [CrossRef]
2. Rodrigues, V.C.; Soares, J.C.; Soares, A.C.; Braz, D.C.; Melendez, M.E.; Ribas, L.C.; Scabini, L.F.S.; Bruno, O.M.; Carvalho, A.L.; Reis, R.M.; et al. Electrochemical and optical detection and machine learning applied to images of genosensors for diagnosis of prostate cancer with the biomarker PCA3. *Talanta* **2021**, *222*, 121444. [CrossRef] [PubMed]
3. Bertoldo, S.A.; PASQUINI, V.Z. Câncer de próstata: Um desafio para saúde do homem. *Revi Enfer UNISA* **2010**, *11*, 138–142.
4. Neto, J.B.M.R.; Taketa, T.B.; Bataglioli, R.A.; Pimentel, S.B.; Santos, D.M.; Fiamingo, A.; Costa, C.A.R.; Campana-Filho, S.P.; Carvalho, H.F.; Beppu, M.M. Tailored chitosan/hyaluronan coatings for tumor cell adhesion: Effects of topography, charge density and surface composition. *Appl. Surf. Sci.* **2019**, *486*, 508–518. [CrossRef]
5. Neto, J.B.M.R.; Soares, A.C.; Bataglioli, R.A.; Carr, O.; Costa, C.A.R.; Oliveira, O.N.; Beppu, M.M.; Carvalho, H.F. Polysaccharide Multilayer Films in Sensors for Detecting Prostate Tumor Cells Based on Hyaluronan-CD44 Interactions. *Cells* **2020**, *9*, 1563. [CrossRef] [PubMed]
6. Swiston, A.J.; Cheng, C.; Um, S.H.; Irvine, D.J.; Cohen, R.E.; Rubner, M.F. Surface functionalization of living cells with multilayer patches. *Nano Lett.* **2008**, *8*, 4446–4453. [CrossRef]
7. Vasconcellos, F.C.; Swiston, A.J.; Beppu, M.M.; Cohen, R.E.; Rubner, M.F. Bioactive Polyelectrolyte Multilayers: Hyaluronic Acid Mediated B Lymphocyte Adhesion. *Biomacromolecules* **2010**, *11*, 2407–2414. [CrossRef]
8. Zöller, M. CD44: Can a cancer-initiating cell profit from an abundantly expressed molecule? *Nat. Rev. Cancer* **2011**, *11*, 254–267. [CrossRef]
9. Chen, C.; Zhao, S.; Karnad, A.; Freeman, J.W. The biology and role of CD44 in cancer progression: Therapeutic implications. *J. Hematol. Oncol.* **2018**, *11*, 1–23. [CrossRef]
10. Stern, R. *Hyaluronan in Cancer Biology*; Academic Press: San Diego, CA, USA, 2009.
11. Sneath, R.J.; Mangham, D.C. The normal structure and function of CD44 and its role in neoplasia. *Mol. Pathol.* **1998**, *51*, 191–200. [CrossRef]
12. Li, W.; Ma, H.; Zhang, J.; Zhu, L.; Wang, C.; Yang, Y. Unraveling the roles of CD44/CD24 and ALDH1 as cancer stem cell markers in tumorigenesis and metastasis. *Sci. Rep.* **2017**, *7*, 13856. [CrossRef]
13. Assis, O.B.G. Caracterização estrutural e da capacidade de absorção de água em filmes finos de quitosana processados em diversas concentrações. *Polímeros* **2003**, *13*, 223–228. [CrossRef]
14. De Moura, C.M.; de Moura, J.M.; Soares, N.M.; de Almeida Pinto, L.A. Evaluation of molar weight and deacetylation degree of chitosan during chitin deacetylation reaction: Used to produce biofilm, Chem. Eng. Process. *Chem. Eng. Process. Process Intensif.* **2011**, *50*, 351–355. [CrossRef]
15. Rinaudo, M. Chitin and chitosan: Properties and applications. *Prog. Polym. Sci.* **2006**, *31*, 603–632. [CrossRef]
16. Ahmed, S.; Ikram, S. *Chitosan: Derivatives, Composites and Applications*; John Wiley & Sons: Hoboken, NJ, USA, 2017; 516p.
17. Peniche, C.; Argüelles-Monal, W.; Goycoolea, F.M. Chitin and Chitosan: Major Sources, Properties and Applications. In *Monomers, Polymers and Composites from Renewable Resources*; Elsevier: Kidlington, UK, 2008; pp. 517–542. [CrossRef]
18. Zamani, A.; Edebo, L.; Sjöström, B.; Taherzadeh, M.J. Extraction and Precipitation of Chitosan from Cell Wall of Zygomycetes Fungi by Dilute Sulfuric Acid. *Biomacromolecules* **2007**, *8*, 3786–3790. [CrossRef] [PubMed]
19. Younes, I.; Rinaudo, M. Chitin and Chitosan Preparation from Marine Sources. Structure, Properties and Applications. *Mar. Drugs* **2015**, *13*, 1133–1174. [CrossRef] [PubMed]
20. Delezuk, J.A.d.M.; Cardoso, M.B.; Domard, A.; Campana-Filho, S.P. Ultrasound-assisted deacetylation of beta-chitin: Influence of processing parameters. *Polym. Int.* **2011**, *60*, 903–909. [CrossRef]
21. Verlee, A.; Mincke, S.; Stevens, C.V. Recent developments in antibacterial and antifungal chitosan and its derivatives. *Carbohydr. Polym.* **2017**, *164*, 268–283. [CrossRef] [PubMed]
22. Park, S.-C.; Nah, J.-W.; Park, Y. pH-dependent mode of antibacterial actions of low molecular weight water-soluble chitosan (LMWSC) against various pathogens. *Macromol. Res.* **2011**, *19*, 853–860. [CrossRef]
23. Krajewska, B.; Wydro, P.; Jańczyk, A. Probing the Modes of Antibacterial Activity of Chitosan. Effects of pH and Molecular Weight on Chitosan Interactions with Membrane Lipids in Langmuir Films. *Biomacromolecules* **2011**, *12*, 4144–4152. [CrossRef]
24. Costa, E.M.; Silva, S.; Pina, C.; Tavaria, F.K.; Pintado, M.M. Evaluation and insights into chitosan antimicrobial activity against anaerobic oral pathogens. *Anaerobe* **2012**, *18*, 305–309. [CrossRef]
25. Sutar, Y.B.; Telvekar, V.N. Chitosan based copolymer-drug conjugate and its protein targeted polyelectrolyte complex nanoparticles to enhance the efficiency and specificity of low potency anticancer agent. *Mater. Sci. Eng. C* **2018**, *92*, 393–406. [CrossRef]
26. Ahsan, S.M.; Thomas, M.; Reddy, K.K.; Sooraparaju, S.G.; Asthana, A.; Bhatnagar, I. Chitosan as biomaterial in drug delivery and tissue engineering. *Int. J. Biol. Macromol.* **2018**, *110*, 97–109. [CrossRef] [PubMed]
27. Suresh, L.; Brahman, P.K.; Reddy, K.R.; Bondili, J.S. Development of an electrochemical immunosensor based on gold nanoparticles incorporated chitosan biopolymer nanocomposite film for the detection of prostate cancer using PSA as biomarker. *Enzym. Microb. Technol* **2018**, *112*, 43–51. [CrossRef]
28. Rodrigues, V.C.; Moraes, M.L.; Soares, J.C.; Soares, A.C.; Sanfelice, R.; Deffune, E.; Oliveira, O.N., Jr. Immunosensors made with layer-by-layer films on chitosan/gold nanoparticle matrices to detect D-dimer as biomarker for venous thromboembolism. *Bull. Chem. Soc. Jpn.* **2018**, *91*, 891–896. [CrossRef]

29. Tamer, T.M.; Valachová, K.; Hassan, M.A.; Omer, A.M.; El-Shafeey, M.; Eldin, M.S.M.; Šoltés, L. Chitosan/hyaluronan/edaravone membranes for anti-inflammatory wound dressing: In vitro and in vivo evaluation studies. *Mater. Sci. Eng. C* **2018**, *90*, 227–235. [CrossRef]
30. Soundarya, S.P.; Menon, A.H.; Chandran, S.V.; Selvamurugan, N. Bone tissue engineering: Scaffold preparation using chitosan and other biomaterials with different design and fabrication techniques. *Int. J. Biol. Macromol.* **2018**, *119*, 1228–1239. [CrossRef] [PubMed]
31. Nezhad-Mokhtari, P.; Akrami-Hasan-Kohal, M.; Ghorbani, M. An injectable chitosan-based hydrogel scaffold containing gold nanoparticles for tissue engineering applications. *Int. J. Biol. Macromol.* **2020**, *154*, 198–205. [CrossRef]
32. Coma, V.; Deschamps, A.; Martial-Gros, A. Bioactive Packaging Materials from Edible Chitosan Polymer—Antimicrobial Activity Assessment on Dairy-Related Contaminants. *J. Food Sci.* **2003**, *68*, 2788–2792. [CrossRef]
33. Zivanovic, S.; Chi, S.; Draughon, A.F. Antimicrobial Activity of Chitosan Films Enriched with Essential Oils. *J. Food Sci.* **2005**, *70*, 45–51. [CrossRef]
34. Arkoun, M.; Ardila, N.; Heuzey, M.-C.; Ajji, A. Chitosan-Based Structures/Coatings With Antibacterial Properties. *Handb. Antimicrob. Coat.* **2018**, 357–389. [CrossRef]
35. Belbekhouche, S.; Bousserrhine, N.; Alphonse, V.; Carbonnier, B. From beta-cyclodextrin polyelectrolyte to layer-by-layer self-assembly microcapsules: From inhibition of bacterial growth to bactericidal effect. *Food Hydrocoll.* **2019**, *95*, 219–227. [CrossRef]
36. Kogan, G.; Šoltés, L.; Stern, R.; Gemeiner, P. Hyaluronic acid: A natural biopolymer with a broad range of biomedical and industrial applications. *Biotechnol. Lett.* **2007**, *29*, 17–25. [CrossRef] [PubMed]
37. Rinaudo, M. Main properties and current applications of some polysaccharides as biomaterials. *Polym. Int.* **2008**, *57*, 397–430. [CrossRef]
38. Fallacara, A.; Baldini, E.; Manfredini, S.; Vertuani, S. Hyaluronic acid in the third millennium. *Polymers* **2018**, *10*, 701. [CrossRef]
39. Bowman, S.; Awad, M.E.; Hamrick, M.W.; Hunter, M.; Fulzele, S. Recent advances in hyaluronic acid based therapy for osteoarthritis. *Clin. Transl. Med.* **2018**, *7*, 6. [CrossRef]
40. Liu, X.-W.; Hu, J.; Man, C.; Zhang, B.; Ma, Y.-Q.; Zhu, S.-S. Insulin-like growth factor-1 suspended in hyaluronan improves cartilage and subchondral cancellous bone repair in osteoarthritis of temporomandibular joint. *Int. J. Oral Maxillofac. Surg.* **2011**, *40*, 184–190. [CrossRef]
41. Bartlett, S.; Lin, K.; Bartlett, S.; Matsuo, K.; Livolsi, V.; Parry, C.; Hass, B.; Whitaker, L. Hyaluronic acid-filled mammary implants: An experimental study. *Plast. Reconstr. Surg.* **1994**, *94*, 306–315.
42. Patterson, J.; Siew, R.; Herring, S.W.; Lin, A.S.P.; Guldberg, R.; Stayton, P.S. Hyaluronic acid hydrogels with controlled degradation properties for oriented bone regeneration. *Biomaterials* **2010**, *31*, 6772–6781. [CrossRef]
43. Prestwich, G.D. Hyaluronic acid-based clinical biomaterials derived for cell and molecule delivery in regenerative medicine. *J. Control. Release* **2011**, *155*, 193–199. [CrossRef]
44. Lu, B.; Luo, D.; Zhao, A.; Wang, H.; Zhao, Y.; Maitz, M.F.; Yang, P.; Huang, N. pH responsive chitosan and hyaluronic acid layer by layer film for drug delivery applications. *Prog. Org. Coat.* **2019**, *135*, 240–247. [CrossRef]
45. Taketa, T.B.; Bataglioli, R.A.; Neto, J.B.M.R.; Beppu, M.M. Probing axial metal distribution on biopolymer-based layer-by-layer films for antimicrobial use. *Colloids Surf. B Biointerfaces* **2021**, *199*, 111505. [CrossRef] [PubMed]
46. Hernández-Montelongo, J.; Nascimento, V.F.; Murillo, D.; Taketa, T.B.; Sahoo, P.; de Souza, A.A.; Beppu, M.M.; Cotta, M.A. Nanofilms of hyaluronan/chitosan assembled layer-by-layer: An antibacterial surface for Xylella fastidiosa. *Carbohydr. Polym.* **2016**, *136*, 1–11. [CrossRef]
47. Swiston, A.J.; Gilbert, J.B.; Irvine, D.J.; Cohen, R.E.; Rubner, M.F. Freely suspended cellular "backpacks" lead to cell aggregate self-assembly. *Biomacromolecules* **2010**, *11*, 1826–1832. [CrossRef] [PubMed]
48. Neto, J.B.M.R.; Lima, G.G.; Fiamingo, A.; Germiniani, L.G.L.; Taketa, T.B.; Bataglioli, R.A.; da Silveira, G.A.T.; da Silva, J.V.L.; Campana-Filho, S.P.; Oliveira, O.N., Jr. Controlling antimicrobial activity and drug loading capacity of chitosan-based layer-by-layer films. *Int. J. Biol. Macromol.* **2021**, *172*, 154–161. [CrossRef] [PubMed]
49. Xu, H.; Ma, L.; Shi, H.; Gao, C.; Han, C. Chitosan–hyaluronic acid hybrid film as a novel wound dressing: In vitro and in vivo studies. *Polym. Adv. Technol.* **2007**, *18*, 869–875. [CrossRef]
50. Lin, Q.-K.; Ren, K.-F.; Ji, J. Hyaluronic acid and chitosan-DNA complex multilayered thin film as surface-mediated nonviral gene delivery system. *Colloids Surf. B Biointerfaces* **2009**, *74*, 298–303. [CrossRef]
51. Soliman, O.Y.; Alameh, M.G.; de Cresenzo, G.; Buschmann, M.D.; Lavertu, M. Efficiency of Chitosan/Hyaluronan-Based mRNA Delivery Systems In Vitro: Influence of Composition and Structure. *J. Pharm. Sci.* **2020**, *109*, 1581–1593. [CrossRef] [PubMed]
52. Tang, Q.; Hu, Z.; Jin, H.; Zheng, G.; Yu, X.; Wu, G.; Liu, H.; Zhu, Z.; Xu, H.; Zhang, C.; et al. Microporous polysaccharide multilayer coated BCP composite scaffolds with immobilised calcitriol promote osteoporotic bone regeneration both in vitro and in vivo. *Theranostics* **2019**, *9*, 1125–1143. [CrossRef] [PubMed]
53. Decher, G. Fuzzy nanoassemblies: Toward layered polymeric multicomposites. *Science* **1997**, *277*, 1232–1237. [CrossRef]
54. Klitzing, R.V. Internal structure of polyelectrolyte multilayer assemblies. *Phys. Chem. Chem. Phys.* **2006**, *8*, 5012–5033. [CrossRef]
55. Ariga, K.; Hill, J.P.; Ji, Q. Layer-by-layer assembly as a versatile bottom-up nanofabrication technique for exploratory research and realistic application. *Phys. Chem. Chem. Phys.* **2007**, *9*, 2319–2340. [CrossRef]

56. Guo, S.; Zhu, X.; Loh, X.J. Controlling cell adhesion using layer-by-layer approaches for biomedical applications. *Mater. Sci. Eng. C* **2017**, *70*, 1163–1175. [CrossRef] [PubMed]
57. Berg, M.C.; Yang, S.Y.; Hammond, P.T.; Rubner, M.F. Controlling Mammalian Cell Interactions on Patterned Polyelectrolyte Multilayer Surfaces. *Langmuir* **2004**, *20*, 1362–1368. [CrossRef]
58. Richert, L.; Lavalle, P.; Payan, E.; Shu, X.Z.; Prestwich, G.D.; Schaaf, P.; Voegel, J.; Picart, C. Layer by Layer Buildup of Polysaccharide Films: Physical Chemistry and Cellular Adhesion Aspects. *Langmuir* **2004**, *20*, 448–458. [CrossRef]
59. Dubas, S.T.; Schlenoff, J.B. Factors Controlling the Growth of Polyelectrolyte Multilayers. *Macromolecules* **1999**, *32*, 8153–8160. [CrossRef]
60. Shiratori, S.S.; Rubner, M.F. pH-Dependent Thickness Behavior of Sequentially Adsorbed Layers of Weak Polyelectrolytes. *Macromolecules* **2000**, *33*, 4213–4219. [CrossRef]
61. Yoo, D.; Shiratori, S.S.; Rubner, M.F. Controlling bilayer composition and surface wettability of sequentially adsorbed multilayers of weak polyelectrolytes. *Macromolecules* **1998**, *31*, 4309–4318. [CrossRef]
62. Li, D.; Dai, F.; Li, H.; Wang, C.; Shi, X.; Cheng, Y.; Deng, H. Chitosan and collagen layer-by-layer assembly modified oriented nanofibers and their biological properties. *Carbohydr. Polym.* **2021**, *254*, 117438. [CrossRef] [PubMed]
63. Manabe, K.; Belbekhouche, S. Construction of low-wettable free-standing layer-by-layer multilayer for fibrinogen adsorption. *Colloids Surf. A Physicochem. Eng. Asp.* **2020**, *604*, 125303. [CrossRef]
64. Ding, C.; Xu, S.; Wang, J.; Liu, Y.; Hu, X.; Chen, P.; Feng, S. Controlled loading and release of methylene blue from LbL polyurethane/poly(acrylic acid) film. *Polym. Adv. Technol.* **2012**, *23*, 1283–1286. [CrossRef]
65. Da Câmara, P.C.F.; Balaban, R.C.; Hedayati, M.; Popat, K.C.; Martins, A.F.; Kipper, M.J. Novel cationic tannin/glycosaminoglycan-based polyelectrolyte multilayers promote stem cells adhesion and proliferation. *RSC Adv.* **2019**, *9*, 25836–25846. [CrossRef]
66. Taketa, T.B. Obtenção e caracterização de recobrimentos de quitosana/ácido hialurônico e quitosana/alginato de sódio pela técnica layer-by-layer para aplicações antimicóticas. Master's Thesis, Universidade Estadual de Campinas, Campinas, Brazil, 2013.
67. Tsaih, M.L.; Chen, R.H. Effects of Ionic Strength and pH on the Diffusion Coefficients and Conformation of Chitosans Molecule in Solution. *J. Appl. Polym. Sci.* **1999**, *73*, 2041–2050. [CrossRef]
68. Nascimento, V.; Franca, C.; Hernández-Montelongo, J.; Machado, D.; Lancellotti, M.; Cotta, M.; Landers, R.; Beppu, M. Influence of pH and ionic strength on the antibacterial effect of hyaluronic acid/chitosan films assembled layer-by-layer. *Eur. Polym. J.* **2018**, *109*, 198–205. [CrossRef]
69. Ranella, A.; Barberoglou, M.; Bakogianni, S.; Fotakis, C.; Stratakis, E. Tuning cell adhesion by controlling the roughness and wettability of 3D micro/nano silicon structures. *Acta Biomater.* **2010**, *6*, 2711–2720. [CrossRef]
70. Zhang, S.; Liu, W.; Liang, J.; Li, X.; Liang, W.; He, S.; Zhu, C.; Mao, L. Buildup mechanism of carboxymethyl cellulose and chitosan self-assembled films. *Cellulose* **2013**, *20*, 1135–1143. [CrossRef]
71. Picart, C.; Mutterer, J.; Richert, L.; Luo, Y.; Prestwich, G.D.; Schaaf, P.; Voegel, J.-C.; Lavalle, P. Molecular basis for the explanation of the exponential growth of polyelectrolyte multilayers. *Proc. Natl. Acad. Sci. USA* **2002**, *99*, 12531–12535. [CrossRef]
72. Neto, J.B.M.R.; Neto, R.J.G.; Bataglioli, R.A.; Taketa, T.B.; Pimentel, S.B.; Baratti, M.O.; Costa, C.A.R.; Carvalho, H.F.; Beppu, M.M. Engineering the surface of prostate tumor cells and hyaluronan/chitosan multilayer films to modulate cell-substrate adhesion properties. *Int. J. Biol. Macromol.* **2020**, *158*, 197–207. [CrossRef]
73. Kumorek, M.; Minisy, I.M.; Krunclová, T.; Voršiláková, M.; Venclíková, K.; Chánová, E.M.; Janoušková, O.; Kubies, D. pH-responsive and antibacterial properties of self-assembled multilayer films based on chitosan and tannic acid. *Mater. Sci. Eng. C* **2020**, *109*, 110493. [CrossRef]

polysaccharides

Article

Conversion of Electrospun Chitosan into Chitin: A Robust Strategy to Tune the Properties of 2D Biomimetic Nanofiber Scaffolds

Natalia Toncheva-Moncheva [1,2], Abdelhafid Aqil [1], Moreno Galleni [3] and Christine Jérôme [1,*]

1 Center for Education and Research on Macromolecules (CERM), University of Liège, Cesam RU, Allée du 6 août 13, Sart-Tilman, B6a, B-4000 Liège, Belgium; ntoncheva@polymer.bas.bg (N.T.-M.); a.aqil@uliege.be (A.A.)
2 Laboratory of Polymerization Processes, Institute of Polymers, Bulgarian Academy of Sciences, Akad. G. Bonchev 103-A, 1113 Sofia, Bulgaria
3 Laboratory of Biological Macromolecules, Center for Protein Engineering (CIP), InBioS, University of Liège, Sart-Tilman, B6a, B-4000 Liège, Belgium; mgalleni@uliege.be
* Correspondence: c.jerome@uliege.be

Abstract: New biomimetic micro- and nano-CsU-based fibrous scaffolds electrospun from solution containing high purity-medical grade chitosan (CsU) of fungus origin (CsU1, Mv ~174,000 and CsU2, 205,000, degree of deacetylation (DDA) ~65%) and polyethylene oxide (PEO, Mv ~ 900,000), in the presence of given amounts of Triton X-100 (from 0.01 to 0.5 wt%) as surfactant were fabricated. We demonstrate that by carefully selecting compositions and surfactant levels, porous mats with CsU content up to 90% (at this molecular weight and DDA) were achieved. Remarkable long-term stability in water or phosphate buffer solution storage were obtained by developing post-electrospinning treatment allowing the complete elimination of the PEO from the CsU-fibers as demonstrated by TGA, DSC and ESEM analysis. Subsequent reacetylation procedure was applied to convert 2D biomimetic chitosan mats to chitin (CsE)-based ones while preserving the nanofiber structure. This innovative procedure allows tuning and modifying the thermal, mechanical properties and more importantly the biodegradation abilities (fast enzymatic biodegradation in some cases and slower on the others) of the prepared nanofibrous mats. The established reproducible method offers the unique advantage to modulate the membrane properties leading to stable 2D biomimetic CsU and/or chitin (CsE) scaffolds tailor-made for specific purposes in the field of tissue engineering.

Keywords: chitosan; chitin; fiber; electrospinning; 2D scaffold

check for updates

Citation: Toncheva-Moncheva, N.; Aqil, A.; Galleni, M.; Jérôme, C. Conversion of Electrospun Chitosan into Chitin: A Robust Strategy to Tune the Properties of 2D Biomimetic Nanofiber Scaffolds. *Polysaccharides* **2021**, *2*, 271–286. https://doi.org/10.3390/polysaccharides2020019

Academic Editor: Azizur Rahman

Received: 26 March 2021
Accepted: 26 April 2021
Published: 2 May 2021

1. Introduction

Chitin (CsE) or poly(β-(1→4)-2-acetamido-2-deoxy-D-glucopyranose] is an abundant and naturally occurring polysaccharide and one of the most popular and studied biopolymers [1]. CsE has a crystalline structure and is commonly found as a constituent of the exoskeleton of invertebrates as crustacean and molluscs [2]. Moreover, it is a main polymer component of the cell wall of some fungi and yeasts [3]. Extracted from mushroom waste at industrial scale, it offers the advantage of being an animal-free and well-controlled renewable material particularly attractive when biomedical applications are foreseen [4]. Being just behind cellulose in the amount of annual biosynthesis production, chitin leads to chitosan by deacetylation yielding to novel biomaterials (CsU) (Scheme 1) [5].

CsU is nontoxic, biodegradable and possesses antimicrobial properties that have led to significant research towards biological applications such as drug delivery, artificial tissue scaffolds for functional tissue engineering, and wound dressings [6]. Depending on the source and deacetylation method, the molecular weight ranges from 100 to over 1000 kDa [7]. The degree of deacetylation (DDA) can vary between 30% and 90%, thus allowing CsU solubilization in acidic media [8]. For application as tissue scaffolds, ability to form highly porous mats of micro and nanometer-sized fibers, such as those fabricated via electrospinning (ESP) is quite attractive [9]. CsU fibers possess an acetamido/amino functionality that imparts many biological properties along with possibilities of chemical modification and has remarkable affinity to proteins [10–12]. If poorly soluble in organic solvents [13], CsU has the advantage over chitin to be soluble in acetic–water mixtures that are not releasing toxic residues from the fibers [14]. Nevertheless, ESP of CsU with high degree of deacetylation is especially challenging because the amine groups protonated under acidic condition [15] make CsU a highly charged cationic polyelectrolyte, resulting in a high solution viscosity [16]. At low polymer concentrations (2–2.5 times above the entanglement concentration), it remains quite difficult to be electrospun due to the highly charged chains [17]. An alternative approach is the preparation of CsU:PEO blends, in which PEO helps the formation of the CsU fibers [18]. PEO was often selected, because it is highly soluble in water, could be used for producing ultra-fine fibers by ESP, and can form hydrogen bonds with polysaccharides [19]. However, a major limitation of CsU ESP is the sensitivity of the method, as well as the stability of the obtained protonated mats in aqueous media [20]. Apart the above described hitches, CsU ESP still remains very attractive for tissue engineering especially for skin repair [21]. In that field, CsU films, hydrogels and sponges appeared less efficient wound dressing as compared to electrospun nanofiber based 2D scaffolds [22–24].

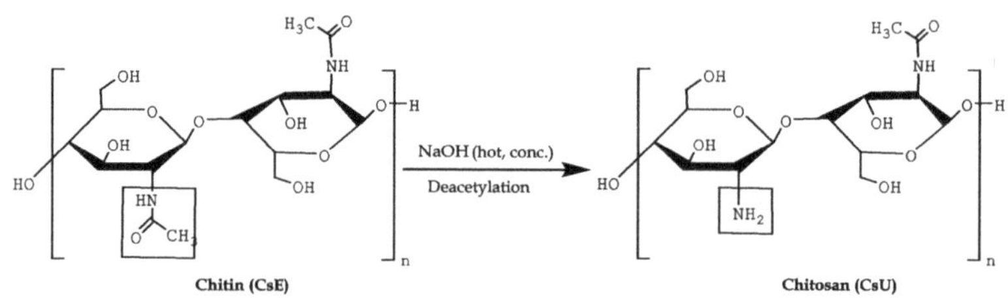

Scheme 1. Production of chitosan (CsU) by deacetylation of chitin (CsE).

The first aim of this paper is to report on a robust process for the formation of nonwoven mats of chitosan nanofibers, stable in water and in phosphate buffer by electrospinning of a high molecular weight and medical grade CsUs of fungus origin with average level of DDA ~65%. Furthermore, we investigated an acetylation procedure to convert the obtained 2D biomimetic chitosan mat to chitin-based one while preserving the nanofiber structure. This innovative process allows the tuning of the chemical, biodegradation, thermal and mechanical properties of the nanofiber nonwoven mats so that it can be tailored to fulfil the targeted tissue specificities.

2. Materials and Methods

2.1. Materials

Sodium ethoxide solution, 21 wt% in Ethanol, (CH_3CH_2ONa, Sigma Aldrich, Overijse, Belgium), Sodium ethoxide, 95%, (CH_3CH_2ONa, Sigma Aldrich, Overijse, Belgium), Ethylene glycol diglycidyl ether, techn. ($C_8H_{14}O_4$, Sigma Aldrich, Overijse, Belgium), 1-4-butanediol diglycidyl ether, $\geq$ 95%, ($C_{10}H_{18}O_4$, Sigma Aldrich, Overijse, Belgium), Calcium hydride, coarse granules, 95% (CaH_2, Sigma Aldrich, Overijse, Belgium), Sodium Hydroxide, pellets for analysis, ACS reagent (NaOH, Merck, Overijse, Belgium), Sodium hydroxide (NaOH), MP Biomedicals, Bruxelles, Belgium, anhydrous, (1534955, Fisher Scientific, Mechelen, Belgium), Sodium carbonate, 99.5%, extra pure, anhydrous (Na_2CO_3, Acros Organics, Gent, Belgium), Acetic acid 100%, ACS reagent, ISO anhydrous GR for analysis (CH_3COOH, Merck, Overijse, Belgium), Poly(ethylene oxide) (PEO)average Mv ~900,000, powder (189456, Sigma Aldrich, Overijse, Belgium), Triton™ X-100 BioXtra (t-Oct-C_6H_4-(OCH_2CH_2)xOH, x = 9–10 Sigma Aldrich, Overijse, Belgium) were used as received. Ethanol-absolute, analytical grade reagent, (CH_3CH_2OH, Fisher Scientific, Mechelen, Belgium) was stored under molecular sieve 3Å. Deionized water was obtained by Millipore MilliQ system and was additionally filtered through a 220 nm PTFE filter. The chitosan samples were purchased from Kitozyme, Liege, Belgium (KiOmedine-CsU®CAS: [9012-76-4] is an ultra-pure chitosan of non-animal origin, produced from white mushrooms (*Agaricus bisporus*)). The main characteristics of both types of used chitosan are given Table 1.

Table 1. General information about the used medical grade CsUs.

Chitosan Code	Molecular Weight (Mv) (g/mol)	Degree of Acetylation (mol %)	Apparent Viscosity (1% sol. in 1% HAc) (mPa·s)
CsU1 (L09306)	174,000	32.3	115
CsU2 (L10204)	205,000	34.0	125

2.2. *Mats Processing*

2.2.1. Electrospinning Conditions

Different solution compositions were prepared by dissolving appropriate amounts of chitosan in deionized water/acetic acid mixture, PEO in ultra-pure deionized water and gently stirring for 18 h (Scheme 2 and Table 2).

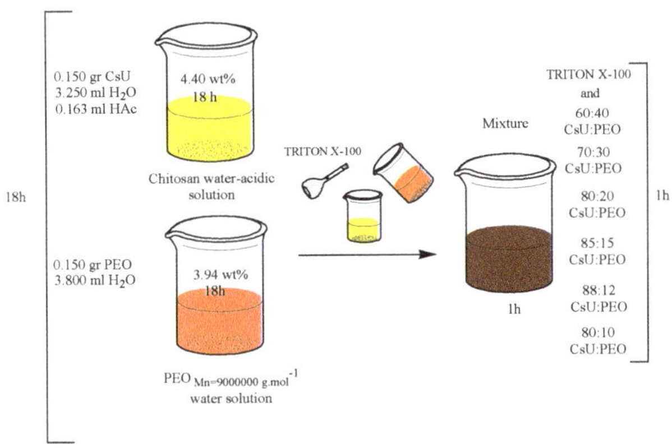

Scheme 2. Schematic representation of the ESP solutions preparation protocol.

Afterwards, solutions were mixed to the desired CsU:PEO ratio and a given amount of Triton X-100 was added (Table 2, Scheme 2). A total of 3 mL of the resulting well-homogenized mixture were transferred into a plastic syringe equipped with orthogonally cut-ended needle (G 21 1½″, K51 Luer-lock, B. Braun Group, Italy). The solution was driven by syringe pump (Razel Scientific pump, Razel Scientific Instruments, Vermont, USA) at 1 mL·h^{-1} debit, and electrospinning voltage (by the use of Spellman SL10 power supply, model is 88906 (A-99)) ranging from 20 to 35 kV was applied between the horizontal needle and a perpendicular fixed aluminum foil used as collector (Figure 1a).

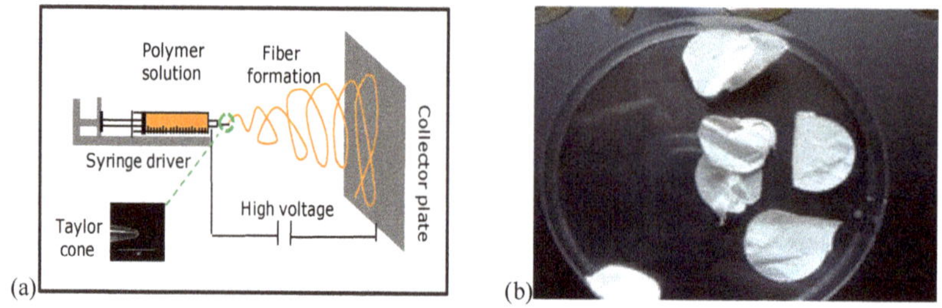

Figure 1. Electrospinning technique sketch (**a**) and collected CsU-based mats (**b**).

The temperature of 25 °C was selected for the ESP. The distance between the needle and the aluminum foil was 15 cm. The electrospinning set-up was placed in a homemade box, equipped with UV lamp for ensuring sterilization. In order to collect easily defect-free nano-fibrous mats (Figure 1b) from the collector surface, the electrospinning was performed during 5–6 h. The different electrospinning conditions are summarized in Tables 2 and 3.

Table 2. Electrospinning conditions: solution composition and voltage.

Mat Samples Code *	Type of CsU Used	Initial CsU:PEO Ratio	Final Solution Conc. (wt%)	Triton X-100 Conc. (wt%)	ΔV (kV)
CN14	CsU1 (L09306)	60:40	4.22	-	27.0
CN36		60:40	4.22	0.05	28.0
CN64_65		85:15	4.32	0.20	30.0
CN73		88:12	4.33	0.20	34.0
CN50		90:10	4.34	0.40	31.0
CN7_8A	CsU2 (L10204)	60:40	4.22	0.05	20.0
CN5_6A		85:15	4.32	0.10	28.0
CN9_10A		88:12	4.33	0.20	30.4

* CsU initial conc. = 4.40 wt% in solvent HAc/H_2O, PEO initial conc. = 3.94 wt% in solvent deionized H_2O, Mixing time of CsU and PEO solutions = 1 h, ESP time = 6 h, Debit 1 mL·h^{-1}, Distance between collector and the needle = 15 cm, Temperature of ESP = 25 °C.

Table 3. Electrospinning conditions for CsU1(L09306):PEO.

Mat Samples Code	Initial CsU:PEO Ratio *	Triton X-100 (wt%)	Voltage (kV)
CN32	60:40	0.5	22
CN36		0.05 **	25
CN34		0.03	26
CN33		0.01	25
CN41	70:30	0.1 **	25
CN48	80:20	0.05	27
CN40		0.1 **	27
CN47	85:15	0.3	27
CN49		0.2 **	28
CN46		0.1	27
CN73	88:12	0.2 **	35
CN43	90:10	0.1	35
CN44		0.3	32
CN50		0.4 **	31
CN45		0.5	27

* CsU initial conc. = 4.40 wt% in solvent HAc/H_2O, PEO initial conc. = 3.94 wt% in solvent deionized H_2O, mixing time of CsU and PEO solutions = 1 h, ESP time = 6 h, debit 1 mL·h^{-1}, distance between collector and the needle = 15 cm, temperature of ESP = 25 °C. ** Mat samples obtained with the most appropriate concentration of Triton X-100.

2.2.2. Stabilization of the Mats

In order to impart stability in water and in PBS buffer solutions of the obtained mats, the as-spun membranes were treated with dry absolute EtOH/NaOH (0.5 M) mixture for several minutes, followed by intensive washing with sterile water until neutral pH was obtained. Subsequently, the membranes were dried under vacuum.

2.2.3. Reacetylation of the Mats

In order to convert chitosan back into chitin, a reacetylation process was performed. The electrospun and stabilized chitosan mats were immersed for 1 h under stirring in a solution of >99% acetic anhydride diluted 32 times in methanol. Then, the mats were rinsed three times in 10 mL of sterile water and dried under air (Scheme 3).

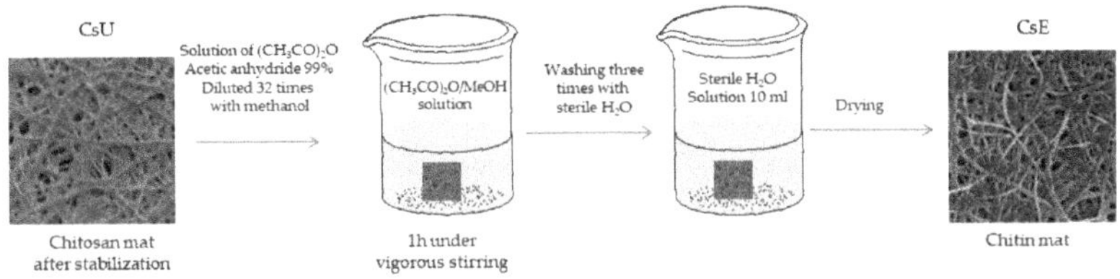

Scheme 3. Reacetylation reaction procedure of chitosan into chitin.

2.3. *Characterization Methods*

2.3.1. Rheology

Viscosity measurements on an ARES G2 Rheometer from TA instruments using a parallel plate geometry (gap diameter 25 mm) equipped with a Peltier plate for temperature control (at 25 °C), soak time 10 s, at a linear shear rate from 0.1 to 100 $1{\cdot}s^{-1}$ were performed. The data were collected with the TRIOS software.

2.3.2. Scanning and Transmission Electron Microscopy

Environmental scanning electron microscopy (ESEM JeolJSM-840A, Tokyo, Japan) to analyze the morphology, of as-spun CsU membranes was used. The produced fibers were metal coated by platinum before ESEM analysis.

2.3.3. Differential Scanning Calorimetry (DSC)

To sense the presence of PEO in the fiber mats, DSC (TA Instrument DSC Q100 V9.0 Build 275) was carried out in the −50–150 °C temperature range, at a heating rate of 20 $°C{\cdot}min^{-1}$ under a nitrogen flow (50 $mL{\cdot}min^{-1}$). Samples (6.0 ± 0.1 mg) were heated up to 150 °C at a rate of 20 $°C{\cdot}min^{-1}$ (first scan) and then quenched to −50 °C at a rate of 100 $°C{\cdot}min^{-1}$. Afterwards, they were heated again up to 150 °C (second scan) at a rate of 20 $°C{\cdot}min^{-1}$. The melting temperature (T_m) was defined as the temperature maximum of the melting endotherm.

2.3.4. Thermal Gravimetric Analysis (TGA)

A TGA Q500 V6.3 build 189 from TA Instruments was used in the range of 0–600 °C under a nitrogen flow of 40 $mL.min^{-1}$. The heating rate chosen was 20 $°C{\cdot}min^{-1}$ for conventional TGA and 40 $°C{\cdot}min^{-1}$ in case of high-resolution TGA with a resolution parameter of 4. Actually, the heating rate was continuously adjusted to track changes in the sample decomposition rate. This parameter was tuned within an eight-step scale to maximize the weight loss resolution. The ±1 °C accuracy on the degradation temperature determined from the derivate of the weight losses temperature curve was established.

2.3.5. Enzymatic Biodegradation

In vitro biodegradation of chitosan and chitin fibrous mats was performed by incubating the testing circular samples (diameter 10 mm) with three different media: PBS, chitotriosidase and lysozyme, at 37 °C, for the period of three weeks. The dilution factor was 9/11. After treatment, the specimens were carefully washed with deionized ultrapure water to stop further enzymatic hydrolysis, and then dried under vacuum at room temperature for two days prior to biodegradation rate estimation. All measurements were performed for three replicates of samples and averaged to obtain the final result.

2.3.6. Mechanical Properties—Tensile Strength on CsU Based Fiber Mats

Tensile testing was performed with Instron equipment following an already described procedure [25]. In all cases the collected samples (before, after the stabilization, and acetylation procedure) were obtained after 6 h ESP (Figure 1b). Next, they were cut to the specimens with the following dimensions, 3 cm length × 0.5 cm width, dried overnight under vacuum without heating and placed between the jaws of the Instron. Young's modulus (MPa) and break strain (%) were automatically calculated by Instron software (Bluehill 2, Elancourt, France). Experiments were performed at 25 °C on the specimens with dimensions 3.0 × 0.5 cm, which were dried overnight under vacuum without heating.

3. Results and Discussion

3.1. Electrospinning of Two-Component Water–Acidic Solutions Containing CsU and PEO

For the preparation of micro- and nano-fibrous mats, two types of high molecular weight and medical grade chitosans (CsU) of fungus origin (CsU1 Mv ~174,000 and CsU2 ~205,000, degree of deacetylation (DDA) ~65%, Table 1) were used for electrospinning (ESP), by using two component solutions containing CsU and PEO (Mv ~900,000). Number of experiments aiming to find appropriate ESP conditions by tuning the water:acid ratio (from 95:5 to 85:15) of the CsU solutions, the concentration of the stock CsU solution (from 4.77 to 3.88 wt%), as well as concentration of the stock PEO water solution (from 4.73 to 3.68 wt%) and CsU:PEO ratio (from 88:12 to 50:50) were carried out. It was estimated that defect-free membranes (CN14 sample code, Figure 2a) can be successfully electrospun at 25 °C, fixed 60:40 CsU/PEO ratio, using the initial concentrations of the stock PEO and CsU solutions as follows: 3.94 wt% (150 mg PEO in 3.8 mL water) and 4.4 wt% (150 mg CsU in 3.25 mL water and 0.163 mL acetic acid). All attempts outside these parameters to produce blended fibers easily detachable from the collector surface were not satisfied or even did not result in fiber formation.

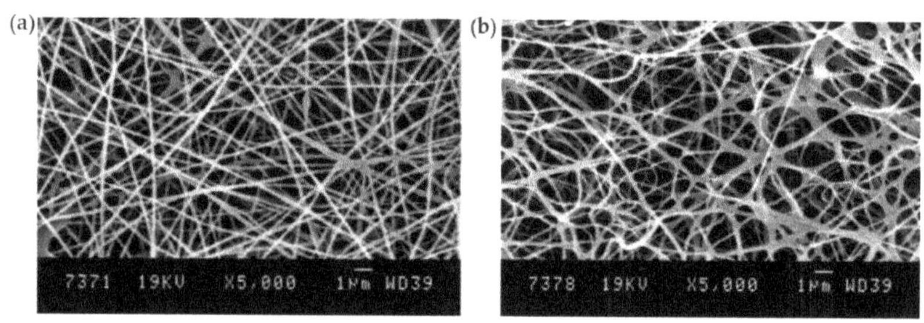

Figure 2. ESEM images of electrospun fibers from a two component solution of the chitosan (4.4 wt% CsU1) and PEO (60:40 ratio) (CN14, Table 2) (**a**) before and (**b**) after stabilization and PEO removal.

Moreover, instability of the jet leading to drops with the time of the ESP, were observed. The detailed literature survey confirmed these findings [26–30]. To overcome these difficulties further, studies were carried out with addition of surfactant to CsU/PEO ESP solution.

3.2. ESP of Three-Component Water–Acidic Solutions Containing CsU, PEO and Triton-X 100

Triton-X 100 is considered as a comparatively mild nondenaturing surfactant, reported in numerous references frequently used for ESP chitosan [26,27]. Addition of such nonionic surfactant to the CsU/PEO mixture is expected to decrease the viscosity of the polymer solution. Moreover, some plasticizing effects could open a prospect for broadening the window of electrospinning possibilities of CsU:PEO water–acidic solutions. It should be also noticed that Triton X-100 has no antimicrobial properties [27].

Following the above described preparation procedures (Scheme 2), Tables 2 and 3 show different three-component solutions containing CsU-PEO-Triton X-100, dissolved in water–acidic media. Solutions of a composition at constant 60:40 CsU:PEO ratio varying only the concentration in the range 0.5 wt% ÷ 0.01 wt% of Triton X-100 were firstly examined (Table 3). The obtained mixtures were subjected to detailed viscosity measurements at 25 °C (Figure 3a). As a rule, increasing the surfactant content in the mixtures decreases the viscosity that dropped reaching its lower level for CN32 ESP sample containing the higher surfactant concentration (0.5 wt%) (Figure 3a). The mixtures were electrospun and high quality mats easily detachable from the collector were obtained in all cases. In order to resolve the morphology of the as prepared fibers ESEM was performed (Figure 3b).

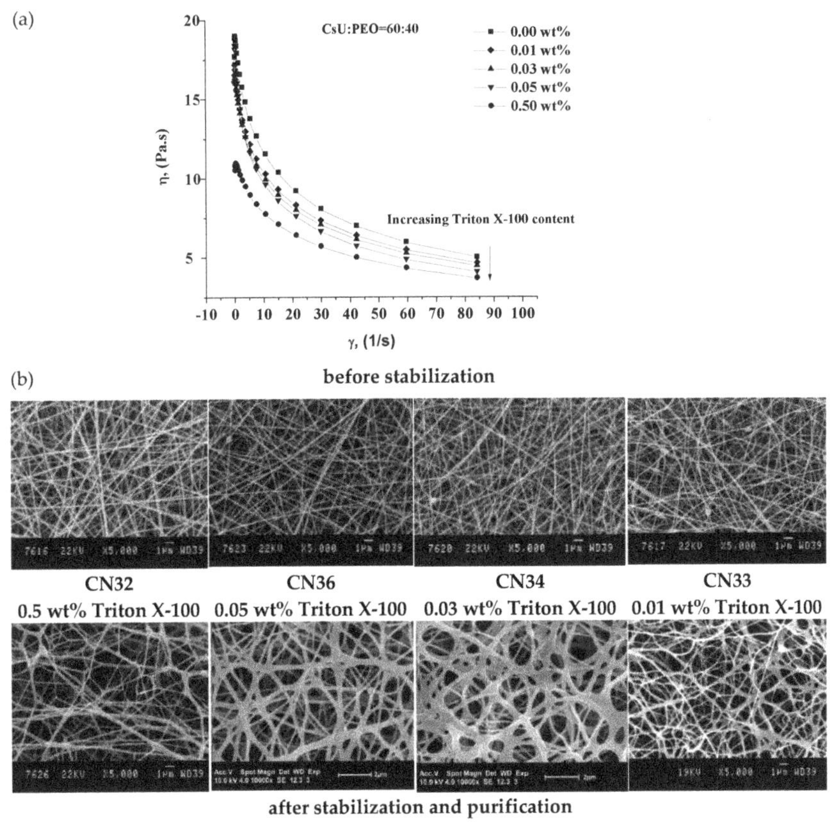

Figure 3. Viscosity vs. shear rate (**a**) and ESEM image of electrospun membranes (**b**) for solutions of the CsU1:PEO 60:40 and various contents of Triton X (see Table 3).

Linear dependence of the mats' defects disappearance on the concentration of Triton X was observed as exemplified by Figure 3b. A lot of bead defects are observed on the CN33 image, i.e., at low concentration of Triton X, the amount of these defects progressively decreases while increasing the concentration (CN34 and CN36 images) and completely disappear at high concentration (CN32). The decrease of the interfacial tension in presence of Triton X, notably at the air/water droplet where the jet is generated favors the electrospinning process leading to the formation of smooth and beadless nanofibrous chitosan-based mats. Moreover, the addition of surfactant decreases the viscosity of the polymer solution, reduces the onset voltage required to induce spinnability allowing polymer solution to remain spinnable over a longer period of time and thus improving reproducibility of the process. Furthermore, the obtained CsU:PEO porous mats keep their fibrillar morphology after subsequent treatment with a mixture of dry absolute ethanol EtOH/0.5 M NaOH for 5 min followed by three times washing with deionized water (for 5 min). This procedure is applied to the collected mats in order to stabilize them in neutral aqueous media. It aims to combine two effects, i.e., the deprotonation of the NH_3^+ group of the chitosan and the dissolution of the PEO and Triton X-100 resulting in pure CsU nano-fiber mats and therefore increasing the membrane stability in neutral or weak alkaline aqueous media as physiological and cell culture media, required for tissue engineering applications.

As can be seen from Figure 3, the structure of the purified and stabilized 2D scaffolds is entirely preserved. They are highly porous and without defects. Taking into account the composition, concentration of Triton X-100, viscosity measurements and quality of the mats, CN36 sample mat was selected as the one with the best characteristics (Figure 3b). The TGA and DSC profiles of the electrospun CN36 fibers before and after their deprotonation and purification were recorded and compared to the profiles of the pure CsU, and pure PEO (as a powder) (Figure 4) [15–17].

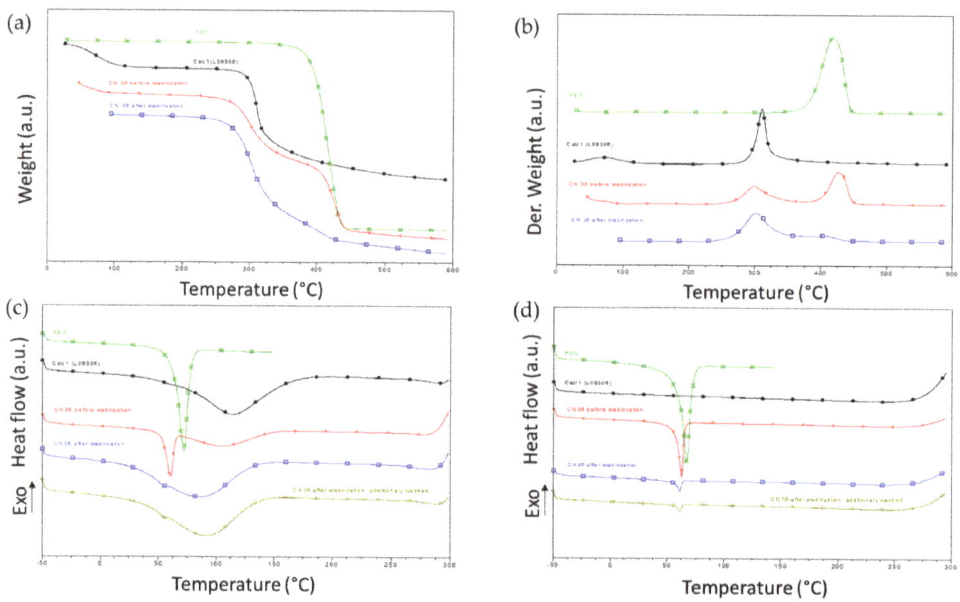

Figure 4. Thermal behavior of CsU1 (black traces), PEO (green curves), CN36 mat before (red traces) and after stabilization and purification (blue traces and grey traces after additional washing). TGA curves (**a**) and TGA derivatives (**b**); DSC curves first (**c**) and second (**d**) heating ramps.

The TGA curves of the CN36 before stabilization shows clearly two degradation steps. The first one corresponds to the degradation of CsU (around 300 °C) and the second (around 400 °C) to the degradation of the PEO part. The TGA curves of the CN36 after stabilization shows also the main degradation step (around 300 °C) connected with degradation of CsU and one very small peak (shoulder like) around 400 °C related to the degradation of PEO traces in the mats. The stabilization process is thus able to preserve the chitosan fiber from the dissolution and to remove the main part of PEO from the electrospun mats, leading to chitosan 2D scaffolds.

The DSC data confirm these observations. On the first heating run of the CN36 sample before stabilization, two distinct endotherms were observed. The first one corresponds to the PEO melting endotherm (around 60 °C). The second broad endotherm could be attributed to the evaporation of the water absorbed by CsU (around 100 °C). Following the fast cooling, the second heating run was started. On the recorded thermogram only one melting peak of PEO around 60 °C appeared. The disappearance of the second peak (around 100 °C) could be connected with the full elimination of the absorbed water [31]. The DSC curves of the CN36 after stabilization also displayed two main peaks at the first heating run and one (around 60 °C) during the second run having very low intensity.

In order to entirely remove PEO, CN36 sample was additionally washed three times with deionized water for 30 min and the DSC measurement was repeated. The persistence of the PEO melting peak was detected but its intensity progressively decreases i.e., the peak slowly disappeared with the washing time. However, PEO elimination from 60:40 CsU:PEO ratio compositions at washing time of 45 min was not complete.

In this context, the major aim of next studied compositions was to reduce the PEO content in the starting CsU:PEO mixture without affecting the produced mats' quality by finding the appropriate Triton X-100 concentration. For each of the following CsU:PEO ratios, 70:30, 80:20, 85:15, 88:12 up to 90:10, the concentration of Triton X-100 was adjusted in order to compensate for the increase of the viscosity due to the decrease of the PEO content and therefore preserve the formation of defect-free fiber mats (Table 3 and Figure 5a).

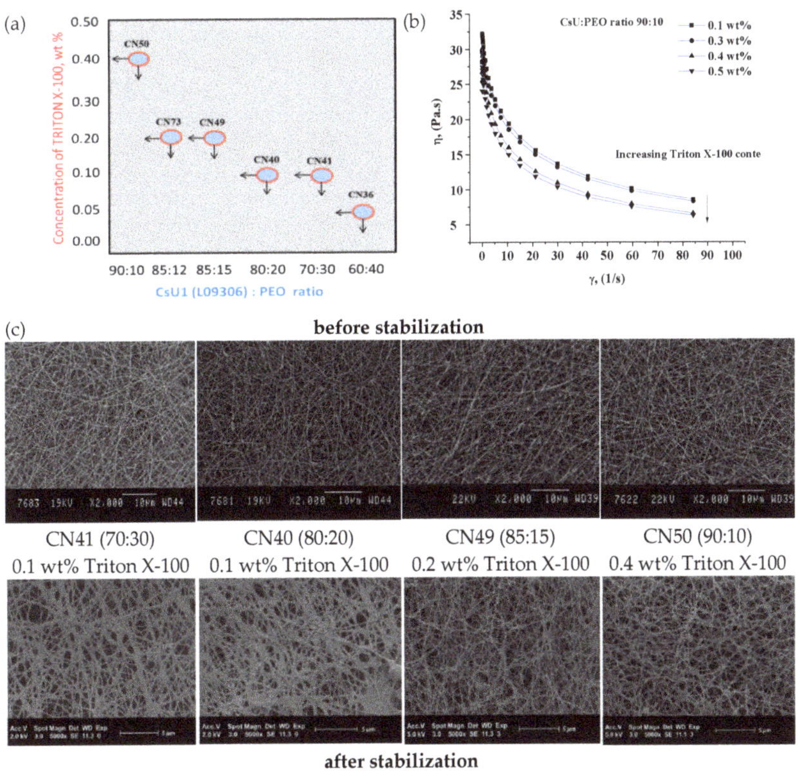

Figure 5. Minimal Triton X-100 concentration for achieving ESP of CsU/PEO solutions of various ratios (**a**); viscosity of CsU1:PEO 90:10 solutions for various Triton X-100 concentration at 25 °C (**b**); ESEM analysis of the collected mat of the corresponding solution (Table 3) (**c**).

Figure 5b shows that increasing the Triton X-100 concentration until 0.4 wt% allows to reach a viscosity around 5 Pa·s at 85 s^{-1} for a solution containing as low as 10% of PEO therefore leading to well-defined fiber mats. The most appropriate concentrations of Triton X-100 for each composition are given in Figure 5a. The criteria used were the stability of the jet during the ESP, the quality of the obtained mats and their porous fiber morphology evidenced by ESEM (Figure 5c).

The mats obtained consist of well-defined fibers and are highly porous before and after the stabilization. Both TGA and DSC measurements of CN49 mat sample (Figure 6) evidenced the decrease of the PEO content in the fiber and its quasi-complete removal after the stabilization step for 15 min.

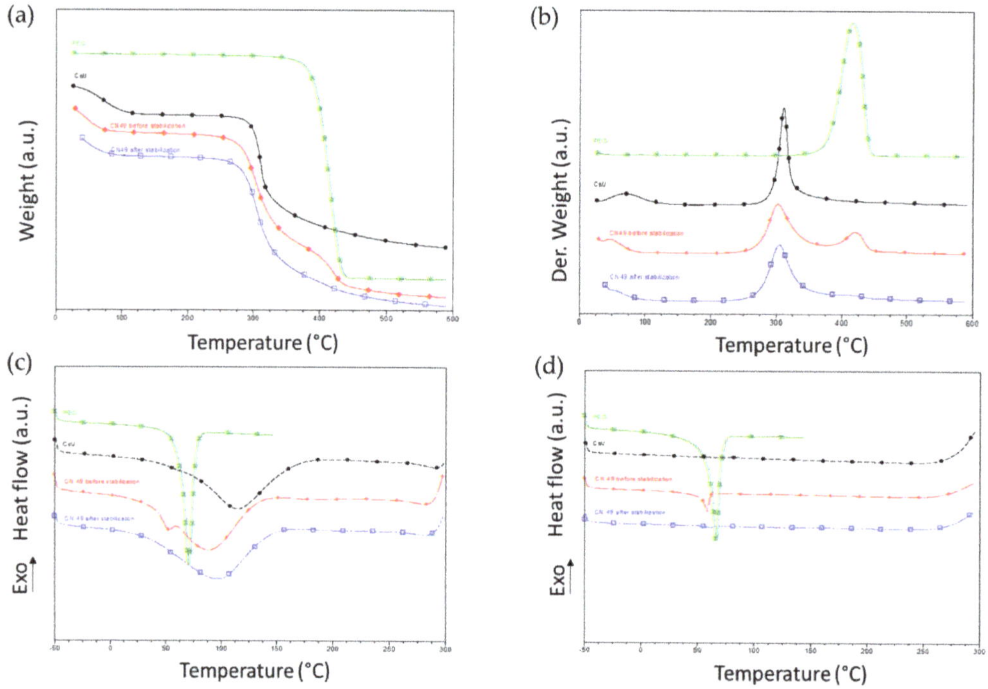

Figure 6. Thermal behavior of CsU1 (black traces), PEO (green curves), CN49 mat before (red traces) and after (blue traces) stabilization and purification. TGA curves (**a**) and TGA derivatives (**b**); DSC curves first (**c**) and second (**d**) heating ramps.

By applying the three-component solution approach and the outlined ESP limits (determined for CsU1 type), well defined nano-fibrous mats based on CsU2 of higher molar mass and viscosity were obtained revealing the robustness of these conditions. The best compositions and ESP parameters are summarized in Table 2. The collected high-quality mats based on CsU2 (L10204) were characterized by ESEM analyses before and after stabilization (Figure 7). The obtained 2D scaffolds before as well as after stabilization are free of any kind of defects as drops, beads, holes, etc.

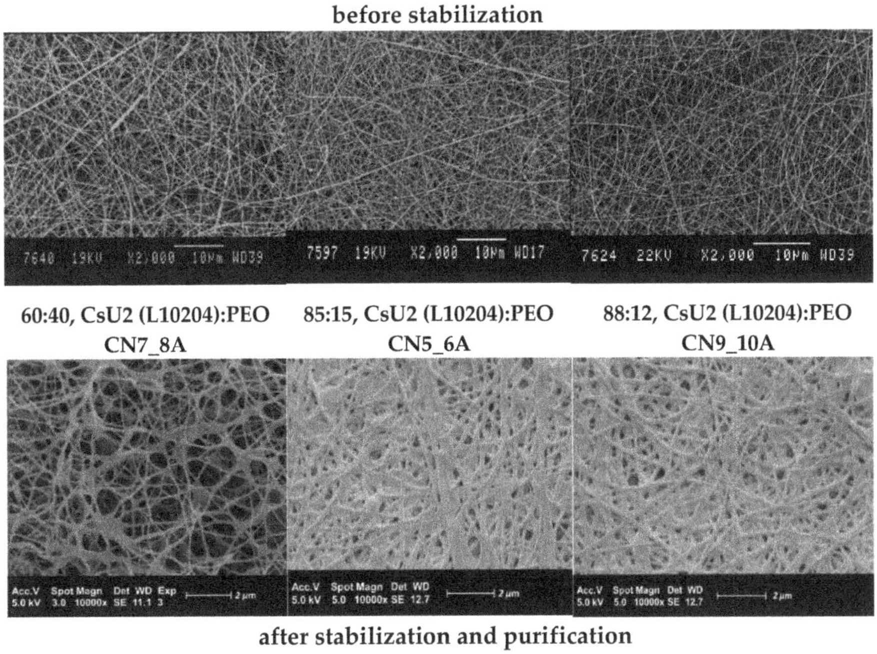

Figure 7. ESEM analysis of CsU2 (L10204) based mats.

The most important advantage of the proposed three-component ESP is that, with small adjustments of the surfactant concentration, CsU of different molecular weights with close DDA can be successfully electrospun with a low content of PEO. Moreover, the PEO can be entirely removed by the described washing/stabilization process.

3.3. *Reacetylation of the Electrospun Chitosan Mats into Chitin Nanofiber Mats*

In order to tune the mechanical and degradation properties of the nanofiber mats, the conversion of chitosan into chitin by reacetylation of the electrospun mats based on high molecular weight CsUs was investigated. Indeed, such reacetylation process gives the opportunity to obtain nanofibers of chitin that are difficult to get directly by ESP since chitin is poorly soluble in most solvents. For that purpose, the CN73 (CsU1 (L09306), 82:12 CsU:PEO ratio, 0.2 wt% Triton X-100) and CN9_10 (CsU2 (L10204), 88:12 CsU:PEO ratio, 0.2 wt% Triton X-100) electrospun and stabilized samples were immersed in a methanol solution of acetic anhydride (see Scheme 3. In this medium, the membranes did not dissolve and the amine groups of the chitosan can be acetylated again and thus converted back from chitosan into chitin. After, applying the acetylation treatment, the membranes were additionally washed and dried by previous established procedure (see Section 2.2.3, Scheme 3). Then, they were analyzed by ESEM that evidenced the successful conservation of their fibrous structure (Figure 8).

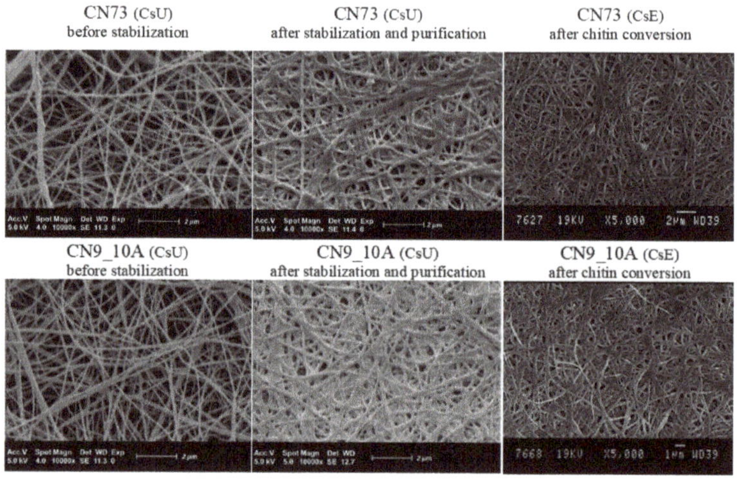

Figure 8. ESEM analysis of electrospun CsU1 and CsU2 mats before, after stabilization and after reacetylation.

First indication of the chitosan-based membranes' reacetylation into chitin is a clear change of their solubility in dilute acidic aqueous conditions. Indeed, while the chitosan mats are quickly dissolved in these conditions, after the chemical modification of the amine groups, the mats resist solubilization, as it is expected for chitin. The reaction occurrence was additionally supported by the improvement of the fiber mats' thermal stability after reacetylation (Figure 9).

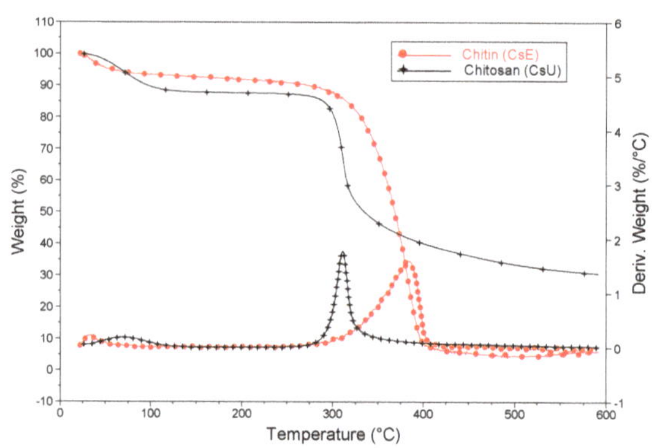

Figure 9. TGA analysis of electrospun CsU1 mat before (black) and after (red) reacetylation.

Since a higher thermal stability is observed when the N-acetyl content increases as well as the crystallinity [10], one may conclude that chitosan nanofibers have been successfully converted back into chitin ones [20,32].

3.4. Biodegradation Properties of the Nanofiber Mats

These polysaccharide mats are expected to be quite stable in PBS buffer but rather sensitive towards enzymatic degradation [33]. Therefore, we tested the hydrolytic stability of the chitin and chitosan mats in PBS buffer and compared this to data obtained in the presence of two enzymes: Human 3-Chitotriosidase (HCHT) and Lysozyme. The biotransformation of chitosan and chitin fibrous mats was observed only in the presence of HCHT enzyme (Figure 10). The detected weight losses were more than 90% for chitin specimens and around 87% for chitosan ones. HCHT is able to hydrolyze both chitin and chitosan via an endoprocessive mechanism. The preferences of HCHT subsites for acetylated (chitin) versus nonacetylated (chitosan) sugars confirmed that the catalytic activity of the enzyme on chitin is major compared to the catalytic activity on chitosan. Nevertheless, the difference of their catalytic efficiency does not exceed a factor of 10. After 3 weeks of incubation, both materials being highly degraded, no major difference between the two polymers is observed. In the presence of Lysozyme, the degradation was not detected for both investigated samples. The latter is expected, as it is well known that the biotransformation of high molecular chitin membranes with Lysozymes proceeds very slowly [34]. In addition, the Lysozymes primarily affect the chitin component in the chitosan-based materials, hence they are poorly degradable, which is with good agreement with the presented data in Figure 10 [35–37].

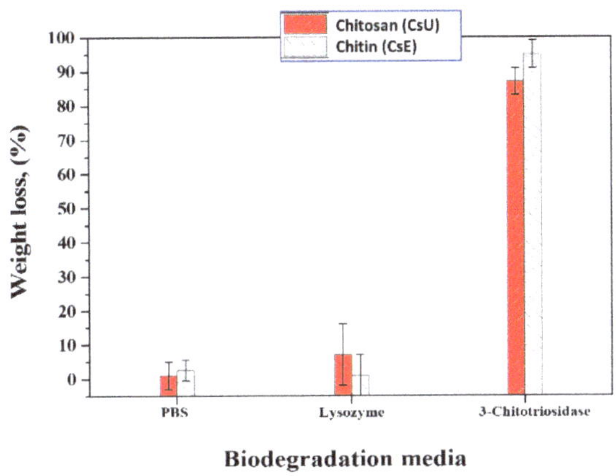

Figure 10. Biodegradation of CsU1 (L09306) based stabilized membranes and its derivate chitin (CsE) based membranes in different media for a period of three weeks.

3.5. Mechanical Properties Evaluation

The mechanical properties of different types of samples, i.e., ESP chitosan mats obtained before, after stabilization and chitin conversion were investigated. Tensile testing data are summarized in Table 4. The Young's modulus is one of the most crucial factors for fiber performance evaluation [18]. As a rule, comparing the mats before and after the stabilization, the Young's modulus increases after the stabilization i.e., the chitosan deprotonation and PEO removal. A slight decrease around 10% of the tensile stress and strain at break was detected (Table 4). Remarkably, the mechanical strength of the chitosan mats obtained after stabilization reaches similar values as those after reacetylation, i.e., made of chitin. These results show that the stabilization process allows getting comparable mechanical properties values for chitosan and chitin 2D biomimetic scaffolds which have made them ideal candidates for their further biomedical application, especially as scaffolds for skin regeneration [26,38].

Table 4. Tensile properties of the electrospun chitosan (CsU) nanofiber mats before and after stabilization, and after reacetylation (CsE conversion).

Sample Code	Type of CsU Used	Initial CsU:PEO Ratio	Young's Modulus * (MPa)	Tensile Stress at Break * (MPa)	Tensile Strain at Break * (%)
Before Stabilization and Purification (bs) CsU					
CN36 $_{bs}$	CsU1	60:40	155± 22	6.99 ± 0.13	8.55 ± 1.63
CN64_65 $_{bs}$	(L09306)	85:15	79 ± 22	2.12 ± 0.23	2.99 ± 0.83
CN7_8A $_{bs}$	CsU2 (L10204)	60:40	114 ± 18	11.27 ± 2.81	20.81 ± 7.8
CN5_6A $_{bs}$		85:15	147 ± 19	10.98 ± 1.47	14.70 ± 1.51
CN9_10A $_{bs}$		88:12	147 ± 21	9.49 ± 0.97	12.32 ± 1.79
After Stabilization and Purification (as) CsU					
CN36 $_{as}$	CsU1	60:40	181 ± 25	7.54 ± 0.97	10.44 ± 4.85
CN64_65 $_{as}$	(L09306)	85:15	125 ± 19	3.36 ± 0.53	3.67 ± 1.36
CN7_8A $_{as}$	CsU2 (L10204)	60:40	191 ± 34	10.70 ± 2.01	16.75 ± 4.37
CN5_6A $_{as}$		85:15	276 ± 31	10.70 ± 0.87	10.87 ± 1.18
CN9_10A $_{as}$		88:12	224 ± 22	9.14 ± 0.22	9.97 ± 1.08
After Reacetylation(ar) CsE					
CN9_10A$_{ar}$	CsU2 (L10204)	88:12	216 ± 23	7.56 ± 0.35	8.44 ± 0.95

* The values presented are averaged from five experiments at standard deviation.

4. Conclusions

Well-defined and stable chitosan based 2D biomimetic scaffolds mats were successfully prepared by reproducible ESP of high molecular weight and medical grades CsUs in the presence of PEO and given amounts of Triton X-100 as surfactant. The Triton X-100 addition combines several advantages. The sensitivity of the ESP was successfully overcome. Thus, the concentration of CsU in the starting CsU:PEO ratio was easily varied. Finally, the surfactant addition allows getting nano-fibrous membranes at high CsU content (up to 90%) from which PEO is easily removed. The proposed stabilization and purification approach provided long-term water and physiological media stability and entirely PEO removal from the as-spun CsU membranes preserving at the same time the initial nanofiber morphology. It was achieved by simple neutralization, without any use of chemical cross-linkers or chlorine-containing organic solvents. Remarkably, the stabilized chitosan nanofiber mats, exempt from PEO, exhibit high Young's modulus and improved mechanical properties, making them suitable for handling in medical applications.

Furthermore, acetylation procedure was investigated and proved to be effective to adjust the solubility, the thermal and the enzymatic stability of the nanofiber mats by converting chitosan mats into chitin ones. In other words, the proposed reaction strategy allowed preparation of well-defined chitosan and chitin nanofibrous mats with predefined mechanical properties and biodegradation abilities (fast enzymatic biodegradation in some cases and slower on the others). The latter can be extremely useful towards a variety of biomedical purposes as in a wound dressing field and/or tissue engineering.

Author Contributions: Conceptualization, C.J. and M.G.; methodology, A.A. and C.J.; validation, N.T.-M., A.A. and C.J.; formal analysis, A.A.; investigation, N.T.-M.; resources, C.J. and M.G.; data curation, A.A., and C.J.; writing—original draft preparation, N.T.-M.; writing—review and editing, A.A., C.J. and M.G.; supervision, A.A. and C.J.; project administration, C.J.; funding acquisition, C.J. All authors have read and agreed to the published version of the manuscript.

Funding: The present work was supported by the Science Policy Office of the Belgian Federal Government (IAP 7/05). CERM thanks the Walloon Region for supporting research on chitosan in the frame of CHITOPOL, TARGETUM, GOCELL and HOMECELL projects. CIP thanks the Walloon Region for supporting research in the frame of GOCELL and HOMECELL.

Institutional Review Board Statement: Not applicable.

Informed Consent Statement: Not applicable.

Data Availability Statement: Not applicable.

Conflicts of Interest: The authors declare no conflict of interest.

References

1. Croisier, F.; Jérôme, C. Chitosan-based biomaterials for tissue engineering. *Eur. Polym. J.* **2013**, *49*, 780–792. [CrossRef]
2. Pillai, C.K.S.; Sharma, C.P. Electrospinning of chitin and chitosan nanofibers. *Trends Biomater. Artif. Organs* **2009**, *22*, 179–201.
3. Majeti, N.V.; Kumar, R. A review of chitin and chitosan applications. *React. Funct. Polym.* **2000**, *46*, 1–27.
4. Kannan, M.; Nesakumari, M.; Rajarathinam, K.; Ranjit Singh, A.J.A. Production and characterization of mushroom chitosan under solid-state fermentation conditions. *Adv. Biol. Res.* **2010**, *4*, 10–13.
5. Madihally, S.V.; Howard, M.W.T. Porous chitosan scaffolds for tissue engineering. *Biomaterials* **1999**, *20*, 1133–1142. [CrossRef]
6. Nisbet, D.R.; Forsythe, J.S.; Shen, W.; Finkelstein, D.I.; Horne, M.K. Review paper: A review of the cellular response on electrospun nanofibers for tissue engineering. *J. Biomater. Appl.* **2009**, *24*, 7–29. [CrossRef]
7. Pillai, C.K.S.; Paul, W.; Sharma, C.P. Chitin and chitosan polymers: Chemistry, solubility and fiber formation. *Prog. Polym. Sci.* **2009**, *34*, 641–678. [CrossRef]
8. Homayoni, H.; Ravandi, S.A.H.; Valizadeh, M. Electrospinning of chitosan nanofibers: Processing optimization. *Carbohydr. Polym.* **2009**, *77*, 656–661. [CrossRef]
9. Ohkawa, K.; Minato, K.-I.; Kumagai, G.; Hayashi, S.; Yamamoto, H. Chitosan Nanofiber. *Biomacromolecules* **2006**, *7*, 3291–3294. [CrossRef] [PubMed]
10. Desai, K.; Kit, K.; Li, J.; Zivanovic, S. Morphological and surface properties of electrospun chitosan nanofibers. *Biomacromolecules* **2008**, *9*, 1000–1006. [CrossRef] [PubMed]
11. Zhang, Y.Z.; Su, B.; Ramakrishna, S.; Lim, C.T. Chitosan Nanofibers from an easily electrospinnable UHMWPEO-Doped chitosan solution system. *Biomacromolecules* **2008**, *9*, 136–141. [CrossRef]
12. Haider, S.; Park, S.-Y. Preparation of the electrospun chitosan nanofibers and their applications to the adsorption of Cu(II) and Pb(II) ions from an aqueous solution. *J. Membr. Sci.* **2009**, *328*, 90–96. [CrossRef]
13. Sill, T.J.; Recum, H.A. Electrospinning: Applications in drug delivery and tissue engineering. *Biomaterials* **2008**, *29*, 1989–2006. [CrossRef]
14. Ignatova, M.; Manolova, N.; Rashkov, I. Novel antibacterial fibers of quaternized chitosan and poly(vinyl pyrrolidone) prepared by electrospinning. *Eur. Polym. J.* **2007**, *43*, 1112–1122. [CrossRef]
15. Jayakumar, R.; Prabaharan, M.; Nair, S.V.; Tamura, H. Novel chitin and chitosan nanofibers in biomedical applications. *Biotechnol. Adv.* **2010**, *28*, 142–150. [CrossRef] [PubMed]
16. Jayakumar, R.; Menon, D.; Manzoor, K.; Nair, S.V.; Tamura, H. Biomedical applications of chitin and chitosan based nanomaterials -A short review. *Carbohydr. Polym.* **2010**, *82*, 227–232. [CrossRef]
17. Sogias, I.A.; Khutoryanskiy, V.V.; Williams, A.C. Exploring the factors affecting the solubility of chitosan in water. *Macromol. Chem. Phys.* **2010**, *211*, 426–433. [CrossRef]
18. Sarmento, B.; Neves, J. *Chitosan-Based Systems for Bbiopharmaceuticals: Delivery, Targeting and Polymer Therapeutics*, 1st ed.; John Wiley & Sons: Hoboken, NJ, USA, 2012.
19. De Vrieze, S.; Van Camp, T.; Nelvig, A.; Hagstrom, B.; Westbroek, P.; De Clerck, K. The effect of temperature and humidity on electrospinning. *J. Mater. Sci.* **2009**, *44*, 1357–1362. [CrossRef]
20. Nam, Y.S.; Park, W.H.; Daewoo, I.; Hudson, S.M. Effect of the degree of deacetylation on the thermal decomposition of chitin and chitosan nanofibers. *Carbohydr. Polym.* **2010**, *80*, 291–295.
21. Bhardwaj, N.; Kundu, S.C. Electrospinning: A fascinating fiber fabrication technique. *Biotechnol. Adv.* **2010**, *28*, 325–347. [CrossRef]
22. Riva, R.; Ragelle, H.; des Rieux, A.; Duhem, N.; Jérôme, C.; Preat, V. Chitosan and chitosan derivatives in drug delivery and tissue engineering. *Adv. Polym. Sci.* **2011**, *244*, 19–44.
23. Sathiyaseelan, A.; Shajahan, A.; Kalaichelvan, P.T.; Kaviyarasan, V. Fungal chitosan based nanocomposites sponges—An alternative medicine for wound dressing. *Int. J. Biol. Macromol.* **2017**, *104*, 1905–1915. [CrossRef] [PubMed]
24. Sathiyaseelan, A.; Kalaichelvan, P.T. Application of tetracycline hydrochloride loaded-fungal chitosan and *Aloe vera* extract based composite sponges for wound dressing. *J. Adv. Res.* **2018**, *14*, 63–71.
25. Dutta, P.K.; Rinki, K.; Dutta, J. Chitosan: A Promising Biomaterial for Tissue Engineering Scaffolds. In *Chitosan for Biomaterials II*; Jayakumar, R., Prabaharan, M., Muzzarelli, R., Eds.; Advances in Polymer Science; Springer: Berlin/Heidelberg, Germany, 2011; Volume 244, pp. 45–80.
26. Ziani, K.; Henrist, C.; Jerome, C.; Aqil, A.; Maté, J.I.; Cloots, R. Effect of nonionic surfactant and acidity on chitosan nanofibers with different molecular weights. *Carbohydr. Polym.* **2011**, *83*, 470–476.
27. Kriegel, C.; Kit, K.M.; McClemens, D.J.; Weiss, J. Electrospinning of chitosan-poly(ethylene oxide) blend nanofibers in the presence of micellar surfactant solutions. *Polymer* **2009**, *50*, 189–200. [CrossRef]

28. Pepić, I.; Filipović-Grčić, J.; Jalšenjak, I. Interactions in a nonionic surfactant and chitosan mixtures. *Colloids Surf. A Physicochem. Eng. Asp.* **2008**, *327*, 95–102. [CrossRef]
29. Sangsanoh, P.; Supaphol, P. Stability improvement of electrospun chitosan nanofibrous membranes in neutral or weak basic aqueous solutions. *Biomacromolecules* **2006**, *7*, 2710–2714. [CrossRef]
30. Luu, Y.K.; Kim, K.; Hsiao, B.S.; Chu, B.; Hadjiargyrou, M. Development of a nanostructured DNA delivery scaffold via electrospinning of PLGA and PLA–PEG block copolymers. *J. Control. Release* **2003**, *89*, 341–353. [CrossRef]
31. Peter, M.G. Chitin and chitosan in fungi. In *Biopolymers*, Online ed.; Steinbüchel, A., Ed.; Wiley-VCH Verlag GmbH: Weinheim, Germany, 2005; Part 6; pp. 123–132.
32. Corazzari, I.; Nistico, R.; Turci, F.; Faga, M.G.; Franzoso, F.; Tabasso, S.; Magnacca, G. Advanced physico-chemical characterization of chitosan by means of TGA coupled on-line with FTIR and GCMS: Thermal degradation and water adsorption capacity. *Polym. Degrad. Stab.* **2015**, *112*, 1–9. [CrossRef]
33. Salas, C.; Thompson, Z.; Bhattarai, N. 15—Electrospun chitosan fibers. In *Electrospun Nanofibers*; Afshari, M., Ed.; Elsevier: Cambridge, UK, 2017; pp. 371–398.
34. Han, T.; Nwe, N.; Furuike, T.; Tokura, S.; Tamura, H. Methods of *N*-acetylated chitosan scaffolds and it's *in vitro* biodegradation by lysozyme. *J. Biomed. Sci. Eng.* **2012**, *5*, 15–23. [CrossRef]
35. Gorzelanny, C.; Pöppelmann, B.; Pappelbaum, K.; Moerschbacher, B.M.; Schneider, S.W. Human macrophage activation triggered by chitotriosidase-mediated chitin and chitosan degradation. *Biomaterials* **2010**, *31*, 8556–8563. [CrossRef] [PubMed]
36. Kean, T.; Thanou, M. Biodegradation, bio distribution and toxicity of chitosan. *Adv. Drug Deliv. Rev.* **2010**, *62*, 3–11. [CrossRef]
37. Croisier, F.; Atanasova, G.; Poumay, Y.; Jérôme, C. Polysaccharide-Coated PCL nanofibers for wound dressing applications. *Adv. Healthc. Mater.* **2014**, *3*, 2032–2039. [CrossRef] [PubMed]
38. Tchemtchoua, V.T.; Atanasova, G.; Aqil, A.; Filee, P.; Garbacki, N.; Vanhooteghem, O.; Deroanne, C.; Noel, A.; Jerome, C.; Nusgens, B.; et al. Development of a Chitosan Nanofibrillar Scaffold for Skin Repair and Regeneration. *Biomacromolecules* **2011**, *12*, 3194–3204. [CrossRef] [PubMed]

polysaccharides

Review

Didymo and Its Polysaccharide Stalks: Beneficial to the Environment or Not?

Hurmat Ejaz [1,2], Esther Somanader [1,2], Uday Dave [1,2], Hermann Ehrlich [1,3,4] and M. Azizur Rahman [1,2,*]

1 Centre for Climate Change Research, Toronto, ON M4P 1J4, Canada; hurmat@ar-environment.ca (H.E.); esther@climatechangeresearch.ca (E.S.); uday@climatechangeresearch.ca (U.D.); Hermann.Ehrlich@esm.tu-freiberg.de (H.E.)
2 A.R. Environmental Solutions, ICUBE-University of Toronto Mississauga, Mississauga, ON L5L 1C6, Canada
3 Institute of Electronic and Sensor Materials, TU Bergakademie Freiberg, 09599 Freiberg, Germany
4 Center for Advanced Technology, Adam Mickiewicz University, 61614 Poznan, Poland
* Correspondence: aziz@climatechangeresearch.ca or mazizur.rahman@utoronto.ca

Abstract: *Didymosphenia geminata* diatoms, or Didymo, was first found to be an invasive species that could have negative impacts on the environment due to the aggressive growth of its polysaccharide-based stalks. The stalks' adhesive properties have prompted park officials to alert the general public to limit further spread and contamination of this algae to other bodies of water. Although the negative effects of Didymo have been studied in the past, recent studies have demonstrated a potential positive side to this alga. One of the potential benefits includes the structural component of the polysaccharide stalks. The origin of the polysaccharides within stalks remains unknown; however, they can be useful in a waste management and agricultural setting. The primary purpose of this study was to describe both the harmful and beneficial nature of Didymo. Important outcomes include findings related to its application in various fields such as medicine and technology. These polysaccharides can be isolated and studied closely to produce efficient solar power cells and batteries. Though they may be harmful while uncontained in nature, they appear to be very useful in the technological and medical advancement of our society.

Keywords: polysaccharides; biomineralization; chitin; algae; diatom; didymo; biosilica; biopolymers; environment

Citation: Ejaz, H.; Somanader, E.; Dave, U.; Ehrlich, H.; Rahman, M.A. Didymo and Its Polysaccharide Stalks: Beneficial to the Environment or Not? *Polysaccharides* **2021**, *2*, 69–79. https://doi.org/10.3390/polysaccharides2010005

Received: 19 December 2020
Accepted: 3 February 2021
Published: 17 February 2021

1. Introduction

Didymosphenia geminata, commonly known as Didymo, is a relatively new alga that has been found to affect stream systems throughout the world. When Didymo grows into stalks, it can produce negative environmental and ecological impacts. In the International Workshop on *D. geminata* in 2008, it was stated that Didymo stalks are composed of proteins, sulfated polysaccharides, and some uronic acid. Figure 1 provides a closer look into the stalks' ability to grow to a length of 500 μm or more [1]. The structure of biomineral components of this species is different from other freshwater [2] and marine algae [3]. For example, polysaccharide-based stalks of Didymo are reinforced with unique calcitic nanofibers [1]; however, the cell is still made of biosilica. These algae tend to form a dense mass through cell division that can be seen as clumps in rivers (as seen in Figure 2). These diatoms can cover up to 100% of a substrate and form a thick layer of more than 20 cm, which can cause physical and biological problems in streams. Moreover, Didymo is able to cover areas over 20 km and can be present in streams for many months. This causes disturbance in the benthic regions and does not allow for the growth of other algae. These stalks also interfere with the fish communities underwater and have been a nuisance to humans due to their ability to clog water filters and hydro plants. Even though Didymo has caused disruptions regarding different underwater communities and human activity, there may now be a promising use of these stalks and their ability to act as reservoirs of mineral ions due to their microcapillary action.

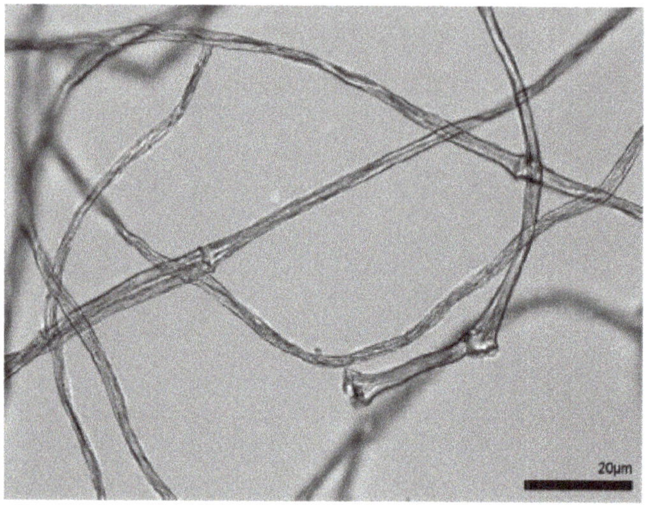

Figure 1. Polysaccharide-based and cell-free *D. geminata* stalks isolated after ultrasonic treatment [1].

Figure 2. Appearance of didymo within rivers [4].

Their structure is considered to be helpful when it comes to metal removal applications due to the large amounts of complex extracellular biopolymers. Due to this characteristic, they are being considered as a fibrous absorbent of lead (II), nickel (II), and cadmium (II) in wastewater systems [5]. As more research was done on this species of algae, it was found that Didymo would be transferred to different parts of the river by sticking to the bottom of fishermen's boots. As the fishing industry grew in Vancouver Island, so did the amount of Didymo sightings. It was during this time that a connection was made between fishermen's boots contaminating more of an area than normal [6]. It was further discovered that Didymo was not only transferred through fishermen's boots, but also through people coming to the rivers for recreational activities. Their swimwear and fishing equipment would be used as a vessel for Didymo transference to other parts of the river, as people themselves explored the area. To limit the spread of Didymo, proper signage was made for public awareness. As Figure 3 shows, certain steps needed to be taken by the visitors [7]. The proper management of sensitive areas can help avoid problems in the future and can help protect the environment.

This paper focuses on the harmful and beneficial nature of Didymo and its stalks. By analyzing both sides of its nature, future measures can be taken according to the overall consensus of its status. For example, if the negative impacts of Didymo exceed its positive impacts, communities with an infestation would need to produce and utilize a permanent solution for its removal. However, if the benefits of Didymo appear to outweigh its disadvantages, more research should be devoted to how to utilize these polysaccharide stalks in industries outside of agriculture.

Figure 3. Didymo information signs posted at the Battenkill River and other streams in New York recommend drying fishing gear or cleaning with household cleaner/disinfectant. Photo credit: Samantha Root [7].

2. Current Status of Didymo in Canada

Although the Vancouver Islands have seen Didymo in their rivers for the last few decades (since the 1990s) [6], on the other side of Canada, in Montreal, they were discovered much later. In 2006, Didymo was first found in the Matapedia River in Montreal [8]. Another river in Quebec, the Causapscal River, remained Didymo-free and was used as a control site to observe the difference in aquatic food webs. Since the Matapedia River is a source of aquatic salmon in the region, there was fear of a lack of availability of salmon which is an important source of food in the area, and therefore studies were done between the different rivers. Comparisons between the different environments have shown that the macroinvertebrate densities have increased in rivers where Didymo was present compared to the control site where Didymo was not found. The structure of the community in the region did change when certain flies (mayflies, stoneflies, and caddisflies) were seen to decline by about 30–50% where Didymo was present, therefore showing an inverse relationship between this particular species of flies and Didymo [8].

As seen in Figure 4, other areas where Didymo was found and surveryed in Canada were the Eastern Rocky Mountains, specifically, the Athabasca River (surveyed 2003–2007), Bow River (surveyed 2002–2007), and Red Deer River (surveyed 2004–2006) [9].

The commonality between these three rivers was that they were in the middle of forests and running downstream. In 18 rivers from the South Saskatchewan River Basin, 80% of 50 sites showed Didymo growth. Dense growth has also been recorded in almost half of the sites in Alberta. In Alberta's Bow River and Red Deer River, a negative relationship was seen between Didymo biomass and mean discharge. This negative relationship could be due to the changes in temperature due to changes in flow at a dam outflow. These changes can allow for various types of nutrient growth and can change the chemistry of the water [9]. More research needs to be done in other provinces within Canada, especially Ontario, where a large amount of the population live around water bodies and could potentially contaminate these areas fairly quickly. As can be seen in Figure 5, there has not been much research in the northern parts of Canada and a future research project is needed in order to properly portray the appearance of Didymo. With the help of GIS techniques and remote sensing, a closer look can be taken at Didymo in certain areas, and a record of past observances can allow us to better monitor potential areas of rapid growth.

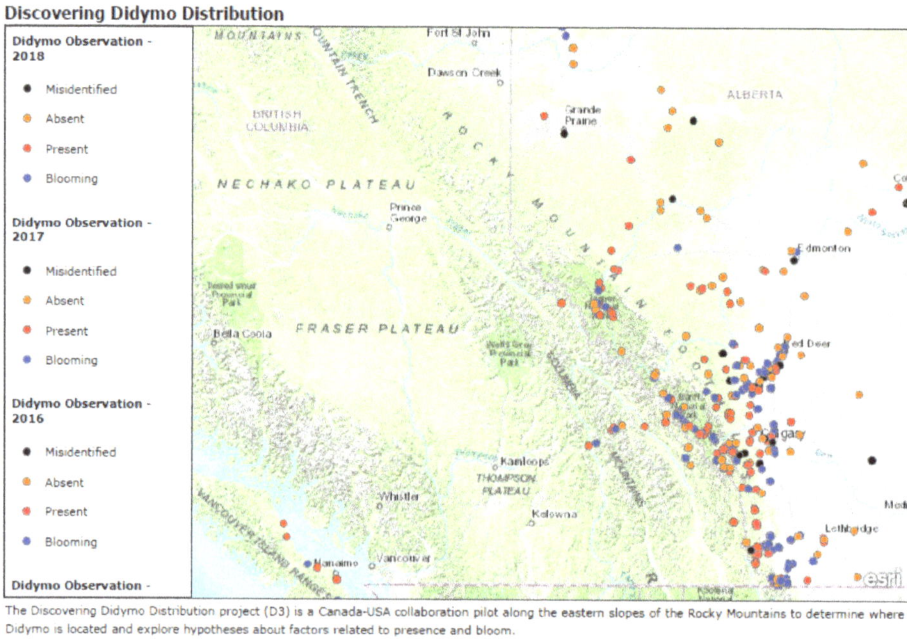

Figure 4. Didymo presence in British Columbia and Alberta [10].

Figure 5. Worldwide distribution of records for *D. geminata* [7].

In Figure 4, we can also notice that some areas (blue) are showing new growth of Didymo and it is essential to make note of these in further studies to see how fast this spreads. There may even be a chance to observe some blooming Didymo that are not currently present, which may lead to further study of the area to learn about what may have stopped the growth. These data can be beneficial to data analysts for modeling future increases or decreases in Didymo within the area. An issue with collecting data within Canada comes when satellites are not able to detect the growth on land due to snow cover. This may be the reason why, in Figure 5, only the southern parts of the provinces are shown. Therefore, it is essential to keep in mind that a project done on Didymo in the northern areas of Canada has to be done during the summer months. This will allow us to properly observe growth on land without natural interruptions.

3. How Harmful Is This Species to the Canadian Environment and Public Health?

In terms of the negative effects that Didymo can have on the environment, several studies have focused on the impact that *D. geminata* has on native species and biodiversity in different regions. Within the last decade, *D. geminata* has spread outside of its native regions to the rivers and lakes of several countries including New Zealand, Argentina, and southern Chile. In the benthic diatom communities in the Chilean rivers, it was found that community heterogeneity decreased in the presence of *D. geminata*, making the site more homogenized. This result was observed through the increase in small stalked diatom density and the decrease in species turnover in the invaded rivers. The authors did not observe species exclusion; however, it was observed that *D. geminata* is inclined towards a particular group of diatom species. This inclination can displace other diatom species and, therefore, cause different types of impacts, such as alterations in fluvial trophic webs [11].

There are several studies that have commented on the possible effects that *D. geminata* could have on fluvial tropic webs. For example, in British Columbia, it was observed that a salmon channel covered with *D. geminata* showed reduced zoobenthos heterogeneity in comparison to the mainstream. Due to the presence of *D. geminata*, the distribution of salmonid food organisms increased in the channel from 57% to 93% Chironomidae [9]. This was a similar observation found in the Matapedia River Valley (located in Quebec). Even though there is currently no scientific evidence that suggests this alga has significant effects on the Atlantic salmon population in Eastern Canada, Gillis and Chalifour [8] reported that the presence of *D. geminata* could act as an additional stressor to this population. They discovered that benthic macroinvertebrates, such as mayflies, stoneflies, and caddisflies, diminished due to the presence of *D. geminata*, while chironomid proportions had increased and become the dominant taxa. This finding confirms that this community has a less homogenous macroinvertebrate distribution due to the presence of this alga [8]. More research would have to be done to determine the long-term impacts that *D. geminata* has on salmonid populations, but there is clear evidence that the presence of *D. geminata* does change the distribution of salmonid food organisms, which, in turn, could hve impacts on the diet and physical condition of the salmonid population.

While it has been generally accepted that *D. geminata* has the potential to alter ecosystems, it is not considered to be directly harmful to human health. However, while there is no direct impact of *D. geminata* on human health, there is a growing literature that demonstrates that the changing of ecosystem structures can have a negative impact on public health [12]. This is an area of research that is at the intersection of health and ecology. While the presence of *D. geminata* in a community would not necessarily lead to a risk in exposure to infectious diseases, a decrease in biodiversity or the abundance of a species could have other significant human health impacts. A lack of biodiversity could contribute the diminishing of marine wildlife and cause a nutritional crisis in certain communities [12]. More research would have to be done to determine whether the change in the ecosystem structure due to *D. geminata* has negative impacts on public health.

As we now aware, *D. geminata*, forms brownish-whites mucilaginous stalks that have been reported to be as thick as 20 cm [13]. As Didymo spans many kilometers, it can monumentally degrade aquatic ecosystems and cause a drastic change in benthic communities [13], such as replacing invertebrate species like mayflies and stoneflies with much smaller midges [13]. Furthermore, shallow cold-water streams, rivers and lake margins with rocky substrates that have constant flows and low nutrients allow Didymo to grow favorably [13]. Didymo has been known to be a significant nuisance for local trout and salmon populations, being the ideal habitat for those species of fish [13]. Previous studies have found that Didymo has an effect on the microenvironment by reducing fish populations [14]. Due to the nature of fish spermatozoa, within the ejaculate, they are immobile, and only after osmotic shock in water can they begin to move or swim [14].

Overall, while there are distinct patterns in how the presence of *D. geminata* affects the biodiversity of certain communities both internationally and locally, there has not been enough research done to demonstrate the longitudinal effects that this species has on trophic food webs and public health. This area of research would have to be explored further.

4. Benefits of Didymo

While most of the research surrounding *D. geminata* has focused on its structure, management, and possible negative impacts, there has been recent research published that demonstrates the possible positive effects that Didymo can have on the environment. In a paper written by Wysokowski et al. [5], the authors state that the polysaccharide stalks of *D. geminata* have the ability to absorb harmful metal ions, such as Pb (II), Ni (II), and Cd (II). These stalks contain large amounts of complex extracellular biopolymers, which allows them to remove metals comparable to other polysaccharides. The authors used a pseudo-second-order kinetic model to describe the absorption kinetics of the stalks and found that *D. geminata* demonstrated an extraordinary sorption capacity for Pb (II) ions of 175.48 mg g^{-1} and a high sorption capacity for Cd (II) ions of 145.86 mg g^{-1}. They also observed a sorption capacity of Ni (II) ions of 130.27 mg g^{-1}. In terms of applicable use, after performing tests on industrial wastewater, the authors suggest that this alga can be used for the purification of industrial wastewater, especially water that has a high content of Pb (II) ions. Due to *D. geminata's* effectiveness of absorption, this biological material can be competitive with manufactured and low-cost adsorbents [1].

The discovery of *D.geminata's* biological absorption ability can have profound effects in the treatment of industrial wastewater; however, its methods for commercial use will have to be further investigated. Industrial wastewater is typically treated through wastewater plants. Due to the sheer volume of wastewater, *D.geminata* could be used within this process, so long as it is contained within a closed system. Another positive effect surrounding this freshwater diatom is its role in nitrogen fixation. Even though little is known about the different microbial species associated with *D. geminata*, it is assumed that many different microbial communities are known to co-exist with the freshwater algae [15]. Found in these communities are nitrogen-fixing bacteria, which are commonly found in habitats that *D. geminata* covers [15]. Furthermore, bottom-feeding cyanobacteria can illustrate the environmental conditions of the freshwater body [15]. Previous studies have stated that *D. geminata* and nitrogen-fixing bacteria have a mutualistic relationship, where nitrogen fixers could provide nitrogen to Didymo in a low-nutrient environment [15]. The study goes on to state that it believes the genus Godleya could be one of the cyanobacteria that favor similar habitats and that fix nitrogen for *D. geminata*. With this in mind, our ability to obtain suitable growth patterns of *D. geminata* can allow for these nitrogen-fixing bacteria to proliferate in areas that are low in nutrients. Studies have concluded that due to 78% of the earth's atmosphere is made up of nitrogen but exists as dinitrogen [13]. Moreover, to allow living organisms to use nitrogen, it must be transformed into ammonia or nitrate via endosymbiotic bacteria [16]. Ultimately, if *D. geminata* is sustained with less environmental impact it can benefit aquatic plants growing in unfavorable low-nitrogen environments by providing enough nitrogen to grow.

$$N_2 + 8e- + 8H^{+} + 16ATP \text{ à } 2NH_3 + H_2 + 16\,ADP + 16\,P_i$$

Not only do the polysaccharide stalks of *D. geminata* have incredible absorption abilities, it was also found that the structure of the stalks allows the algae to be both stable and flexible under various flow conditions. Ehrlich et al. [2] discovered that the stalks contained nanostructured calcite-based scaffolds. This structural attribute could further provide opportunities for biometric approaches in developing unique and innovative biomaterials with adhesive properties. While *D. geminata* has been termed as a "nuisance species," there have been reports that show the possible positive effects on the environment and in the development of biomaterials.

This group has been extensively studied by researchers since the 18th century because of its unique and complicatedly patterned frustules (silica cell walls). The mechanical structure of frustules was well investigated, revealing interesting characterizations and patterns [17–19]. Interestingly, the frustule is mainly composed of chitin [20] in biogenic silica, which is different than other chitin-based algae [1,21] in nature. Because of their unique mechanical properties, they have a large variety of applications, such as in the modeling of mechanical properties [19,22], drug delivery [23,24], electronics [25], the modifier of resin [26], and biomimetics [27].

5. Management of *Didymosphenia geminata*

D. geminata has the potential to have catastrophic effects on local freshwater environments, due to the high isolation and endemism found in species living in rivers, streams, lakes and ponds [28]. In recent years, since this discovery, many different research programs have investigated different potential control products to manage the spread of Didymo [28]. A previous study performed rigorous testing to identify the best chemical control compound [14]. Such testing looked at the toxicity of chemicals such as chelated copper compounds known as Germex, EDTA, Hydrothol 191, and Organic Interceptor, otherwise known as pine oil formulation [17]. Numerous factors went into the testing process to determine which of these chemicals had the best effect on Didymo [17]. These factors included effectiveness, non-target species impacts, stalk removal, degradation profile, risks to health and safety, ease of application, neutralization potential, cost, and local regulatory requirements [14]. The study found that both Gemex and Organic Interceptor were the best in terms of biocidal efficacy on Didymo [28]. However, when Organic Interceptor was exposed to species of fish, the researchers noticed a higher mortality rate under laboratory conditions [28]. However, Germex was noted to have negative effects on invertebrates, allowing stakeholders to approve the chemical as the best option to tackle this freshwater diatom, but further testing is required. Other household products have also been discussed for controlling the spread of *D. geminata*. Among several common decontamination treatments that were tested, the study by Root and O'Reilly [7] found that dish liquid detergent was the most effective, followed by bleach, Virkon, and salt. The above decontaminants were highly effective on the *D. geminata* cells still attached to their stalks [7].

However, due to the nature of Germex and other toxins which could have adverse effects on nontarget algae by disrupting the base of the food web, many researchers have looked at novel biological control methods [29]. Studies surrounding the use of Germex have proven that whole-stream chemical poisoning can have major implications for the surrounding environment, causing more ecological costs than benefits. Short-term applications of Gemex, repeated weekly, can cause major damage to local trout populations [29]. Currently, most management policies to prevent *D. geminata* from uncontrollable blooms have adapted the "Check, Clean, Dry" marketing campaign used in New Zealand [29]. While many of these decontamination programs are elected to be the first line of defense against unwanted organisms, their efficacy has not been proven to slow down the spread of free-living microorganisms such as *D. geminata*. With this in mind, scientists studying *D. geminata* in New Zealand have noted that diatoms do not develop in streams and rivers with higher than normal concentrations of inorganic phosphorus not exceeding 2 micrograms per liter [18], thus suggesting an alternative control method for *D. geminata* through phosphorus augmentation [29]. Phosphorus augmentation may only provide short-term control, because it needs continuous application to control the spread of bloom. If that is the case, increasing phosphorus concentrations can be counter-intuitive, as most freshwater bodies would like to keep phosphorus concentrations low in order to prevent cultural eutrophication. Moreover, increasing phosphorus concentrations can be costly to maintain a continuous flow of inorganic phosphorus. Such practices can only be suggested if the stream that *D. geminata* inhabits is threatening other species living in that freshwater body, as additional phosphorus can only be recommended for a short period [29].

The economic impacts of Didymo are currently unknown. With its ability to be used in wastewater management and agricultural practices, we can see it potentially having great prospects in the coming years. However, it is hard to say if its effects on certain species in river ecosystems will reduce the amount of economic growth that region could have without access to essentials like salmon. Further research on these topics is necessary to make an informed prediction.

Diverse aspects of the structural biology of *D. geminata* have been recently described [17,18,30,31]. Thus, *D. geminata* stalks are examples of unique biocomposites, which contain calcite [1], some proteins, and polysaccharides, which are still poorly investigated. According to Bothwell and Spaulding [32], these stalks are "composed primarily of sulfated polysaccharides with significant uronic acid content, and protein. Monosaccharide analysis has revealed predominately galactosyl and xylosyl residues and linkage analysis has shown predominantly 3,4-Gal and 4-Xyl. The polysaccharide portion of the stalk therefore appears to be primarily sulfated xylogalactan, which has been reported for stalks of related diatoms *Gomphonema* and *Cymbella*, where it was shown to be intrinsically hydrophilic and linked by ionic cross-bridging".

Preliminary experiments which have been carried out in our lab for the identification of the nature and origin of the polysaccharides within Didymo stalks showed an absence of chitin. However, possible localization within siliceous cells, similar to the phenomenon reported in *T. pseudonana* diatoms [10,33], has yet to be studied. We also suggest that, due to the confirmed presence of xylose within *D. geminata* stalks [32], corresponding analytical studies should be carried out in the near future with the aim of identifying xylosamine as a structural alternative compound to chitin.

As seen in Figures 6 and 7, the scanning electron microscope (SEM) images show the Didymo stalks in higher resolution, allowing for the intricate details of the structures to be made clear. It is these structures that are important for the future study and management of this species. These images can help us obtain information about the surface topology and inner structural peculiarities.

Figure 6. SEM image of Didymo siliceous cell tightly bound to the biomineralized polysaccharides-based stalk.

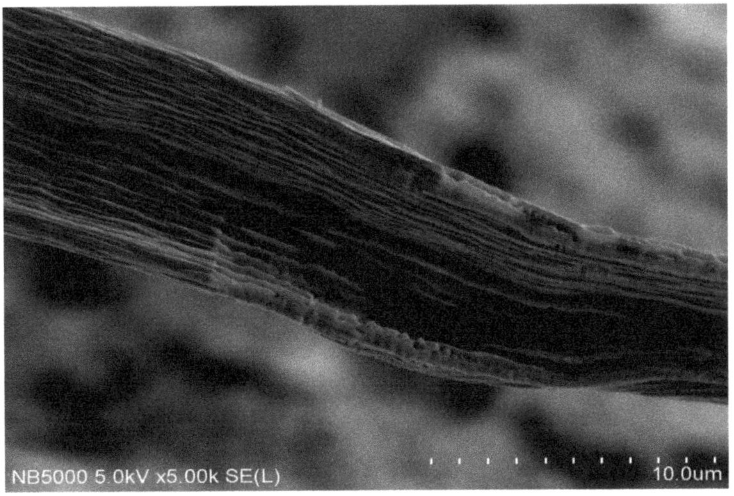

Figure 7. SEM image of the Didymo stalk cross-section confirms its multilayered structure with capillary-like morphology.

6. Conclusions

The unique nature of Didymo has been proven to cause environmental and ecological damage. Higher mortality rates in fish, affecting water filtration systems and hydro-plants, are major areas of concern for streams affected by Didymo. This is spreading from coast to coast; it not only occurs in Canada, but is a worldwide problem. Very few benefits come to mind when talking about *D. geminata*, like its ability, when controlled, to absorb harmful metals and co-exist with nitrogen-fixing bacteria. However, managing the species can be expensive using synthetic chemical agents and can further damage the environment. Other solutions, like management policies and using inorganic phosphorus augmentation, can control the spread of the algae. We also applied GIS tools to identify the location and current condition of this alga which could help researchers in modeling and management to either control or collect the beneficial applications. With all of the information provided above, it can be confirmed that *D.geminata* can, in fact, be a "nuisance species", but, due to its structure (e.g., polysaccharide stalks, chitin-based frustule), scientists have the ability to manage the spread, and possibly use this species to our benefit. Therefore, permanent removal of its polysaccharide stalks is not needed, but its growth should be isolated from the environment so that its negative effects in rivers and streams are minimized while it helps maximize our technological growth in all sectors.

Author Contributions: Conceptualization, M.A.R. and H.E. (Hermann Ehrlich); software, H.E. (Hurmat Ejaz); validation, M.A.R., H.E. (Hermann Ehrlich) and H.E. (Hurmat Ejaz); resources, M.A.R. and H.E. (Hurmat Ejaz); data curation, H.E. (Hurmat Ejaz), M.A.R., E.S., U.D.; writing—original draft preparation, M.A.R., H.E. (Hurmat Ejaz), E.S. and U.D.; writing—review and editing, M.A.R. and H.E. (Hermann Ehrlich); visualization, M.A.R. and H.E. (Hurmat Ejaz); supervision, M.A.R.; project administration, M.A.R.; funding acquisition, M.A.R. All authors have read and agreed to the published version of the manuscript.

Funding: This research was funded by Eco Canada, Innovation Assistant Program of IRAP-NRC (Project No.: 950567, 959949), Canada.

Institutional Review Board Statement: Not applicable.

Informed Consent Statement: Not applicable.

Data Availability Statement: Not applicable.

Conflicts of Interest: The authors declare no conflict of interest.

References

1. Ehrlich, H.; Motylenko, M.; Sundareshwar, P.V.; Ereskovsky, A.; Zgłobicka, I.; Noga, T.; Płociński, T.; Tsurkan, M.V.; Wyroba, E.; Suski, S.; et al. Multiphase Biomineralization: Enigmatic Invasive Siliceous Diatoms Produce Crystalline Calcite. *Adv. Funct. Mater.* **2016**, *26*, 2503–2510. [CrossRef]
2. Žuljević, A.; Kaleb, S.; Peña, V.; Despalatović, M.; Cvitković, I.; De Clerck, O.; Le Gall, L.; Falace, A.; Vita, F.; Braga, J.C.; et al. First freshwater coralline alga and the role of local features in a major biome transition. *Sci. Rep.* **2016**, *6*, 19642. [CrossRef] [PubMed]
3. Rahman, M.A.; Halfar, J. First evidence of chitin in calcified coralline algae: New insights into the calcification process of *Clathromorphum compactum*. *Sci. Rep.* **2014**, *4*, 6162. [CrossRef]
4. Invasive Species: Didymo or "Rock Snot". 21 August 2019. Available online: https://www.nps.gov/yose/learn/nature/didymo.htm (accessed on 16 November 2020).
5. Wysokowski, M.; Bartczak, P.; Żółtowska-Aksamitowska, S.; Chudzińska, A.; Piasecki, A.; Langer, E.; Bazhenov, V.V.; Petrenko, I.; Noga, T.; Stelling, A.L.; et al. Adhesive Stalks of Diatom *Didymosphenia geminataas* a Novel Biological Adsorbent for Hazardous Metals Removal. *CLEAN Soil Air Water* **2017**, *45*, 1600678. [CrossRef]
6. Bothwell, M.L.; Taylor, B.W.; Kilroy, C. The Didymo story: The role of low dissolved phosphorus in the formation of Didymosphenia geminata blooms. *Diatom Res.* **2014**, *29*, 229–236. [CrossRef]
7. Root, S.; O'Reilly, C.M. Didymo Control: Increasing the Effectiveness of Decontamination Strategies and Reducing Spread. *Fisheries* **2012**, *37*, 440–448. [CrossRef]
8. Gillis, C.-A.; Chalifour, M. Changes in the macrobenthic community structure following the introduction of the invasive algae Didymosphenia geminata in the Matapedia River (Québec, Canada). *Hydrobiologia* **2009**, *647*, 63–70. [CrossRef]
9. Whitton, B.A.; Ellwood, N.T.W.; Kawecka, B. Biology of the freshwater diatom Didymosphenia: A review. *Hydrobiologia* **2009**, *630*, 1–37. [CrossRef]
10. ArcGIS. Available online: https://www.arcgis.com/home/webmap/viewer.html?webmap=44af8ce71bc844b8be2386b6bdea6f59 (accessed on 8 December 2020).
11. Pereira, J.S.; Pérez, A.O. Community signals of the effect of *Didymosphenia geminata* (Lyngbye) M. Schmidt on benthic diatom communities in Chilean rivers. *Rev. Chil. Hist. Nat.* **2019**, *92*, 4. [CrossRef]
12. Myers, S.S.; Gaffikin, L.; Golden, C.D.; Ostfeld, R.S.; Redford, K.H.; Ricketts, T.H.; Turner, W.R.; Osofsky, S.A. Human health impacts of ecosystem alteration. *Proc. Natl. Acad. Sci. USA* **2013**, *110*, 18753–18760. [CrossRef] [PubMed]
13. Beville, S.T.; Kerr, G.N.; Hughey, K.F. Valuing impacts of the invasive alga *Didymosphenia geminata* on recreational angling. *Ecol. Econ.* **2012**, *82*, 1–10. [CrossRef]
14. Olivares, P.; Orellana, P.; Guerra, G.; Peredo-Parada, M.; Chavez, V.; Ramirez, A.; Parodi, J. Water contaminated with *Didymosphenia geminata* generates changes in Salmo salar spermatozoa activation times. *Aquat. Toxicol.* **2015**, *163*, 102–108. [CrossRef] [PubMed]
15. Novis, P.M.; Schallenberg, M.; Smissen, R.D. Aquatic nitrogen-fixing cyanobacteria associated with blooms of *Didymosphenia geminata*: Insights from a field study. *Hydrobiologia* **2015**, *770*, 37–52. [CrossRef]
16. Dalton, D.A.; Kramer, S. Nitrogen-fixing bacteria in non-legumes. In *Plant-Associated Bacteria*; Springer Nature: Berlin/Heidelberg, Germany, 2007; pp. 105–130.
17. Zgłobicka, I.; Li, Q.; Gluch, J.; Płocińska, M.; Noga, T.; Dobosz, R.; Szoszkiewicz, R.; Witkowski, A.; Zschech, E.; Kurzydłowski, K.J. Visualization of the internal structure of *Didymosphenia geminata* frustules using nano X-ray tomography. *Sci. Rep.* **2017**, *7*, 1–7. [CrossRef] [PubMed]
18. Zglobicka, I.; Chmielewska, A.; Topal, E.; Kutukova, K.; Gluch, J.; Krüger, P.; Kilroy, C.; Swieszkowski, W.; Kurzydlowski, K.J.; Zschech, E. 3D Diatom–Designed and Selective Laser Melting (SLM) Manufactured Metallic Structures. *Sci. Rep.* **2019**, *9*, 1–9. [CrossRef]
19. Topal, E.; Rajendran, H.; Zgłobicka, I.; Gluch, J.; Liao, Z.; Clausner, A.; Kurzydłowski, K.J.; Zschech, E. Numerical and Experimental Study of the Mechanical Response of Diatom Frustules. *Nanomaterials* **2020**, *10*, 959. [CrossRef]
20. Brunner, E.; Richthammer, P.; Ehrlich, H.; Paasch, S.; Simon, P.; Ueberlein, S.; Van Pée, K.-H. Chitin-Based Organic Networks: An Integral Part of Cell Wall Biosilica in the Diatom Thalassiosira pseudonana. *Angew. Chem. Int. Ed.* **2009**, *48*, 9724–9727. [CrossRef]
21. Rahman, M.A.; Halfar, J.; Adey, W.H.; Nash, M.; Paulo, C.; Dittrich, M. The role of chitin-rich skeletal organic matrix on the crystallization of calcium carbonate in the crustose coralline alga Leptophytum foecundum. *Sci. Rep.* **2019**, *9*, 1–8. [CrossRef]
22. Hamm, C.E.; Merkel, R.; Springer, O.; Jurkojc, P. Architecture and material properties of diatom shells provide effective me-chanical protection. *Nature* **2003**, *421*, 841–843. [CrossRef]
23. Aw, M.S.; Simovic, S.; Yu, Y.; Addai-Mensah, J.; Losic, D. Porous silica microshells from diatoms as biocarrier for drug delivery applications. *Powder Technol.* **2012**, *223*, 52–58. [CrossRef]
24. Medarević, Đ.; Losic, D.; Ibrić, S. Diatoms—Nature materials with great potential for bioapplications. *Chem. Ind.* **2016**, *70*, 613–627. [CrossRef]
25. Jeffryes, C.; Campbell, J.; Li, H.; Jiao, J.; Rorrer, G.L. The potential of diatom nanobiotechnology for applications in solar cells, batteries, and electroluminescent devices. *Energy Environ. Sci.* **2011**, *4*, 3930. [CrossRef]
26. Zglobicka, I. Frustules of Didymosphenia geminata as a modifier of resins. *Inżynieria Mater.* **2018**, *1*, 10–16. [CrossRef]
27. Gordon, R.; Losic, D.; Tiffany, M.; Nagy, S.; Sterrenburg, F. The Glass Menagerie: Diatoms for novel applications in nanotechnology. *Trends Biotechnol.* **2009**, *27*, 116–127. [CrossRef]

28. Jellyman, P.G.; Clearwater, S.J.; Clayton, J.S.; Kilroy, C.; Blair, N.; Hickey, C.W.; Biggs, B.J.F. Controlling the Invasive Diatom Didymosphenia geminata: An Ecotoxicity Assessment of Four Potential Biocides. *Arch. Environ. Contam. Toxicol.* **2011**, *61*, 115–127. [CrossRef] [PubMed]
29. Taylor, B.W.; Bothwell, M.L. The Origin of Invasive Microorganisms Matters for Science, Policy, and Management: The Case of Didymosphenia geminata. *Bioscience* **2014**, *64*, 531–538. [CrossRef] [PubMed]
30. Zgłobicka, I. Aspects of Structural Biology of *Didymosphenia geminata* (Lyngb.) M. Schmidt (Bacillariophyta). *Int. J. Algae* **2013**, *15*, 291–310. [CrossRef]
31. Zglobicka, I. Exploratory Study of the Use of Didymosphenia geminata Stalks as a Functional Biomaterial. The Warsaw University of Technology, Poland, Ph.D. Thesis, 2015, p. 82.
32. Bothwell, M.L.; Spaulding, S.A. (Eds.) *Proceedings of the 2007 International Workshop on Didymosphenia geminata*; Canadian Technical Report of Fisheries and Aquatic Sciences 2795; Fisheries and Oceans Canada, Science Branch, Pacific Region, Pacific Biological Station: Nanaimo, BC, Canada, 2008; p. 58.
33. Wustmann, M.; Poulsen, N.; Kröger, N.; Van Pée, K.-H. Chitin synthase localization in the diatom *Thalassiosira pseudonana*. *BMC Mater.* **2020**, *2*, 1–7. [CrossRef]

MDPI AG
Grosspeteranlage 5
4052 Basel
Switzerland
Tel.: +41 61 683 77 34

Polysaccharides Editorial Office
E-mail: polysaccharides@mdpi.com
www.mdpi.com/journal/polysaccharides

www.ingramcontent.com/pod-product-compliance
Lightning Source LLC
Chambersburg PA
CBHW061547210926
51850CB00009B/2024